Unimodality, Convexity, and Applications

This is a volume in
PROBABILITY AND MATHEMATICAL STATISTICS
Z. W. Birnbaum and E. Lukacs, editors
A list of titles in this series appears at the end of this volume.

Unimodality, Convexity, and Applications

Sudhakar Dharmadhikari

Department of Mathematics
Southern Illinois University
at Carbondale
Carbondale, Illinois

Kumar Joag-dev

Department of Mathematics
University of Illinois
Urbana, Illinois

ACADEMIC PRESS, INC.
Harcourt Brace Jovanovich, Publishers
Boston San Diego New York
Berkeley London Sydney
Tokyo Toronto

ACADEMIC PRESS, INC.
1250 Sixth Avenue, San Diego, CA 91202

United Kingdom Edition published by
ACADEMIC PRESS, INC. (LONDON) LTD.
24–28 Oval Road, London NW1 7DX

Library of Congress Cataloging-in-Publication Data

Dharmadhikari, S. W. (Sudhakar Waman), Date-
Unimodality, convexity, and applications.

(Probability and mathematical statistics)
Bibliography: p.
Includes index.
1. Distribution (Probability theory) 2. Convex domains. I. Joag-dev, Kumar, Date- . II. Title. III. Series.
QA273.6.D49 1987 519.2 87-18715
ISBN 0-12-214690-5

88 89 90 91 9 8 7 6 5 4 3 2 1
Printed in the United States of America

To Pradnya and Chitra

Contents

Preface

The purpose of this monograph is to develop the notion of unimodality as it appears in the field of probability and statistics and to present some of its applications. In most textbooks, unimodality of distributions is mentioned only in passing. Concepts of multivariate unimodality are generally not mentioned at all. During the last thirty years, and especially since 1970, the subject of multivariate unimodality has received a lot of attention in the literature. The present monograph is an attempt to bring together the basic notions and tools of unimodality for their own sake. We hope that, at the very least, this account will show that there is much more to the concept of unimodality than the casual description of a curve going up and then down.

As the title indicates, arguments involving convexity play a crucial role throughout this monograph. Since the turn of the century, properties of convex sets and functions have been extensively applied in several branches of mathematics. The minimax theorems in game theory and the simplex method in linear programming are two of the better known instances of such applications. The main reason for the popularity of convexity arguments is that the basic ideas involved are easy to describe and have intuitive geometric appeal. In probability and statistics, the use of convexity to prove Jensen's inequality and the Rao–Blackwell theorem is standard textbook material.

Properties of convex sets also play an important role in decision theory because of its close connection with game theory.

The relevance of convexity to the subject matter of this monograph is easily explained. Khintchine's definition of a unimodal distribution involves the convexity of the distribution function. His characterization of unimodal distributions is a theorem of the Krein–Milman type. Chung discovered that the property of unimodality is not preserved under convolutions, and this discovery led to Ibragimov's characterization of strongly unimodal distributions through the logconcavity of their densities. Olshen and Savage developed Choquet-type representation theorems in the context of generalized unimodality. The Brunn–Minkowski inequality was the basic tool in Anderson's work on multivariate unimodality. This work and its generalizations have found extensive applications in statistical inference. The monograph by Marshall and Olkin on majorization contains several applications of Schur convexity in statistics. We believe that convexity is the thread that unifies the material of this monograph.

We have tried to make our material accessible to graduate students in probability and statistics. A graduate-level course on probability and familiarity with the most elementary properties of convex sets would, in our opinion, be an adequate background. We have also attempted to simplify the proofs as much as possible. The short appendix contains some material on convexity and weak convergence, which will be useful at a few places in the monograph.

The core material on classes of unimodal distributions on the line and on higher dimensional spaces is contained in the first two chapters. The third chapter presents useful and important generalizations of some of the multivariate results of Chapter 2. Chapter 4, on the unimodality of discrete distributions, is essentially independent of the rest of the monograph. The unimodality of univariate and multivariate infinitely divisible distributions is discussed in Chapter 5. The last four chapters consist of applications of unimodality. The diversity of these applications should convince the reader that the theoretical results on convexity and unimodality form an important tool in the hands of a researcher in probability and statistics.

This is not an encyclopedic book, and we realize that most readers will find that some favorite item of theirs is not mentioned by us. While we have tried our best to research the literature and acknowledge original sources of results, we may have inadvertently made some lapses. For all such errors of omission and commission, we apologize in advance.

The project of writing this monograph has benefited from input by several of our professional colleagues. During 1977, Kumar Joag-dev was fortunate to be a participant in a seminar on multivariate methods at Stanford University. At that time, Ingram Olkin kindly made available to Kumar the galley proofs of some of the chapters of the Marshall-Olkin monograph on majorization. During and after Kumar's lectures at Stanford, many useful comments were made by Ted Anderson, Michael Perlman, and Persi Diaconis. We have received several useful suggestions and enthusiastic comments from Don Jensen, Peter Nuesch, Pranab Sen, Yung Tong, and Brian Young. We are indebted to Frank Proschan for his interest in this project and for giving a careful reading to early versions of some chapters. We were also greatly encouraged by the constructive comments from Professor Birnbaum, Professor Lukacs, and the referees. To all these individuals and to others who have helped in small and big ways, we express our heart-felt thanks.

We thank the National Science Foundation for supporting our research during the late 1970s and for providing support during Kumar's 1977 visit to Stanford. We also thank the Air Force Office of Scientific Research for supporting our research during the last six years, initially through the Reliability Center at the Florida State University and later through the grant AFOSR 84-0208 at the University of Illinois. We gratefully acknowledge sabbatical support from Virginia Tech and EPF at Lausanne, Switzerland, during the final stages of the work on the manuscript.

Finally, we are indebted to Linda Macak for the efficient typing of the manuscript and to Pascha Gerlinger and William Sribney of Academic Press for excellent editorial support.

Sudhakar Dharmadhikari
Kumar Joag-dev

1 Properties of Univariate Unimodal Distributions

1.0. Summary

This chapter brings together some of the basic results involving unimodal distributions on the line. Khintchine's general definition of unimodality is given in Section 1. Section 2 presents results on the convex structure of the set of unimodal distributions and also gives some characterizations of such distributions. Convolution properties are discussed in Section 3 and these lead to a discussion of strong unimodality in Section 4. Section 5 contains a detailed treatment of the Gauss inequality for unimodal distributions. The final section discusses the mean-median-mode inequality.

1.1. A General Definition of Unimodality on the Line

By the term unimodal density on R one usually understands a density f which has a maximum at a unique point $x = v$ and *decreases* as x goes away from v in either direction. The normal and the Cauchy distributions are unimodal in this sense. If we want to include distributions whose support is only a part of the real line, like the beta or the gamma distributions, it becomes evident that one will have to allow the density $f(x)$ to be just *nonincreasing* as x goes away from the mode v. Furthermore, densities are not everywhere unique. Thus, it is desirable to have a definition in terms of distribution functions. Such a definition was given by Khintchine (1938) and is stated as follows.

Definition 1.1. A real random variable X or its distribution function F is called *unimodal* about a *mode* (or *vertex*) v if F is convex on $(-\infty, v)$ and concave on (v, ∞).

Since the definition involves convexity, standard results on convex functions can be used to prove some properties of unimodal distributions. In particular, the following proposition is useful; see Royden (1968), p. 109.

Proposition. *Let φ be a convex function on (a, b). Then φ is absolutely continuous on every closed subinterval of (a, b). The right and left derivatives of φ exist at each point of (a, b) and are equal to each other except on a countable set. The left and right derivatives are nondecreasing functions and at each point, the left derivative is less than or equal to the right derivative.*

We now note a few simple consequences of Definition 1.1.

(i) If F is unimodal about v, then apart from a possible mass at v, F is absolutely continuous.
(ii) If F is unimodal about v, then the left and right derivatives of F exist everywhere except possibly at v.
(iii) If F is absolutely continuous, then the unimodality of F about v is equivalent to the existence of a density f, which is nondecreasing on $(-\infty, v)$ and nonincreasing on (v, ∞). This is the definition usually adopted in textbooks and mentioned at the beginning of this section.
(iv) The distribution degenerate at v is unimodal about v. Such degenerate distributions are the only discrete distributions which are unimodal according to Definition 1.1. In Chapter 3, unimodality of discrete distributions will be discussed through a different definition.
(v) If F corresponds to the uniform distribution on an interval (a, b), then F is unimodal about every v in $[a, b]$. Thus a unimodal distribution may have several modes. In general, if F is unimodal, then its modes form a bounded closed interval.
(vi) If F_1 and F_2 are both unimodal about the *same* mode v, then $\alpha F_1 + (1-\alpha)F_2$ is also unimodal about v for every $\alpha \in [0, 1]$. This result clearly extends to mixtures involving more than two components.

Under minimal measurability assumptions the result also extends to "generalized mixtures" $F(x)$ of the type

$$F(x) = \int_T F_t(x)\, dQ(t),$$

where $\{F_t, t \in T\}$ is a family of unimodal distribution functions having the same mode and Q is a probability measure on T.

The following theorem gives an important property of unimodal distributions.

Theorem 1.1. *The class of all unimodal distributions on R is closed under weak limits. The same holds for the class of all distributions which are unimodal about a fixed mode* v.

Proof. Let F_n be unimodal about v_n and let $F_n \to F$ weakly. Let v be any limit point of $\{v_n\}$. We show that F is unimodal about v. This will establish both the assertions of the theorem.

Let x_1, x_2 be continuity points of F in the interval $(-\infty, v)$. Then $\alpha x_1 + (1-\alpha)x_2$ is also a continuity point of F for almost all $\alpha \in (0, 1)$. By going to a subsequence, if necessary, we may assume that $v_n \to v$. Assuming $x_1 < x_2 < v$, we see that F_n is convex on $(-\infty, x_2]$ for all sufficiently large n. Therefore, for almost all $\alpha \in (0, 1)$,

$$\begin{aligned} F[\alpha x_1 + (1-\alpha)x_2] &= \lim_{n\to\infty} F_n[\alpha x_1 + (1-\alpha)x_2] \\ &\le \lim_{n\to\infty} [\alpha F_n(x_1) + (1-\alpha)F_n(x_2)] \\ &= \alpha F(x_1) + (1-\alpha)F(x_2). \end{aligned}$$

Now, by the right continuity of F, we see that

$$F[\alpha x_1 + (1-\alpha)x_2] \le \alpha F(x_1) + (1-\alpha)F(x_2)$$

for all $x_1 < x_2 < v$ and $\alpha \in [0, 1]$. This means that F is convex on $(-\infty, v)$. Similarly F is concave on (v, ∞). Finally, v cannot be infinite. For, if $v = \pm\infty$, then F would be convex or concave on the whole real line. But such an F would be either constant or unbounded and hence cannot be a distribution function. Thus v is finite and F is unimodal about v. ■

1.2. The Convex Structure of the Set of Unimodal Distributions

Let $\mathscr{U}$ denote the set of all distribution functions on R which are unimodal about a mode v. It was seen in the last section [see property (vi) and Theorem 1.1] that $\mathscr{U}$ is convex under mixtures and closed under weak limits. It turns out that the convex set $\mathscr{U}$ has extreme points and $\mathscr{U}$ is the closed convex hull of the set of these extreme points. This is, of course, a result of the Krein–Milman type and it can be further strengthened to yield a Choquet-type representation for unimodal distributions. These facts are proved in Theorem 1.2 below.

We take $v=0$ because this involves no loss of generality. The symbol W will denote the uniform distribution function on (0, 1). Similarly W_a will denote the uniform distribution function on $(0, a)$ or $(a, 0)$ according as $a>0$ or $a<0$. Naturally, W_0 will correspond to the point mass at 0. As in Section 1.1, we use the topology of weak convergence.

Theorem 1.2. *The closed convex hull $\mathscr{C}$ of the set $\{W_a, 0<|a|<\infty\}$ coincides with the set $\mathscr{U}$ of all distribution functions on R which are unimodal about* 0. *Moreover every distribution function in $\mathscr{U}$ is a generalized mixture of the distribution functions W_a, $-\infty<a<\infty$.*

Proof. The distribution functions W_a, $-\infty<a<\infty$ are all in $\mathscr{U}$. Since $\mathscr{U}$ is closed and convex, the closed convex hull $\mathscr{C}$ of $\{W_a, 0<|a|<\infty\}$ must be a subset of $\mathscr{U}$.

Conversely, suppose F is unimodal about 0. Then F is a convex combination of three distribution functions F_0, F_1 and F_2, where F_0 corresponds to the point mass at zero and F_1, F_2 are continuous distribution functions such that $F_1(0)=0$ and $F_2(0)=1$. We note that F_1 is concave on $(0,\infty)$ and that F_2 is convex on $(-\infty, 0)$. Let n be a *fixed* positive integer. For $m=0, 1, 2, \ldots,$ let $a_{nm}=m/n$. Let G_n be the distribution function which agrees with F_1 at the points a_{nm} and which is linear between the successive a_{nm}'s. Then G_n can be expressed as an infinite convex combination of uniform distribution functions. In symbols,

$$G_n=\sum_{m=1}^{\infty}\alpha_{nm}W_{a_{nm}}, \tag{1.1}$$

where the weights α_{nm} can be obtained by equating the derivatives of both sides of (1.1) at points $x\in(a_{n,i-1}, a_{ni})$. For $i\geqslant 1$, we get

$$\sum_{m=i}^{\infty}\frac{\alpha_{nm}}{a_{nm}}=\frac{F_1(a_{ni})-F_1(a_{n,i-1})}{a_{ni}-a_{n,i-1}}. \tag{1.2}$$

The concavity of F_1 shows that the right side of (1.2) is nonincreasing in i and so the α_{nm} are nonnegative. Multiplying both sides of (1.2) by $(a_{ni} - a_{n,i-1})$ and summing over i, one easily shows that $\sum_{m=1}^{\infty} \alpha_{nm} = 1$. Thus $G_n \in \mathscr{C}$. But $G_n \to F_1$ weakly as $n \to \infty$. Therefore $F_1 \in \mathscr{C}$. Similarly $F_2 \in \mathscr{C}$. Finally $F_0 \in \mathscr{C}$ because it is the weak limit of W_{b_n} with $b_n \to 0$. This shows that $F \in \mathscr{C}$ and completes the proof of the first assertion.

To prove the second assertion, consider equation (1.1) again. Let H_n be the distribution function on R which puts mass α_{nm} at the point a_{nm}, $m \geqslant 1$. Then (1.1) can be written as

$$G_n(x) = \int_{0-}^{\infty} W_z(x)\, dH_n(z), \qquad x \in \mathbf{R}.$$

Let $x > 0$ be fixed. By Helly's theorem, there is a subsequence $\{n_k\}$ and a nondecreasing function $H^{(1)}$ on R such that $H_{n_k}(z) \to H^{(1)}(z)$ at all continuity points z of $H^{(1)}$. Further $W_z(x) = x/z$ for $z > x$. Therefore $W_z(z) \to 0$ as $z \to \infty$. We also know that $G_n(x) \to F_1(x)$ for all x. Thus,

$$F_1(x) = \int_{0-}^{\infty} W_z(x)\, dH^{(1)}(z).$$

Now, by the monotone convergence theorem,

$$1 = F_1(\infty) = \int_{0^-}^{\infty} W_z(\infty)\, dH^{(1)}(z) = \int_{0^-}^{\infty} dH^{(1)}(z).$$

Therefore, $H^{(1)}$ is a distribution function. Since it is clear that $H^{(1)}(z) = 0$ for $z < 0$, we can write

$$F_1(x) = \int_{-\infty}^{\infty} W_z(x)\, dH^{(1)}(z).$$

This last formula holds trivially for $x < 0$ because $W_z(x) = 0$ for $z \geq 0$ and $H^{(1)}$ puts zero mass on $(-\infty, 0)$. We can treat F_2 in a similar fashion and obtain a distribution function $H^{(2)}$ with $H^{(2)}(0+) = 1$. Finally,

$$F_0(x) = \int_{-\infty}^{\infty} W_z(x)\, dH^{(0)}(z),$$

where $H^{(0)}$ is the degenerate distribution function at 0. Combining everything, we get a distribution function H on $(-\infty, \infty)$ such that

$$F(x) = \int_{-\infty}^{\infty} W_z(x)\, dH(z). \tag{1.3}$$

Thus F is a generalized mixture of the distributions W_z, $z \in R$. The proof of the theorem is now complete. ■

The Choquet representation (1.3) has a very interesting and useful probabilistic interpretation. This interpretation is due to Shepp (1962) and is given in the next theorem.

Theorem 1.3. *A distribution function F on R is unimodal about 0 if, and only if, there exist independent random variables U and Z such that U is uniform on $(0, 1)$ and the product UZ has distribution function F.*

Proof. Suppose F is unimodal about 0. By Theorem 1.2, F has the representation (1.3). Suppose Z is a random variable with distribution function H. Let U be independent of Z and be uniform on $(0, 1)$. Now from (1.3), we get

$$1 - F(x) = \int_{-\infty}^{\infty} [1 - W_z(x)]\, dH(z). \tag{1.4}$$

Suppose $x > 0$. Then $W_z(x) = 1$ for $z \leq x$ and $W_z(x) = x/z = W(x/z) = P(U \leq x/z)$ for $z > x$. Therefore, (1.4) shows that

$$\begin{aligned} 1 - F(x) &= \int_{x}^{\infty} [1 - W_z(x)]\, dH(z) \\ &= \int_{x}^{\infty} P\left[U > \frac{x}{z}\right] dH(z) \\ &= P[UZ > x]. \end{aligned}$$

Similarly, we show that, for $x < 0$,

$$F(x) = P[UZ \leq x].$$

It follows that UZ has distribution function F. The "only if" part of the theorem is thus proved.

Conversely, suppose U and Z are as in the statement of the theorem and let Z have an arbitrary distribution function H. Suppose further that UZ has distribution function F. Then, for $x > 0$, we reverse the above steps to obtain

$$\begin{aligned} 1 - F(x) &= P[UZ > x] \\ &= \int_{x}^{\infty} [1 - W_z(x)]\, dH(z) = \int_{-\infty}^{\infty} [1 - W_z(x)]\, dH(z). \end{aligned}$$

Similarly, for $x < 0$,

$$F(x) = \int_{-\infty}^{\infty} W_z(x)\, dH(z).$$

We therefore conclude that F is a generalized mixture of the distribution functions W_z. Since each W_z is unimodal about 0, we see that F is also unimodal about 0. The proof of the theorem is now complete. ■

Corollary. *A distribution with characteristic function φ is unimodal about* 0 *if, and only if, there is a characteristic function ψ such that*

$$\varphi(t) = \int_0^1 \psi(tu)\, du, \qquad t \in R. \tag{1.5}$$

Proof. If φ and ψ, respectively denote the characteristic functions of F and Z, then the corollary is just a restatement of Theorem 1.3 in terms of characteristic functions. ■

Remark. The above corollary was proved by Khintchine (1938) who identified the distribution function H in (1.4) in a different way (to be discussed in the next remark). While the interpretation of H as the distribution function of the random variable Z was not given by Khintchine, we know that Theorem 1.3 is equivalent to its corollary. For this reason, we refer to the results of Theorems 1.2 and 1.3 as *Khintchine's Representation* for unimodal distributions. A good discussion of the ideas of Theorems 1.2 and 1.3 is given by Olshen and Savage (1970).

Remark. Suppose F is unimodal about 0. Then the mixing distribution function H given in the representation (1.3) is uniquely determined by F. There are two ways of proving this claim. The simpler way is to use (1.5). A change of variables $v = tu$ shows that

$$t\varphi(t) = \int_0^t \psi(v)\, dv.$$

Therefore

$$\frac{d}{dt}\,[t\varphi(t)] = \psi(t),$$

and so φ determines ψ. This calculation also shows that the characteristic function of a unimodal distribution is differentiable on $R - \{0\}$.

Another approach to proving the uniqueness of H is to use (1.3). The unimodality of F implies that F is absolutely continuous on $R - \{0\}$. So let

$f(x)$ be a density of $F(x)$ on $R - \{0\}$. Proceeding formally, we get from (1.3)

$$f(x) = \int_x^\infty \frac{1}{z}\, dH(z), \qquad x > 0.$$

This last formula shows that the Radon–Nikodym derivative of H w.r.t. f is $-x$. Therefore

$$1 - H(x) = \int_x^\infty dH(x) = -\int_x^\infty z\, df(z).$$

Thus

$$H(x) = 1 + \int_x^\infty z\, df(z), \qquad x > 0.$$

Integrating by parts and noting that $xf(x) \to 0$ as $x \to \infty$, we get, for all continuity points x of f on $(0, \infty)$,

$$H(x) = F(x) - xf(x).$$

This last formula is also valid for all continuity points x on f on $(-\infty, 0)$. At the remaining values of x, we determine H by right continuity. Thus, we again see that F determines H. In particular, if $f'(x) = F''(x)$ exists and is continuous everywhere, then H has the density h given by

$$h(x) = -xf'(x).$$

In summary, the representation (1.3) for a unimodal distribution function is unique.

As before, let $\mathscr{U}$ denote the set of all distribution functions on R which are unimodal about 0. Let $\mathscr{H}$ be the set of all distribution functions on R. If $H \in \mathscr{H}$, then (1.3) gives a distribution function $F \in \mathscr{U}$ and we can write $F = \theta(H)$. The map θ is clearly affine. That is, for $0 \le \alpha \le 1$ we have

$$\theta[\alpha H_1 + (1-\alpha)H_2] = \alpha\theta(H_1) + (1-\alpha)\theta(H_2).$$

Theorem 1.3 and the subsequent remarks show that θ is one-to-one and onto $\mathscr{U}$. Thus, under the map θ, an extreme point of $\mathscr{H}$ must go into an extreme point of $\mathscr{U}$. Moreover, all the extreme points of $\mathscr{U}$ must be obtained in this fashion. Now, it is well known that the extreme points of $\mathscr{H}$ are the degenerate distribution functions; see Theorem A.5 in the Appendix. But if $H \in \mathscr{H}$ is degenerate at z, then $\theta(H) = W_z$. Thus, the extreme points of $\mathscr{U}$ are the uniform distributions W_z, $z \in R$ and (1.3) expresses a unimodal distribution as an integral over the set of extreme points of $\mathscr{U}$.

We now mention a simple application of Khintchine's representation. Suppose X is a unimodal random variable with mean μ, variance σ^2 and mode M. Assume first that $M = 0$. Then X is distributed as UZ where U,Z are as stated in Theorem 1.3. Let Z have mean v and variance τ^2. Since $E(U) = (\frac{1}{2})$ and $E(U^2) = (\frac{1}{3})$, we easily get

$$v = 2\mu \quad \text{and} \quad v^2 + \tau^2 = 3(\mu^2 + \sigma^2).$$

Eliminating v, we get $\mu^2 + \tau^2 = 3\sigma^2$ and so $\mu^2 \leqslant 3\sigma^2$. For a general mode M, we therefore get the inequality

$$(\mu - M)^2 \leq 3\sigma^2. \tag{1.6}$$

The inequality (1.6) was given by Johnson and Rogers (1951) and rediscovered by Vysochanskii and Petunin (1982). It is clear that equality holds in (1.6) if, and only if, Z is degenerate, that is, if, and only if, X is uniformly distributed.

We now present another useful characterization of unimodality due to Olshen and Savage (1970).

Theorem 1.4. *A random variable X has a unimodal distribution about 0 if, and only if, $tE[g(tX)]$ is nondecreasing in $t > 0$ for every bounded, nonnegative, Borel-measurable function g.*

Proof. Let X have distribution function F and let F be unimodal about 0. Then, by Theorem 1.3, F is also the distribution function of UZ, where U is uniform on (0, 1) and Z is independent of U. Therefore,

$$\begin{aligned} tE[g(tX)] &= tE[g(tUZ)] \\ &= t\int_0^1 E[g(tuZ)]\,du \\ &= \int_0^t E[g(vZ)]\,dv. \end{aligned}$$

Clearly, the last integral is nondecreasing in t whenever g is bounded, nonnegative and measurable. This proves the "only if" part.

Conversely, suppose that $tE[g(tX)]$ is nondecreasing in $t > 0$ for every nonnegative, bounded, measurable g. Suppose $0 < a < b$ and let g be the indicator of the interval $(a, b]$. Then $t[F(b/t) - F(a/t)]$ is nondecreasing in $t > 0$. Write $u = 1/t$. Then

$$K(u) = \frac{F(bu) - F(au)}{u}$$

is nonincreasing in $u > 0$. First, F must be continuous on $(0, \infty)$. For, if $x_0 > 0$ is a discontinuity point of F, then taking a to be a continuity point of F and $b = x_0$, we see that K will increase as u increases through 1. Now we claim that F is concave on $(0, \infty)$. If this is not the case, then the continuity of F shows that we can find a and b in $(0, \infty)$ such that, $a < b$ and for all $t \in (0, 1)$,

$$F[ta + (1 - t)b] < tF(a) + (1 - t)F(b). \tag{1.7}$$

Let $c = \sqrt{ab}$. Then $(c/a) = (b/c) = u$, say. Clearly $u > 1$. Therefore the non-increasing character of K shows that

$$F(c) - F(a) \geq \frac{F(cu) - F(au)}{u} = \frac{F(b) - F(c)}{u}$$

or

$$F(c) \geq \frac{uF(a) + F(b)}{1 + u}. \tag{1.8}$$

Since $c = (ua + b)/(1 + u)$, (1.8) contradicts (1.7). Thus, F is concave on $(0, \infty)$. Similarly F is convex on $(-\infty, 0)$. This shows that F is unimodal and completes the proof of the theorem. ■

Remark. If we put $a = 1$ in the definition of K above, we see that the function F satisfies the condition that $[F(bu) - F(u)]/u$ is nonincreasing in $u > 0$ for every $b > 1$. Such functions are called pre-concave by Olshen and Savage (1970). The proof above that a nondecreasing pre-concave functon is continuous and concave is taken, with minor changes, from their paper.

For distributions on R, the concept of symmetry is easily defined. A distribution function F will be called symmetric if $F(x) + F(-x) = 1$ for all continuity points x of F. If a symmetric distribution function F is unimodal, then 0 is a mode of F. Further, the proofs given in this section can be easily modified to establish the following theorem.

Theorem 1.5. *Given a random variable X with a distribution function F, the following statements are equivalent to one another.*

(a) *F is symmetric and unimodal.*
(b) *F belongs to the closed convex hull of the set of all uniform distributions on symmetric intervals $(-a, a)$ with $a > 0$.*
(c) *There exist independent random variables U and Z such that U is uniform on $(0, 1)$, Z is symmetric and the product UZ has distribution function F.*
(d) *There exist independent random variables V and Z' such that V is uniform on $(-1, 1)$, Z' is nonnegative, and the product VZ' has distribution function F.*

(e) *For every function g which is bounded, nonnegative and Borel-measurable,* $|t|\,E[g(tX)]$ *is symmetric in t and nondecreasing on* $(0, \infty)$.

Remark. If F is symmetric, unimodal, and if F has a density f which is continuously differentiable, then, as observed before, $h(x) = -xf'(x)$ is the density of Z in statement (c) above. The positive random variable Z' of statement (d) has the density $2h(x)$, $0 < x < \infty$.

1.3. Convolution Properties of Unimodal Distributions

Convolution is an important operation in probability and statistics. Therefore, it is important to know whether the property of unimodality is preserved under the operation of convolution.

Let F_1 and F_2 be unimodal about 0. By using the representation (1.3), we can write

$$F_i(x) = \int_{-\infty}^{\infty} W_z(x)\, dH_i(z), \qquad i = 1, 2.$$

Therefore, the convolution $F_1 * F_2$ is given by

$$(F_1 * F_2) = \int_{-\infty}^{\infty} \int_{-\infty}^{\infty} (W_{z_1} * W_{z_2})\, dH_1(z_1)\, dH_2(z_2).$$

Now while $W_{z_1} * W_{z_2}$ is unimodal, its mode depends on the pair (z_1, z_2). For instance, the unique mode of $W_z * W_z$ is z. Thus, $F_1 * F_2$ is obtained as a mixture of distributions with different modes. Consequently, $F_1 * F_2$ may not be unimodal. While this last statement needs to be substantiated by an example, it should be mentioned that it was believed at first that convolutions of unimodal distributions are again unimodal. The first counterexample to this belief was given by Chung [see the discussion on pp. 254–255 of Gnedenko and Kolmogorov (1954)]. Another example is given by Feller (1971, p. 168). We give here a simple example which avoids detailed calculations.

Example 1.1. Suppose D_a is the distribution function corresponding to the point mass at a. As before, let W denote the uniform distribution function on (0, 1). Consider the distribution function $F_a = \frac{1}{2}[W + D_a]$ with $0 \le a \le 1$. The convolution $F_0 * F_0$ puts mass $\frac{1}{4}$ at 0 and has a density on (0, 2) with a unique maximum at 1. Consequently $F_0 * F_0$ is not unimodal. More generally, we show that $F_a * F_a$ is not unimodal unless $a = \frac{1}{2}$. Clearly, F_a is unimodal about a. Further,

$$4(F_a * F_a)(x) = D_{2a}(x) + 2W(x - a) + (W * W)(x).$$

If possible, suppose that $(F_a * F_a)$ is unimodal. Since $(F_a * F_a)$ has a mass point at $2a$, its unique mode must be at $2a$. Now the density of $(W * W)$ is triangular with a unique mode at 1, whereas the density of $W(x-a)$ is constant on $(a, a+1)$. Therefore the unique mode of $(F_a * F_a)$ must also equal 1. Thus $(F_a * F_a)$ is unimodal if, and only if, $2a = 1$. More generally, one easily sees that $F_a * F_b$ is unimodal if, and only if, $a + b = 1$.

We present another example which requires more calculations but has some interesting features. This example is due to Wolfe (1971).

Example 1.2. Let G be the distribution function of the exponential distribution with mean 1. That is, $G(x) = 1 - e^{-x}$, $x \geq 0$. Let G_n be the n-fold convolution of G with itself. We interpret G_0 as the degenerate distribution at 0. For $a > 0$, write $\alpha_n = a^n/(n!)$ and let

$$H_a = e^{-a} \sum_{n=0}^{\infty} \alpha_n G_n.$$

Then H_a is the distribution function of a nonnegative random variable. If ψ and φ_a denote the characteristic functions of G and H_a, respectively, then

$$\varphi_a = \exp[a(\psi - 1)].$$

Consequently, $H_a * H_{a'} = H_{a+a'}$. We also note that H_a puts mass e^{-a} at 0. Therefore, if H_a is unimodal then zero must be its unique mode. We claim that H_a is unimodal if, and only if, $0 < a \leq 2$. To see this, observe that, for $n \geq 1$, a density g_n of G_n is given by

$$g_n(x) = \frac{e^{-x} x^{n-1}}{(n-1)!}, \qquad x > 0.$$

Therefore, for $x > 0$, a density h_a of H_a is given by

$$e^a h_a(x) = \sum_{n=1}^{\infty} \alpha_n g_n(x).$$

Now $g_1'(x) = -g_1(x)$ and $g_n'(x) = g_{n-1}(x) - g_n(x)$ for $n \geq 2$. Therefore,

$$\begin{aligned} e^a h_a'(x) &= -\alpha_1 g_1(x) + \sum_{n=2}^{\infty} \alpha_n [g_{n-1}(x) - g_n(x)] \\ &= \sum_{n=1}^{\infty} (\alpha_{n+1} - \alpha_n) g_n(x) \\ &= \sum_{n=1}^{\infty} \frac{\alpha_{n+1}(a - n - 1) g_n(x)}{a}. \end{aligned}$$

Now $a \le 2 \Rightarrow a \le (n+1)$ for all $n \ge 1 \Rightarrow h_a'(x) \le 0$ for $x > 0$. Thus H_a is unimodal for $a \le 2$. Now $g_1(0+) = 1$ and $g_n(0+) = 0$ for $n \ge 2$. Therefore $h_a'(0+) = ae^{-a}(a-2)/2$. It follows that, for $a > 2$, H_a is not unimodal about 0. But we have noted that the only point about which H_a can be unimodal is zero. Thus H_a is not unimodal if $a > 2$. Now if $1 < a \le 2$, then H_a is unimodal but the convolution $H_a * H_a = H_{2a}$ is not unimodal. We note in passing that each H_a is infinitely divisible. Thus, there exist infinitely divisible unimodal distributions whose convolution is not unimodal.

Remark. Some results from the preceding example can be extended to a more general situation. Suppose G is a distribution function with a nonzero mean μ, variance σ^2 and characteristic function ψ. For $a > 0$, let H_a be the distribution function with characteristic function φ_a given by

$$\varphi_a = \exp[a(\psi - 1)].$$

Then H_a has mean $a\mu$ and variance $a(\mu^2 + \sigma^2)$. Again, H_a puts mass e^{-a} at 0. Therefore, if H_a is unimodal, then its unique mode must be zero and so, by (1.6)

$$(a\mu)^2 \le 3a(\mu^2 + \sigma^2)$$

or

$$a \le \frac{3(\mu^2 + \sigma^2)}{\mu^2}.$$

Thus H_a is not unimodal if $a > 3(\mu^2 + \sigma^2)/\mu^2$. This simpler proof of the nonunimodality of H_a for all sufficiently large a is due to Wolfe (1978a).

We now present some positive results on the convolutions of unimodal distributions. It is easy to see that the convolution of two *symmetric uniform* distributions is unimodal with mode zero. This fact and Theorem 1.5 yield the following theorem due to Wintner (1938).

Theorem 1.6. *The convolution of two symmetric unimodal distributions on R is unimodal.*

Consider a nonnegative function g on R which is not necessarily a probability density. If g is nondecreasing on $(-\infty, v)$ and nonincreasng on (v, ∞), one may say that g is unimodal about v. A symmetric unimodal function is defined in an obvious way. Theorem 1.6 has the following immediate corollary, which will form the basis of an important definition in the next chapter.

Corollary. *A distribution F on R is symmetric unimodal if, and only if, for every $\delta > 0$, the function*

$$g(x) = F(x + \delta) - F(x + \delta)$$

is symmetric unimodal.

Proof. Observe that $g/(2\delta)$ is just the density of the convolution of F with the uniform distribution on $(-\delta, \delta)$. If F is symmetric unimodal, then g is symmetric unimodal by Theorem 1.6. Conversely, if g is symmetric unimodal for every $\delta > 0$ then F must be symmetric unimodal because it is the weak limit, as $\delta \to 0$, of the distribution with density $g/(2\delta)$. ■

While we have seen that the convolution of two unimodal distributions is not always unimodal, there are situations in which the convolution *is* unimodal even if the distributions are not symmetric. The next theorem identifies one such situation.

Theorem 1.7. *A distribution function F is unimodal if, and only if, the convolution $F * W_z$ is unimodal for every z.*

Proof. Let F be unimodal and let v be a mode of F. Write $h(x) = F(x + z) - F(x)$. We have to show that h is a unimodal function. Since F is convex on $(-\infty, v)$, h is nondecreasing on $(-\infty, v - z)$. Similarly h is nonincreasing on $[v, \infty)$. So we need only consider the behavior of h on $(v - z, v)$. Let F' denote the left or right derivative of F on the set $R - \{v\}$. Let $y = \inf\{x \in (v - z, v): F'(x) > F'(x + z)\}$. Then, for $v - z < x < x' < y$,

$$\begin{aligned} h(x') - h(x) &= F(x' + z) - F(x + z) - [F(x') - F(x)] \\ &= \int_x^{x'} [F'(t + z) - F'(t)]\, dt \end{aligned} \tag{1.9}$$

The right side of (1.9) is nonnegative because, in the range of integration, $t < y$ and so $F'(t) \leq F'(t + z)$. It follows that h is nondecreasing on $(v - z, y)$. Similarly h is nonincreasing on (y, v). This shows that $F * W_z$ is unimodal.

Conversely, if $F * W_z$ is unimodal for every z, then we can let $z \to 0$ to conclude that F is unimodal. The theorem is thus proved. ■

Remark. Let F be unimodal and, for fixed z, write $h(x) = F(x + z) - F(x)$. The proof of Theorem 1.7 shows that for $h(x)$ to be maximized, the interval $[x, x + z]$ must contain a mode. Minor modifications of the proof show that, if x_0 is not a mode of F, then we can find an interval I_1 containing x_0 and an interval I_2 containing a mode such that I_1 and I_2 have the same length and F assigns a strictly larger mass to I_2 than to I_1.

Another positive result on convolutions is that symmetrizations of unimodal distributions are unimodal. This result was proved by Hodges and Lehmann (1954). Their proof is somewhat complicated. We use Khintchine's representation (Theorems 1.2 and 1.3) to give a simpler proof. This proof is taken from Dharmadhikari and Joag-dev (1983a). First, we need a lemma.

Lemma 1.1. *Let $c_1, \ldots, c_m$ and $d_1, \ldots, d_n$ be positive real numbers. Then*

$$\sum_{1 \le i \le j \le m} \sum c_i c_j + \sum_{1 \le k \le l \le n} \sum d_k d_l > \sum_{i=1}^{m} \sum_{k=1}^{n} c_i d_k. \tag{1.10}$$

Proof. The left side of (1.10) equals

$$\tfrac{1}{2}[(\sum c_i)^2 + \sum c_i^2 + (\sum d_k)^2 + \sum d_k^2]$$

$$= \tfrac{1}{2}[\sum c_i - \sum d_k]^2 + \sum_i \sum_k c_i d_k + \tfrac{1}{2}[\sum c_i^2 + \sum d_k^2],$$

which clearly implies the required result. ■

Theorem 1.8. *Let X_1, X_2 be independent random variables having the same unimodal distribution. Then $X_1 - X_2$ is unimodal.*

Proof. Let F be the common distribution function of X_1 and X_2. It is sufficient to prove the theorem in the case where F is a finite mixture of uniform distributions on intervals with 0 as one end point.

Denote by u_a (respectively, v_b) the density of the uniform distribution on $(0, a)$ [respectively, $(-b, 0)$], where $a > 0$ and $b > 0$. Suppose the density f of F is

$$f = \sum_{i=1}^{m} \alpha_i u_{a_i} + \sum_{k=1}^{n} \beta_k v_{b_k},$$

with $\alpha_i > 0$, $\beta_k > 0$, and $\sum \alpha_i + \sum \beta_k = 1$. The density of $-X_2$ is then

$$\sum_{i=1}^{m} \alpha_i v_{a_i} + \sum_{k=1}^{n} \beta_k u_{b_k}.$$

Therefore the density g of $X_1 - X_2$ is

$$g(x) = \sum_i \sum_j \alpha_i \alpha_j (u_{a_i} * v_{a_j}) + \sum_k \sum_l \beta_k \beta_l (v_{b_k} * u_{b_l})$$

$$+ \sum_i \sum_k \alpha_i \beta_k [(u_{a_i} * u_{b_k}) + (v_{a_i} * v_{b_k})]. \tag{1.11}$$

Now it is easy to check (see Figures 1.1–1.3) that

$$\left.\begin{aligned}
&(u_a * v_b)'(x) \le 0 && \text{if } x > 0,\\
&(u_a * v_b)'(x) = -(ab)^{-1} && \text{if } 0 < x < a < b,\\
&(u_a * u_b)'(x) = (ab)^{-1} && \text{if } 0 < x < \min(a,b),\\
\text{and}\quad &(u_a * u_b)'(x) \le 0 && \text{if } x > \min(a,b).
\end{aligned}\right\} \tag{1.12}$$

Assume now that $a_1 < a_2 < \cdots < a_m$ and $b_1 < b_2 < \cdots < b_n$. Write $a_0 = b_0 = 0$ and $a_{m+1} = b_{n+1} = \infty$. Suppose $x > 0$ is such that x does not equal any one of the a_i's or the b_j's. Then we can find unique integers r and s such that $a_r < x < a_{r+1}$ and $b_s < x < b_{s+1}$. Using (1.12) and the fact that $(v_a * v_b)'(x) = 0$, we get

$$\begin{aligned}
g'(x) \le &- \sum_{r+1 \le i \le j \le m} \sum \frac{\alpha_i \alpha_j}{a_i a_j}\\
&- \sum_{s+1 \le k \le l \le n} \sum \frac{\beta_k \beta_l}{b_k b_l}\\
&+ \sum_{i \ge r+1} \sum_{k \ge s+1} \frac{\alpha_i \beta_k}{a_i b_k}.
\end{aligned}$$

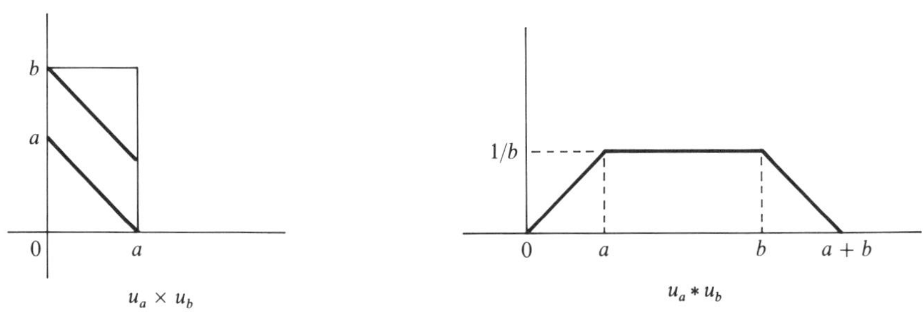

Figure 1.1

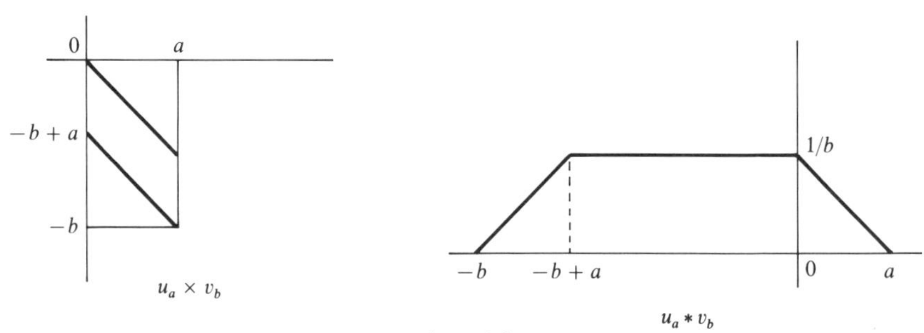

Figure 1.2

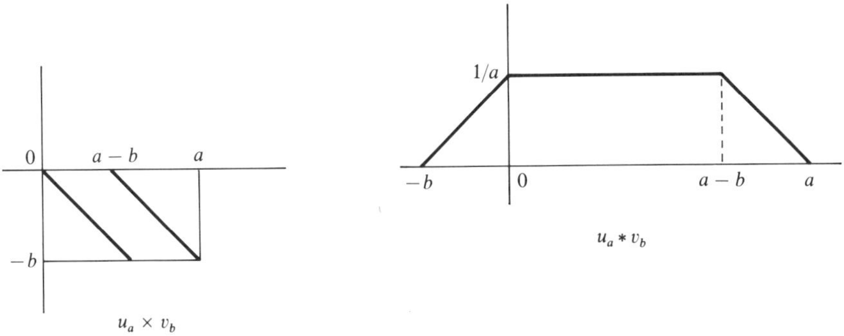

Figure 1.3

Writing $c_i = \alpha_i/a_i$ and $d_k = \beta_k/b_k$, we see from Lemma 1.1 that $g'(x) \le 0$. This proves that g is unimodal and completes the proof of the theorem. ■

1.4. Strong Unimodality

Since convolutions of unimodal distributions are, in general, not unimodal, Ibragimov (1956) called a distribution function G *strongly unimodal* if the convolution $G * F$ is unimodal for every unimodal F. It follows immediately from this definition that

(a) the set of all strongly unimodal distributions is closed under convolutions and weak limits;
(b) a strongly unimodal distribution must be unimodal; and
(c) the degenerate distributions are strongly unimodal.

The next theorem was proved by Ibragimov. The proof given here is different from the one given by Ibragimov (1956).

Theorem 1.9. *A nondegenerate, strongly unimodal distribution is continuous.*

Proof. Suppose F is not continuous and not degenerate. We show that F is not strongly unimodal. If F is not unimodal, then the required result is immediate. So, assume that F is unimodal. Then F has exactly one point of discontinuity, namely, its mode, which we denote by v. Write $F = \alpha G + (1-\alpha)D_v$, where $(1-\alpha)$ is the mass at v and G is the continuous part of F. Clearly $0 < \alpha < 1$. Let g be right or left derivative of G on the set $R - \{v\}$.

Choose a number δ satisfying $0 < \delta < \max\{g(v_+), g(v_-)\}$. Let

$$c = \inf\{x > v : g(x) < \delta\}$$
$$d = \sup\{x < v : g(x) < \delta\}.$$

Then $c \le v \le d$ and $c \ne d$. So either $c < v$ or $v < d$. Without loss of generality, assume that $v < d$. Let $b > \max\{d - v, v - c\}$ and consider the distribution function $H = \beta W_b + (1 - \beta)D_0$. We show that $F * H$ is not unimodal if β is sufficiently close to 1.

If possible, suppose that $F * H$ is unimodal. Since $F * H$ has a mass point at v, the unique mode of $F * H$ must be at v and $(F * H)'$ must be nonincreasing on (v, ∞). Now

$$(F * H)(x) = \alpha\beta(G * W_b)(x) + \alpha(1 - \beta)G(x) + (1 - \alpha)\beta W_b(x - v) + (1 - \alpha)(1 - \beta)D_v.$$

Therefore, for $v < x < b$,

$$(F * H)'(x) = \alpha\beta(G * W_b)'(x) + \alpha(1 - \beta)g(x) + \frac{(1 - \alpha)\beta}{b}. \tag{1.13}$$

Let $\varepsilon_1 = b + c - v$, $\varepsilon_2 = d - v$ and $\varepsilon = \min\{\varepsilon_1, \varepsilon_2\}$. Then from the definitions of c and d, one easily sees that $(G * W_b)'(x)$ is strictly increasing on $(v, v + \varepsilon)$. Therefore, we see from (1.13) that, if β is sufficiently close to 1, then $(F * H)'(v + \varepsilon) > (F * H)'(v_+)$. More precisely, if $k(x) = (G * W_b)'(x)$, then $(F * H)'(v + \varepsilon) > (F * \mathrm{H})'(v_+)$, as soon as $\beta > \beta_0$, where

$$\beta_0 = \frac{g(v_+) - g(v + \varepsilon)}{g(v_+) - g(v + \varepsilon) + k(v + \varepsilon) - k(v)}.$$

Thus, for $\beta_0 < \beta < 1$, $(F * H)$ is not unimodal. But H is unimodal for all $\beta \in (0, 1)$. Therefore F cannot be strongly unimodal. The theorem is thus proved. ■

A nonnegative function g is called *logconcave* if $\log g$ is concave, or, equivalently, if

$$g[\theta x + (1 - \theta)y] \ge [g(x)]^\theta [g(y)]^{1-\theta},$$

for all x, y and all $\theta \in (0, 1)$. It is easy to see that a logconcave density is unimodal. Ibragimov (1956) proved that a nondegenerate distribution is strongly unimodal if, and only if, its density is logconcave. Our proof differs, in some details, from the one given by Ibragimov. In particular, we have utilized the ideas of Keilson and Gerber (1971) who have given the discrete analog of Ibragimov's result. The steps in the proof are as follows. If Ibragimov's characterization is to be valid, then the density of a continuous limit of distributions with logconcave densities must again be logconcave. This is done in Lemma 1.4, which requires the (fairly elementary) results of Lemmas 1.2 and 1.3. The proof of Theorem 1.10 is somewhat long. But most of the steps are straightforward, although the verification of the assertion

$$\text{strong unimodality} \Rightarrow \text{logconcave density}$$

requires some tedious calculations. We use right or left derivatives F', F_n', etc., which exist because of convexity.

Lemma 1.2. *Let F be unimodal about a. Then* $|x| > |a| + \varepsilon \Rightarrow F'(x) \leq 1/\varepsilon$.

Proof. If $x_0 \geq |a| + \varepsilon$ and $F'(x_0) > 1/\varepsilon$ then

$$\int_a^{x_0} F'(x)\, dx > \frac{1}{\varepsilon} \cdot \varepsilon = 1,$$

which is a contradiction. ■

Lemma 1.3. *Let* $\{F_n\}$ *be a sequence of unimodal distribution functions such that* $F_n \to F$ *weakly. Then there is a subsequence* $\{n_k\}$ *such that* $F_{n_k}'(x) \to F'(x)$ *almost everywhere.*

Proof. Let a_n be a mode of F_n and let F be unimodal about 0. By going to a subsequence, we may assume that $a_n \to 0$. Lemma 1.2 shows that, for every $b > 0$, the $F_n'(x)$ are uniformly bounded on $(-\infty, -b]$ and $[b, \infty)$. By Helly's theorem, we may therefore extract a subsequence $\{n_k\}$ such that $F_{n_k}' \to$ some $g(x)$ for almost every $x \neq 0$. Clearly g is nonincreasing on $(0, \infty)$ and nondecreasing on $(-\infty, 0)$. Let a, b be continuity points of F such that $[a, b]$ does not contain 0. Then

$$\begin{aligned} F(b) - F(a) &= \lim_{k\to\infty} \{F_{n_k}(b) - F_{n_k}(a)\} \\ &= \lim_{k\to\infty} \int_a^b F_{n_k}'(x)\, dx = \int_a^b g(x)\, dx. \end{aligned}$$

It follows that $F'(x) = g(x)$ almost everywhere. This proves the lemma. ■

Lemma 1.4. *Suppose* F_n *has support* $(-\infty, \infty)$ *and has a logconcave density* f_n. *Let* $F_n \to F$ *weakly, where F is continuous. Then F has a logconcave density.*

Proof. In this proof we replace a given sequence by a subsequence whenever convenient. Since f_n is logconcave, each F_n is unimodal and hence F is also unimodal. We may assume that 0 is a mode of F. Write $\psi = \log F'$ and $\psi_n = \log f_n$. By Lemma 1.3, $\psi_n(x) \to \psi(x)$ for almost all x. Since ψ_n is concave, we see that, for every fixed $\delta > 0$,

$$\psi(x) \geq \tfrac{1}{2}[\psi(x+\delta) + \psi(x-\delta)] \tag{1.14}$$

for almost all x. We want to show that this last result (1.14) holds for all x. Since ψ is monotone on $(-\infty, 0)$ and on $(0, \infty)$, the only discontinuities of ψ are jumps. If possible, suppose y_0 is such that $\psi(y_0+) - \psi(y_0-) = \alpha > 0$. Let $0 < 4\varepsilon < \alpha$. Choose $\delta > 0$ so that

$$y_0 < x < y_0 + 2\delta \Rightarrow |\psi(x) - \psi(y_0+)| < \varepsilon$$

and

$$y_0 - 2\delta < x < y_0 \Rightarrow |\psi(x) - \psi(y_0-)| < \varepsilon.$$

Let $x_1 \in (y_0 - \delta, y_0)$ be arbitrary. Then,

$$\psi(x_1) - \psi(x_1 - \delta) = [\psi(x_1) - \psi(y_0-)] + [\psi(y_0-) - \psi(x_1 - \delta)] < 2\varepsilon < \alpha - 2\varepsilon.$$

However,

$$\alpha - 2\varepsilon = [\psi(y_0+) - \varepsilon] - [\psi(y_0-) + \varepsilon] < \psi(x_1 + \delta) - \psi(x_1).$$

Thus, $2\psi(x_1) < \psi(x_1 - \delta) + \psi(x_1 + \delta)$ for all $x_1 \in (y_0 - \delta, y_0)$. This contradicts (1.14) and shows that ψ is continuous. The result (1.14) must therefore hold everywhere. This shows that ψ is concave. The lemma is thus proved. ■

We are now ready to present Ibragimov's characterization of strong unimodality.

Theorem 1.10. *A nondegenerate distribution function G is strongly unimodal if, and only if, G is continuous and its density g is logconcave (i.e., log g is concave).*

Proof. (a) Suppose that $\log g$ is concave. Assume also, for the moment, that g is never zero. We show that the convolution $(p * g)$ is unimodal for every smooth unimodal density p. Write $q = p * g$. Assume, without loss of generality, that p is unimodal about 0. We have

$$q(x) = (p * g)(x) = \int_{-\infty}^{\infty} p(x-y)g(y)\,dy.$$

Therefore,

$$q'(x) = \int_{-\infty}^{\infty} p'(x-y)g(y)\,dy$$
$$= \int_{-\infty}^{\infty} p'(y)g(x-y)\,dy. \tag{1.15}$$

Choose x, z in R such that $x < z$. Then (1.15) shows that

$$q'(z) = \int_{-\infty}^{\infty} p'(y)\,\frac{g(z-y)}{g(x-y)}\,g(x-y)\,dy. \tag{1.16}$$

Now g is logconcave. Therefore $g(x_0+\delta)/g(x_0)$ is nonincreasng in x_0 if $\delta > 0$ and nondecreasing in x_0 if $\delta < 0$. Therefore,

$$\frac{g(z-y)}{g(z)} \le \frac{g(x-y)}{g(x)} \qquad \text{if} \quad y \le 0$$

and

$$\frac{g(z-y)}{g(z)} \ge \frac{g(x-y)}{g(x)} \qquad \text{if} \quad y \ge 0.$$

But we also know that $p'(y) \ge 0$ for $y \le 0$ and $p'(y) \le 0$ for $y \ge 0$. Consequently (1.16) shows that

$$q'(z) \le \frac{g(z)}{g(x)} \int_{-\infty}^{\infty} p'(y)g(x-y)\,dy = \frac{g(z)}{g(x)}\,q'(x).$$

Thus $q'(x) \le 0 \Rightarrow q'(z) \le 0$ for $z > x$ and q is unimodal.

The condition that g is never zero can be easily removed. Suppose g has support $[a, b]$. For integers $m \ge 2/(b-a)$, let $\alpha > 0$ be larger than $(\log g)'(a + 1/m)$ and let $\beta < 0$ be smaller than $(\log g)'(b - 1/m)$. Define a density g_m as follows.

$$g_m(x) = \begin{cases} c \cdot g(x), & x \in \left(a + \dfrac{1}{m},\, b - \dfrac{1}{m}\right); \\[2ex] c \cdot g\left(a + \dfrac{1}{m}\right)\exp\left[\alpha\left(x - a - \dfrac{1}{m}\right)\right], & x < a + \dfrac{1}{m}; \\[2ex] c \cdot g\left(b - \dfrac{1}{m}\right)\exp\left[\beta\left(x - b + \dfrac{1}{m}\right)\right], & x > b - \dfrac{1}{m}; \end{cases}$$

where c is a normalizing constant. Then g_m is logconcave and never zero.

Therefore g_m is strongly unimodal by the above proof. Letting $m \to \infty$, we see that g is strongly unimodal. This proves the "if" part.

(b) We first prove the "only if" part in the special case where G has support R and a continuous derivative g. Assume that G is strongly unimodal. If possible, suppose there exist x_0 and δ in R such that

$$g^2(x_0) < g(x_0 + \delta)g(x_0 - \delta).$$

We find a unimodal density h such that $g * h$ is not unimodal. Without loss of generality, we may assume that $\delta = 1$. Define the density h as follows.

$$h(x) = \begin{cases} \dfrac{\beta}{N+1}, & -N < x \leq 0, \\[2ex] \dfrac{\beta+\gamma}{N+1}, & 0 < x \leq 1, \\[2ex] \dfrac{\gamma}{N+1}, & 1 < x \leq (N+1), \\[2ex] 0, & \text{otherwise.} \end{cases}$$

Here N is a nonnegative integer and β, γ are positive numbers such that $\beta + \gamma = 1$. We note that h is unimodal. If $k = g * h$, then

$$(N+1)k(x) = \gamma \int_{x-N-1}^{x-1} g(y)\, dy + (\beta + \gamma) \int_{x-1}^{x} g(y)\, dy + \beta \int_{x}^{x+N} g(y)\, dy.$$

Therefore, one easily gets

$$(N+1)k'(x) = \gamma g(x) - \beta g(x-1) + \beta g(x+N) - \gamma g(x-N-1). \quad (1.17)$$

Since $g^2(x_0) < g(x_0+1)g(x_0-1)$, we can find a $\beta \in (0, 1)$ such that

$$\frac{g(x_0)}{g(x_0-1)} < \frac{\beta}{1-\beta} < \frac{g(x_0+1)}{g(x_0)}.$$

For this value of β, γ is fixed by the condition $(\beta + \gamma) = 1$. Now (1.17) shows that

$$\lim_{N \to \infty} \frac{(N+1)k'(x_0)}{\gamma g(x_0-1)} = \frac{g(x_0)}{g(x_0-1)} - \frac{\beta}{1-\beta} < 0$$

and

$$\lim_{N \to \infty} \frac{(N+1)k'(x_0+1)}{\gamma g(x_0)} = \frac{g(x_0+1)}{g(x_0)} - \frac{\beta}{1-\beta} > 0.$$

Thus, for large N, k is not unimodal. This contradicts the strong unimodaility of G and so g must be logconcave.

(c) To complete the proof of the "only if" part we have to remove the smoothness and the support conditions on G. Let Φ denote the standard normal distribution function. Then Φ is strongly unimodal because it has a logconcave density. Let G be a nondegenerate strongly unimodal distribution function. Write $F_n = G * \Phi_n$, where $\Phi_n(x) = \Phi(nx)$ and let $f_n = F_n'$. Then F_n is strongly unimodal, has a continuous density, and has support R. Therefore, by part (b) of the proof, f_n is logconcave. But $F_n \to G$ weakly and, by Theorem 1.9, G is continuous. Therefore, by Lemma 1.4, G has a logconcave density. This proves the "only if" part and completes the proof of the theorem. ■

Theorem 1.10 enables us to identify several standard distributions that are strongly unimodal. Some of these are:

(i) The normal distributions $N(\mu, \sigma^2)$,
(ii) The uniform distributions on intervals (a, b),
(iii) The gamma distributions with shape parameters $p \geq 1$,
(iv) The beta distributions with parameters (p, q) with $p \geq 1$ and $q \geq 1$.

There are also a few standard unimodal distributions that are not strongly unimodal. This follows if we note that a strongly unimodal density g always possesses a moment generating function. The reason is that the logconcavity and integrability of g easily imply that $g(x) \leq \exp[-c|x|]$ for some $c > 0$ and for all sufficiently large $|x|$. In particular, a strongly unimodal distribution must have all its moments finite. As a consequence, a distribution having an infinite moment of some order will fail to be strongly unimodal. Some examples of such distributions are:

(a) the t distributions,
(b) the F distributions, and
(c) the stable distributions of index $\alpha < 2$.

1.5. The Gauss Inequality and the Three-Sigma Rule for Unimodal Distributions

The usual inequality due to Chebyshev is a special case of Markov's inequality, which states that

$$P(|X| \geq k) \leq \frac{E(|X|^r)}{k^r}, \tag{1.18}$$

where X is a real random variable and k and r are arbitrary positive numbers.

If X has a distribution which is unimodal about 0, then the bound on the right side of (1.18) can be reduced by a factor which depends on r. This is made precise by Theorem 1.11 below, which is due to Camp (1922) and Meidell (1922). It is interesting to note that, for the special case $r = 2$, Theorem 1.11 goes back to Gauss (1821). Thus, the strengthening of Chebyshev's result under the additional assumption of unimodality came much before Chebyshev's time. We begin with a simple lemma.

Lemma 1.5. *Let $r > 0$ and let X be a real random variable with $E(|X|^r) < \infty$. Then we can find a sequence of random variables X_n such that each X_n takes only a finite number of values and $E(|X_n - X|^r) \to 0$. Moreover, if $r \geq 1$, then we can choose the X_n in such a way that $E(X_n) = E(X)$ for all n.*

Proof. It is a standard result that we can find random variables X_n taking only a finite number of values such that $|X_n| \leq |X|$ and $X_n \to X$ everywhere. Now $|X_n - X| \leq |X_n| + |X| \leq 2|X|$ and $E|X|^r < \infty$. Therefore $E|X_n - X|^r \to 0$, by the dominated convergence theorem. Now, if $r \geq 1$, then $E(X_n) \to E(X)$ and so $Y_n = X_n - E(X_n) + E(X)$ is such that $E|Y_n - X|^r \to 0$ and $E(Y_n) = E(X)$. The lemma is thus proved. ■

Theorem 1.11. *Let X have a distribution which is unimodal about zero. Then, for every $r > 0$ and $k > 0$,*

$$P(|X| \geq k) \leq \left(\frac{r}{r+1}\right)^r \frac{E(|X|^r)}{k^r}. \tag{1.19}$$

Moreover, this bound is sharp in the sense described at the end of the proof.

Proof. Since (1.19) is trivialy true if $E|X|^r = \infty$, we assume that $E|X|^r < \infty$. Since X is unimodal about zero, by Theorem 1.3, X has the same distribution as UZ, where U is uniform on $(0, 1)$ and U, Z are independent. Now $E|X|^r = E(|Z|^r)/(r+1)$. Therefore $E|Z|^r < \infty$. Lemma 1.5 shows that it is sufficient to establish (1.19) in the case where Z takes only a fnite number of values. So, if F is the distribution function of X, then we may assume that the density of F takes only a finite number of values. Let $\mathscr{U}^*$ denote the set of all distribution functions which are unimodal about 0 and whose densities take only a finite number of values. We note that $\mathscr{U}^*$ is the convex hull of the set of all uniform distributions W_z, $0 < |z| < \infty$. Now let $\pi_r = [r/(r+1)]^r$ and write (1.19) as

$$k^r \int_{|x| \geq k} dF(x) \leq \pi_r \int_{-\infty}^{\infty} |x|^r \, dF(x). \tag{1.20}$$

We can consider (1.20) as a property of F. It is clear that if (1.20) holds for $F = F_i$, $i = 1, 2$, and if $0 < \alpha < 1$, then it also holds for $F = \alpha F_1 + (1 - \alpha)F_2$. Thus it is sufficient to prove (1.20) for the case where F is an extreme point of $\mathscr{U}^*$. So we can assume that F is one of the uniform distributions W_z. Again, if (1.19) holds for X, it also holds for aX, where $a \in R$. Thus we have reduced the theorem to the case where X is uniform on (0, 1). In this case $E|X|^r = 1/(r+1)$ and

$$P(|X| \geq k) = \begin{cases} (1-k) & \text{if } 0 < k \leq 1 \\ 0 & \text{if } k \geq 1. \end{cases}$$

Therefore,

$$k^r P(|X| \geq k) = \begin{cases} k^r(1-k) & \text{if } 0 < k \leq 1 \\ 0 & \text{if } k \geq 1. \end{cases}$$

For fixed r, the last quantity becomes maximum when $k = r/(r+1)$. The maximum value is $r^r/(r+1)^{r+1}$. Therefore,

$$k^r P(|X| \geq k) \leq \left(\frac{r}{r+1}\right)^r \cdot \frac{1}{r+1} = \left(\frac{r}{r+1}\right)^r E|X|^r,$$

which proves (1.19). The above calculation further shows that this bound is sharp in the following sense. Given positive numbers k and r, there is a unimodal random variable X with mode 0 [indeed, uniformly distributed on the interval $(0, k(r+1)/r)$] for which equality holds in (1.19). This proves the theorem. ■

The special case $r = 2$ gives the Gauss inequality.

Corollary. (Gauss). *If X has a distribution which is unimodal about zero, then, for all $k > 0$,*

$$P(|X| \geq k) \leq \tfrac{4}{9} \cdot \frac{E(X^2)}{k^2}.$$

If X has a distribution which is symmetric and unimodal about μ and $\operatorname{Var}(X) = \sigma^2$, then the above corollary shows that

$$P(|X - \mu| \geq 3\sigma) \leq \tfrac{4}{81} < .05.$$

This last result is the three-sigma rule for symmetric unimodal distributions. Without the assumption of unimodality, the bound $\frac{4}{81}$ has, of course, to be replaced by $\frac{1}{9}$. Recently, Vysochanskii and Petunin (1979) established the three-sigma rule for arbitrary unimodal distributions without the assumption of symmetry. We generalize their result and also simplify their proof below (Theorem 1.12); see Dharmadhikari and Joag-dev (1985b). We need a lemma which is a special case of a theorem of Hoeffding (1955). See also Theorem A.6 in the Appendix.

Lemma 1.6. *For a Borel measurable function g on R and for $c \in R$, let $\mathscr{A}_c$ denote the set of all probability distributions P on R such that P* has finite support and $\int g(x)\, dP(x) = c$. *Then every distribution in $\mathscr{A}_c$ is a finite mixture of distributions in $\mathscr{A}_c$, which assign positive mass to at most two points.*

Proof. Without loss of generality, let $c = 0$. Let $P \in \mathscr{A}_0$ and let v be the size of the support of P. The lemma holds if $v \leq 2$. Suppose the lemma holds for $v \leq n$, where $n \geq 2$. Let Y be a random variable with distribution P and support S, where S has exactly $(n+1)$ points. If $g(y) = 0$ for every $y \in S$, then P is a mixture of $(n+1)$ degenerate distributions in $\mathscr{A}_0$. So, assume that g is not constant over S. Since $E[g(Y)] = 0$, we can find $a \in S$ and $b \in S$ such that $g(a) = -\alpha < 0$ and $g(b) = \beta > 0$. Let $\xi = P(Y = a)$ and $\eta = P(Y = b)$. Without loss of generality, let $\xi\alpha \geq \eta\beta$. Consider the two-point distribution P_0 which puts mass $\alpha/(\alpha+\beta)$ at b and mass $\beta/(\alpha+\beta)$ at a. Then $P_0 \in \mathscr{A}_0$ and we can write

$$P = \theta P_0 + (1-\theta)P_1, \tag{1.21}$$

where $\theta = \eta(\alpha+\beta)/\alpha$. It is clear that $\theta > 0$. Since S has at least three points, we must have $\xi + \eta < 1$. Therefore,

$$\eta(\alpha+\beta) = \eta\alpha + \eta\beta \leq \eta\alpha + \xi\alpha = \alpha(\xi+\eta) < \alpha.$$

That is, $\theta < 1$. Observe that, in the representation (1.21) the mass at b is accounted for by θP_0. Therefore, P_1 is a distribution in $\mathscr{A}_0$ whose support has $\leq n$ points. By the induction hypothesis, P_1 is a finite mixture of distributions in $\mathscr{A}_0$ which put their mass at most two points. Therefore, by (1.21), P is also a mixture of the required type. This proves the lemma. ∎

For ease in writing, we denote the uniform distribution on the interval (b, c) by $W(b, c)$. Further, for $r > 0$, we denote by r^* the unique number satisfying

$$r^* > (r+1) \quad \text{and} \quad r^*(r^* - r - 1)^r = r^r. \tag{1.22}$$

We need two more lemmas before we can present our generalization of the result of Vysochanskii and Petunin (1979).

Lemma 1.7. *Let X have a unimodal distribution with mode M. For $a \in R$ and $r > 0$, let $\tau_r = E(|X - a|^r)$. Then*

$$|a - M|^r \leq r^* \tau_r, \tag{1.23}$$

where r^* is given by (1.22).

Proof. We note that in the case where X has several modes, we must establish (1.23) for all possible choices of M, that is, when M is the mode farthest away from a.

Let F be the distribution function of X. Then (1.23) can be written as

$$|a - M|^r \leq r^* \int |x - a|^r \, dF(x). \tag{1.24}$$

As in the proof of Theorem 1.11, we may assume that F has a density which takes only a finite number of values. Further, the set of all F for which (1.24) is valid is clearly convex. Therefore, it is sufficient to establish (1.24) for the case where $F = W(b, c)$, and $|M - a| = \max(|b - a|, |c - a|)$. Two cases arise.

Case 1. Let $a \in (b, c)$ and suppose, for the sake of definiteness, that a is closer to b than to c. We may then take

$$b = -1 < 0 = a < 1 < c = M.$$

Here $\tau_r = (c^{r+1} + 1)/[(c + 1)(r + 1)]$. Therefore, for $s > (r + 1)$,

$$\begin{aligned}(r + 1)(c + 1)[s\tau_r - |M|^r] &= s(1 + c^{r+1}) - (r + 1)(c + 1)c^r \\ &= (s - r - 1)c^{r+1} - (r + 1)c^r + s \\ &= h(c), \quad \text{say.}\end{aligned}$$

It is easy to check that h(c) becomes minimum when $c = r/(s - r - 1)$ and that the minimum value of $h(c)$ is $s - [r/(s - r - 1)]^r$. This minimum value is nonnegative whenever $s \geq r^*$. Thus, $|M|^r < r^* \tau_r$.

Case 2. Suppose $a \notin (b, c)$. We may assume that

$$a = 0 < b = 1 < c = M.$$

Here $\tau_r = (c^{r+1} - 1)/[(c - 1)(r + 1)]$. Therefore, for $s \geq (r + 1)$,

$$\begin{aligned}(r + 1)(c - 1)[s\tau_r - |M|^r] &= s(c^{r+1} - 1) - (r + 1)(c - 1)c^r \\ &= (s - r - 1)c^{r+1} - s + (r + 1)c^r \\ &= g(c).\end{aligned}$$

Now $g(c)$ is clearly increasing in c and $g(1) = 0$. Therefore, $g(c) \geq 0$ for all $c \geq 1$. Thus, in this case $s\tau_r \geq |M|^r$ for every $s \geq (r+1)$.

The proof of the lemma is now complete. ∎

Lemma 1.8. *Let X have a distribution which is unimodal about M. For $r > 0$ and $a \in R$, let $\tau_r = E(|X - a|^r)$. If $k > 0$ is such that either $a + k \leq M$ or $a - k \geq M$, then*

$$P(|X - a| \geq k) \leq \frac{r^*\tau_r - k^r}{(r^* - 1)k^r}, \tag{1.25}$$

where r^* is given by (1.22).

Proof. We may assume that $a = 0$, $M > 0$ and $0 < k < M$. Let A_k denote the set $\{x : |x| \geq k\}$. We denote the probability distribution of X by P. Consider two cases.

Case 1. Suppose $P(A_k) = 1$. Then $\tau_r \geq k^r$. Therefore,

$$r^*\tau_r - k^r \geq r^*k^r - k^r = (r^* - 1)k^r.$$

Noting that $r^* > 1$, we get

$$\frac{r^*\tau_r - k^r}{(r^* - 1)k^r} \geq 1 = P(A_k).$$

So, (1.25) follows.

Case 2. Suppose $P(A_k) < 1$. Then we can express P as a convex mixture $\theta_1 P_1 + \theta_2 P_2 + \theta_3 P_3$, where P_1 is unimodal with mode k and support $[-k, k]$, P_2 is unimodal with mode M and support $[k, \infty)$ and P_3 is unimodal with mode $(-k)$ and support $(-\infty, -k]$. Write $\tau_{ri} = \int |x|^r \, dP_i(x)$, $i = 1, 2, 3$. By Lemma 1.7, $r^*\tau_{r1} \geq k^r$ and so

$$P_1(A_k) = 0 \leq \frac{r^*\tau_{r1} - k^r}{(r^* - 1)k^r}. \tag{1.26}$$

Further, if $i = 2, 3$, then $P_i(A_k) = 1$ and so Case 1 above shows that

$$P_i(A_k) \leq \frac{r^*\tau_{ri} - k^r}{(r^* - 1)k^r}. \tag{1.27}$$

Now (1.25) follows by taking a convex combination of (1.26) and (1.27). The lemma is thus proved. ∎

We now state the result of Vysochanskii and Petunin (1979).

Theorem 1.12. *Let X be a unimodal random variable with mean μ and variance σ^2. Then for all $k > 0$,*

$$P(|X-\mu| \geq k) \leq \max\left[\frac{4\sigma^2 - k^2}{3k^2}, \frac{4\sigma^2}{9k^2}\right]. \tag{1.28}$$

We generalize Theorem 1.12 in two directions. First, we allow deviations to be taken from an arbitrary $a \in R$. Second, we take account of moments of order other than 2.

Theorem 1.13. *Let X be a random variable with a unimodal distribution. Let $a \in R$, $r > 0$ and $\tau_r = E(|X-a|^r)$. Then, for every $k > 0$,*

$$P(|X-a| \geq k) \leq \max\left[\frac{r^*\tau_r - k^r}{(r^*-1)k^r}, \frac{\pi_r \tau_r}{k^r}\right], \tag{1.29}$$

where r^* is given by (1.22) *and* $\pi_r = [r/(r+1)]^r$.

Before we present the proof of Theorem 1.13, let us observe that $r^* = 4$ when $r = 2$. This explains the expression, $(4\sigma^2 - k^2)/(3k^2)$ in (1.28). But even for this case, Theorem 1.13 is more general than Theorem 1.12 because a need not be the same as μ. It is not known whether (1.29) can be improved if $a = \mu$. But we refer the reader to Ulin (1953) who has made a detailed analysis of the quantity $P(|X-\mu| \geq k\sigma)$ as a function of k when X is unimodal with mean μ and standard deviation σ.

Proof. (of Theorem 1.13). We may assume that $a = 0$ and $\tau_r = 1$. Let M be a mode of X. By Theorem 1.3, X is distributed as $M + UZ$, where U is uniform on $(0, 1)$ and U, Z are independent. Let P and Q be the probability distributions of X and Z, respectively. Since r, r^* and τ_r are all fixed, we can denote the right side of (1.29) by $B(k)$ and write (1.29) as

$$\int_{|x| \geq k} dP(x) \leq B(k). \tag{1.30}$$

In turn, (1.30) is equivalent to

$$\int_0^1 \left\{\int_{D(u,k)} dQ(z)\right\} du \leq B(k), \tag{1.31}$$

where $D(u, k) = \{z : |M + uz| \geq k\}$. We see that if (1.31) holds for $Q = Q_i$, $i = 1, 2$ and $0 < \alpha < 1$, then it also holds for $Q = \alpha Q_1 + (1-\alpha)Q_2$. Moreover, if $g(z) = E(|M + zU|^r)$, then Q must satisfy the condition

$$\int_{-\infty}^{\infty} g(z)\, dQ(z) = E|X|^r = \tau_r = 1.$$

Therefore, Lemmas 1.6 and 1.7 show that we only need to prove (1.31) for the case where the support of Q has exactly two points. In this case the density of P takes exactly two nonzero values. So suppose P has the density f given by

$$f(x) = \begin{cases} \alpha & \text{if } b < x < c \\ \beta & \text{if } c < x < d \\ 0 & \text{elsewhere.} \end{cases}$$

We may assume that $\alpha < \beta$. We also write A_k for the set $\{x: |x| \geq k\}$. Several cases arise.

Case 1. Suppose $0 < c$ and $|b| < d$. Since f is unimodal about d, Lemma 1.8 shows that, if $0 < k < d$, then

$$P(A_k) \leq \frac{r^*\tau_r - k^r}{(r^* - 1)k^r}. \tag{1.32}$$

On the other hand, if $k > d$, then $P(A_k) = 0$ and so,

$$P(A_k) \leq \frac{\pi_r \tau_r}{k^r}. \tag{1.33}$$

Case 2. Suppose $0 < c$ and $|b| > d$. Here we must have $b < 0 < c$. Again Lemma 1.8 shows that (1.32) holds for $0 < k < d$. So, let $k > d$. Define a new distribution P_1 with density f_1 given by

$$f_1(x) = \begin{cases} \gamma & \text{if } 0 < x < d \\ f(x) & \text{elsewhere.} \end{cases}$$

We write $\tau_{r1} = \int |x|^r f_1(x)\, dx$. Using the relation $\gamma d = \alpha c + \beta(d - c)$, we get

$$\begin{aligned} (r+1)[\tau_r - \tau_{r1}] &= (r+1)\left[\int_0^d t^r f(t)\, dt - \int_0^d t^r f_1(t)\, dt\right] \\ &= \alpha c^{r+1} + \beta(d^{r+1} - c^{r+1}) - \gamma d^{r+1} \\ &= (\alpha - \beta)c^{r+1} + \beta d^{r+1} - [\alpha c + \beta(d - c)]d^r \\ &= c(\beta - \alpha)(d^r - c^r). \end{aligned}$$

Since $\alpha < \beta$ and $0 < c < d$, we see that $\tau_r \geq \tau_{r1}$. Now f_1 is unimodal about 0.

So, by Theorem 1.11,

$$P_1(A_k) \le \frac{\pi_r \tau_{r1}}{k^r}.$$

But, since $k > d$, $P_1(A_k) = P(A_k)$ and we know that $\tau_r \ge \tau_{r1}$. Therefore, (1.33) holds.

Case 3. Suppose $c < 0 < d$. Then f is unimodal about 0. Therefore, by Theorem 1.11, (1.33) holds for all $k > 0$.

Case 4. Suppose $d < 0$. Since f is unimodal about c, Lemma 1.8 shows that (1.32) holds when $0 < k < |c|$. Also, if $k \ge |b|$, then (1.33) holds because $P(|X| \ge k) = 0$. So, suppose that $|b| > k > |c|$. Define a new distribution P_2 with density f_2 given by

$$f_2(x) = \begin{cases} \delta & \text{if} \quad c < x < 0 \\ f(x) & \text{elsewhere.} \end{cases}$$

As in Case 2, writing $\tau_{r2} = \int |x|^r f_2(x)\,dx$, we get

$$(r+1)(\tau_r - \tau_{r2}) = \beta |d| (|c|^r - |d|^r),$$

which is positive. Now, if $\delta < \alpha$, then b is a mode of f_2 and so Lemma 1.8 shows that

$$P_2(A_k) \le (r^* \tau_{r2} - k^r)[(r^* - 1)k^r].$$

But $P_2(A_k) = P(A_k)$ and $\tau_{r2} \le \tau_r$. Therefore, (1.32) holds. On the other hand, if $\delta \ge \alpha$, then f is unimodal about 0 and so, by Theorem 1.11,

$$P_2(A_k) \le \frac{\pi_r \tau_{r2}}{k^r}.$$

Again (1.33) follows because $P_2(A_k) = P(A_k)$ and $\tau_{r2} \le \tau_r$.

We have thus shown, in all cases, that either (1.32) or (1.33) holds. Therefore, (1.29) is established and the theorem is proved. ∎

Corollary. *With the same notation as in Theorem* 1.13, *let* $\lambda_r = (\tau_r)^{1/r}$ *and*

$$k_r = \frac{[r^*(r+1)^r - (r^*-1)r^r]^{1/r}}{r+1}.$$

Then, for all $k \ge k_r$, *we have*

$$P[|X - a| \ge k\lambda_r] \le \frac{\pi_r}{k^r}. \tag{1.34}$$

Proof. If we substitute $k\lambda_r$ for k in (1.29), we see that (1.34) holds as soon as

$$\frac{r^* - k^r}{(r^* - 1)} \le \left[\frac{r}{r+1}\right]^r.$$

This last inequality is in turn equivalent to $k \ge k_r$. The corollary is thus proved. ■

We conclude this section by giving two examples. Note first that $k_2 = \sqrt{8/3}$ and $\pi_2 = \frac{4}{9}$. Example 1.3 will show that if a is arbitrary, then (1.34) may fail for k around 1.52. Example 1.4 will show that even if we require that $a = E(X)$, (1.34) may still fail for k as high as $\sqrt{4/3}$. Thus, some condition of the type $k \ge k_r$ seems to be required for (1.34) to hold.

Example 1.3. Let $t > 1$, $P(X = t) = t/(2 + 3t)$ and $P(X \le x) = 2(1 + x)/(2 + 3t)$, $-1 < x < t$. Then $\tau_2 = (2 + 5t^3)/[3(2 + 3t)]$. Further $P(|X| \ge t)$ exceeds $4\tau_2/(9t^2)$ as soon as $t^3 > (8/7)$. In particular, if $t = 1.525$, we get $\tau_2 = 1.0004$ and $t/\sqrt{\tau_2} = 1.524695$.

Example 1.4. Suppose X has the density f given by

$$f(x) = \begin{cases} \dfrac{\alpha}{3} & \text{if} \quad -2 < x < 1 \\[2ex] \dfrac{1-\alpha}{c-1} & \text{if} \quad 1 < x < c \\[2ex] 0 & \text{elsewhere.} \end{cases}$$

Here $1 < c < 2$ and $\alpha = (c + 1)/(c + 2)$. It is easy to check that f is a bonafide density. Further $E(X) = 0$ and $\sigma^2 = \operatorname{Var}(X) = (c + 2)/3$. We take $k = \sigma$ and show that

$$P[|X| \ge k\sigma] > \frac{4}{9k^2}. \tag{1.35}$$

We have

$$P[|X| \ge k\sigma] = P[|X| \ge \sigma^2] = \frac{(1-\alpha)(c-\sigma^2)}{c-1} + \frac{\alpha(2-\sigma^2)}{3}.$$

Substituting the values of α and σ^2, we easily get

$$P[|X| \ge k\sigma] = \frac{5-c}{9}.$$

Now (1.35) is equivalent to $(5-c) > 12/(c+2)$, which is in turn equivalent to $(c-1)(c-2) < 0$. The last result clearly holds for $1 < c < 2$. When c runs through the values in $(1, 2)$, we get values of k from 1 to $\sqrt{4/3}$ for which (1.34) fails.

1.6. The Mean-Median-Mode Inequality

Let X be a real random variable with mean μ. A number m is said to be a *median* of X if

$$P(X \geq m) \geq \tfrac{1}{2} \quad \text{and} \quad P(X \leq m) \geq \tfrac{1}{2}. \tag{1.36}$$

It is easy to show that such a number m always exists. However, a random variable may, in general, have several medians. Suppose now that X is unimodal with a mode M. Then (1.36) easily shows that the median m is unique. Moreover, under suitable conditions, we have either $M \leq m \leq \mu$ or $M \geq m \geq \mu$. This result is usually referred to as the *mean-median-mode inequality*. Conditions for the validity of this inequality have been given by Groenveld and Meeden (1977), Runnenberg (1978) and van Zwet (1979). Here we use the concept of stochastic ordering and present a result (Theorem 1.14), which includes the results of the above-mentioned authors; see Dharmadhikari and Joag-dev (1983c).

In addition to X, let Y be another real random variable. Let F and G, respectively, denote the distribution functions of X and Y. We say that X is *stochastically larger* than Y and write $X >_s Y$ if $F(x) \leq G(x)$ for all x.

Remark. If X and Y have finite expectations and $X >_s Y$, then

$$E(X - Y) = \int_{-\infty}^{\infty} [G(x) - F(x)]\, dx \geq 0. \tag{1.37}$$

Moreover, equality holds in (1.37) if, and only if, X and Y are identically distributed.

Lemma 1.9. *Suppose X and Y are nonnegative random variables with distribution functions F and G. Let $F(0) = G(0)$ and let F, G have densities f, g on $(0, \infty)$. Then either of the following conditions implies that $X >_s Y$.*

(a) *The density g crosses f only once and from above.*

(b) *For all* $t \in (F(0), 1)$,

$$\frac{d}{dt}[F^{-1}(t) - G^{-1}(t)] \geq 0,$$

or equivalently, $f[F^{-1}(t)] \leq g[G^{-1}(t)]$.

Proof. Easy. ■

In the next theorem, we use the standard notation $x^+ = \max(x, 0)$ and $x^- = -\min(x, 0)$.

Theorem 1.14. *Let X be a unimodal random variable. Let m be the median of X and write $\mu = E(X)$. If $(X - m)^+ >_s (X - m)^-$, then X has a mode M satisfying $M \leq m \leq \mu$.*

Proof. We may assume that $m = 0$. *Let* $X^+ >_s X^-$. Then the above remark shows that $\mu = E(X^+ - X^-) \geq 0$. Further, $\mu = 0$ only if X^+ and X^- have the same distribution that happens when the distributon of X is symmetric and then $M = m = \mu$.

Let $[M_1, M_2]$ be the set of all modes of X. We show that $M_1 \leq 0$. Contrary to this, suppose that $M_1 > 0$. Then the median $m = 0$ is a point of continuity of F. Therefore $P[X < 0] = P[X > 0] = \frac{1}{2}$. Further, since 0 is not a mode of X, the remark following Theorem 1.7 shows that there is a number $b > M_1$ such that $P(-b < X < 0) < P[0 < X < b]$. This means that $P(X^- \geqslant b) > P(X^+ \geq b)$, which contradicts the assumption that $X^+ >_s X^-$. Thus $M_1 \leq 0$ and we have found a mode M satisfying $M \leq m$. The theorem is thus proved. ■

The following corollary follows immediately from Theorem 1.14 and Lemma 1.9.

Corollary. *Suppose X is unimodal with mean μ and median m. If G and H are the distribution functions of $(X - m)^+$ and $(X - m)^-$, respectively, assume that $G(0) = H(0) = \frac{1}{2}$ and that G, H have densities g, h on $(0, \infty)$. Then either of the following two conditions implies that $M \leq m \leq \mu$ for some mode M of X.*

(a) *h crosses g only once and from above.*
(b) $g[G^{-1}(t)] \leq h[H^{-1}(t)]$, *for all* $t \in (\frac{1}{2}, 1)$.

Suppose X is a continuous unimodal random variable with density f. Let M, m, μ be as above. MacGillivray (1981) has remarked that if $f(\mu + x) -$

$f(\mu - x)$ changes sign only once from negative to positive (as x goes from 0 to ∞), then, in addition to $M < \mu$, we also have $E(X-\mu)^3 > 0$. Thus, for such types of densities, both the sign of the third central moment and the relative positions of the mean and the mode indicate positive skewness. Of course, the result is reversed if $f(\mu + x) - f(\mu - x)$ changes sign from positive to negative. Extending this argument, MacGillivray has shown that the mean-median-mode inequality holds for all distributions of the Pearson family whose densities are continuous at the boundaries of their supports.

We conclude this discussion of the mean-median-mode inequality by giving two examples.

Example 1.5. This example shows that the median does not necessarily fall between the mean and the mode even if the distribution is unimodal. Consider the density f given by

$$f(x) = \begin{cases} 0 & \text{if } x \le 0, \\ x & \text{if } 0 \le x \le c \\ c \exp[-\lambda(x-c)] & \text{if } x \ge c. \end{cases}$$

For f to be a bonafide density, c must satisfy

$$\frac{c^2}{2} + \frac{c}{\lambda} = 1. \tag{1.38}$$

It is easily checked that

$$\mu = \frac{c^3}{3} + \frac{c^2}{\lambda} + \frac{c}{\lambda^2}. \tag{1.39}$$

Now (1.38) shows that $\lambda \to 2$ as $c \to 1$. Therefore, by (1.39), $\mu \to (13/12) > 1$ as $c \to 1$. Thus, if c is sufficiently close to 1 but $c > 1$, then $\mu > c = M > 1 = m$.

Example 1.6. This is an example of a unimodal distribution for which $\mu < M$ but the third central moment is positive. Thus the relative positions of the mean and the mode indicate negative skewness whereas the sign of the third central moment indicates positive skewness.

Let X have the density

$$f(x) = \begin{cases} x & \text{if } 0 < x \le 1 \\ \dfrac{c}{2} \exp[-c(x-1)] & \text{if } x \ge 1 \\ 0 & \text{elsewhere.} \end{cases}$$

Then $E(X) = \frac{1}{3} + \frac{1}{2}(1 + c^{-1}) = \frac{5}{6} + 1/(2c)$, and

$$E(X-1)^3 = \frac{1}{2}\left[\int_1^\infty (x-1)^3 ce^{-c(x-1)}\,dx + \int_0^1 (x-1)^3 x\,dx\right.$$

$$= \frac{3}{c^3} - \frac{1}{20}.$$

If $c \to 3$, then $\mu \to 1$ and the third central moment $\mu_3 \to \frac{1}{9} - \frac{1}{20} > 0$. Therefore, if c is close to 3 but $c > 3$, then $\mu < 1 = M$ but $\mu_3 > 0$.

2 Concepts of Multivariate Unimodality

2.0. Summary

It was seen in Chapter 1 that there is just one natural definition of unimodality for distributions on R. In this chapter, we show that, for distributions on R^n, unimodality can be defined in several nonequivalent ways. Ideas of convexity play an important role in almost all such definitions. After introducing the necessary notation in Section 1, we discuss a few defnitions of unimodality in Section 2. The important class of logconcave distributions is studied in Section 3. In Section 5, we show that the class of "symmetric unimodal" distributions studied by Kanter (1977) coincides with the class of "central convex unimodal" distributions considered earlier in Section 2. The inter-relationships amongst the various notions of unimodality are discussed in Section 6. The remaining sections of the chapter (namely, Sections 4, 7 and 8) study the closure properties of the various classes of unimodal distributions under the operations of taking weak limits, marginals and convolutions.

2.1 Notation

We mention here some notation which wll be used throughout the chapter. A set $S \subset R^n$ is said to be *star-shaped* about $\boldsymbol{\xi} \in S$ if, for every $\mathbf{x} \in S$, the line segment joining ξ to $\mathbf{x}$ is completely contained in S. The set of points ξ about which a set S is star-shaped is called the *kernel* of S. It is well known that the kernel of a set is always a convex set (possibly empty). A set $S \subset R^n$ will be called *centrally symmetric* if $\mathbf{x} \in S \Rightarrow -\mathbf{x} \in S$. For $A \subset R^n$, we write $-A = \{\mathbf{x} \in R^n\colon\ -\mathbf{x} \in A\}$. A distribution P on R^n is said to be *centrally symmetric* if $P(A) = P(-A)$ for all Borel sets A in R^n. If A and B are subsets of R^n, then $A + B$ denotes the set $\{\mathbf{x} + \mathbf{y} : \mathbf{x} \in A \text{ and } \mathbf{y} \in B\}$. We write $A + \mathbf{b}$ in place of $A + \{\mathbf{b}\}$. Similarly, for $A \subset R^n$ and $c \in R$, cA denotes the set $\{c\mathbf{x} : \mathbf{x} \in A\}$. The indicator function of a set A is denoted by I_A. If $\mathbf{x} \in R^n$, then $\|\mathbf{x}\|$ is the usual Euclidean norm. The usual δ-neighborhood of $\mathbf{x} \in R^n$ is $N_\delta(\mathbf{x}) = \{\mathbf{y} \in R^n\colon \|\mathbf{y} - \mathbf{x}\| < \delta\}$. The symbols $\mathscr{B}_n$ and λ_n, respectively, denote the Borel σ-algebra in R^n and the Lebesgue measure on $\mathscr{B}_n$. Finally, V_δ will stand for the volume $\lambda_n[N_\delta(\mathbf{x})]$ of $N_\delta(\mathbf{x})$.

2.2. Some Definitions of Unimodality for Distributions on R^n

Theorem 1.1 says that the set of distributions on R which are unimodal about zero is just the closed convex hull of the set of all uniform distributions on intervals which include 0. Since intervals in R can be generalized to star-shaped sets in R^n, we may adopt the following definition.

Definition 2.1. A distribution on R^n is called *star unimodal* about $\mathbf{0}$ if it belongs to the closed convex hull of the set of all uniform distributions on sets in R^n which are star-shaped about $\mathbf{0}$.

This defnition is natural in the sense that, if a distribution P has a continuous density f, then requiring P to be star unimodal about $\mathbf{0}$ is the same as requiring that f is nonincreasing along rays going away from $\mathbf{0}$. For ease of reference we isolate this result as a criterion.

Criterion. *Suppose a distribution P has a continuous density f on R^n. Then P is star unimodal about $\mathbf{0}$ if, and only if, for every $s > 0$, the set*

$$C_s = \{\mathbf{x} \in R^n : f(\mathbf{x}) \geq s\}$$

is star-shaped about $\mathbf{0}$ or, equivalently, if, and only if,

$$0 < t < u < \infty \quad \textit{and} \quad \mathbf{x} \neq \mathbf{0} \Rightarrow f(u\mathbf{x}) \leq f(t\mathbf{x}).$$

Proof. Suppose that each C_s is star-shaped about $\mathbf{0}$. Write W_{C_s} for the uniform distribution on C_s and let $g(s)=\lambda_n(C_s)$. Now

$$f(\mathbf{x})=\int_0^{f(\mathbf{x})} ds=\int_0^{\infty} I_{C_s}(\mathbf{x})\, ds.$$

Therefore, for $B\in\mathscr{B}_n$,

$$\begin{aligned} P(B)=\int_B f(\mathbf{x})\,d\mathbf{x}&=\int_B\int_0^{\infty} I_{C_s}(\mathbf{x})\,ds\,d\mathbf{x}\\ &=\int_0^{\infty}\lambda_n(B\cap C_s)\,ds\\ &=\int_0^{\infty} W_{C_s}(B)g(s)\,ds.\end{aligned}$$

Setting $B=R^n$, we see that g is a probability density on $(0,\infty)$. Since each W_{C_s} is clearly star unimodal we see that P is also star unimodal.

To see the converse, suppose that Q is the uniform distribution on a set S which is open, bounded and star-shaped about $\mathbf{0}$. Then, for $0<t<u<\infty$ and $\mathbf{x}\neq\mathbf{0}$,

$$\begin{aligned}\mathbf{y}\in[S\cap N_{u\delta}(u\mathbf{x})]&\Rightarrow\left(\frac{t}{u}\right)\mathbf{y}\in[S\cap N_{t\delta}(t\mathbf{x})]\\ &\Rightarrow\mathbf{y}\in\left(\frac{u}{t}\right)[S\cap N_{t\delta}(t\mathbf{x})].\end{aligned}$$

Therefore,

$$\lambda_n[S\cap N_{u\delta}(u\mathbf{x})]\leqq\left(\frac{u}{t}\right)^n\lambda_n[S\cap N_{t\delta}(t\mathbf{x})].$$

Dividing by $\lambda_n(S)$ and rearranging, we have

$$u^{-n}Q[N_{u\delta}(u\mathbf{x})]\leq t^{-n}\mathrm{Q}[\mathrm{N}_{t\delta}(t\mathbf{x})].$$

This last relation can be easily extended, by taking convex mixtures and weak limits, to all distributions on R^n which are star unimodal about $\mathbf{0}$. So, if P is star unimodal about $\mathbf{0}$, then

$$u^{-n}P[N_{u\delta}(u\mathbf{x})]\leq t^{-n}P[N_{t\delta}(t\mathbf{x})].$$

Dividing by the volume V_δ of $N_\delta(\mathbf{x})$, letting $\delta\to 0$ and noting that P has a continuous density f, we get $f(u\mathbf{x})\leq f(t\mathbf{x})$ as claimed. The criterion is thus proved. ■

We will see in Chapter 3 that Definition 2.1 is a special case of the defnition of "generalized unimodailty" given by Olshen and Savage (1970).

Theorem 1.3 can be generalized to yield a characterization of star unimodal distributions. For use in this generalization, we consider the polar transformation under which a vector $\mathbf{x} \in R^n$ is written as $\mathbf{x} = l\mathbf{d}$, where $l = \|\mathbf{x}\|$ is the length of $\mathbf{x}$ and $\mathbf{d}$ is the direction vector of $\mathbf{x}$. It is well known that the Jacobian of this transformation is $l^{n-1}h(\mathbf{d})$, where h is an easily calculated function. We write $\mathscr{D}$ for the set of all possible direction vectors $\mathbf{d}$.

Theorem 2.1. *A random n-vector* $\mathbf{X}$ *has a star unimodal distribution about* $\mathbf{0}$ *if, and only if,* $\mathbf{X}$ *is distributed as* $U^{1/n}\mathbf{Z}$, *where* U *and* $\mathbf{Z}$ *are independent and* U *is uniformly distributed on* $(0, 1)$.

Proof. (a) Suppose $\mathbf{X}$ is uniformly distributed on a set S which is star-shaped about $\mathbf{0}$ and which can be described in polar form as $0 \le l \le g(\mathbf{d})$, $\mathbf{d} \in \mathscr{D}$, where g is a smooth positive function. Write $\mathbf{X}$ also in polar form as $\mathbf{X} = L\mathbf{D}$. Then the density of $(L, \mathbf{D})$ is

$$p(l, \mathbf{d}) = \frac{l^{n-1}h(\mathbf{d})}{\lambda_n(S)}, \qquad 0 \le l \le g(\mathbf{d}),\ \mathbf{d} \in \mathscr{D}. \tag{2.1}$$

The conditional density of L given $\mathbf{D} = \mathbf{d}$ is easily seen to be

$$q(l \mid \mathbf{d}) = \frac{nl^{n-1}}{[g(\mathbf{d})]^n}, \qquad 0 \le l \le g(\mathbf{d}). \tag{2.2}$$

Let $Y = L/g(\mathbf{D})$. Then (2.2) shows that the conditional density of Y given $\mathbf{D} = \mathbf{d}$ does not involve $\mathbf{d}$. Thus Y and $\mathbf{D}$ are independent and Y has the density

$$r(y) = ny^{n-1}, \qquad 0 \le y \le 1.$$

If $U = Y^n$, then U is uniform on $(0, 1)$ and is independent of $\mathbf{D}$. Writing $\mathbf{Z} = g(\mathbf{D})\mathbf{D}$, we get

$$X = L\mathbf{D} = \left[\frac{L}{g(\mathbf{D})}\right] \cdot g(\mathbf{D})\mathbf{D} = Y\mathbf{Z} = U^{1/n}\mathbf{Z}.$$

The "only if" part of the theorem now follows by taking convex mixtures and weak limits.

(b) From (2.1) above the marginal density of $\mathbf{D}$ is obtained as

$$\xi(\mathbf{d}) = k[g(\mathbf{d})]^n \cdot h(\mathbf{d}), \qquad \mathbf{d} \in \mathscr{D}, \tag{2.3}$$

where k is a normalizing constant. The steps in part (a) of the proof can now be retraced as follows. Suppose g is any smooth positive function on $\mathscr{D}$ and let $\mathbf{D}$ have the density ξ given by (2.3). Let U be independent of $\mathbf{D}$ and be uniform on $(0, 1)$. Write $\mathbf{Z}(g) = g(\mathbf{D})\mathbf{D}$. Then the distribution of $U^{1/n}\mathbf{Z}(g)$ is star unimodal because it is just the uniform distribution on the star-shaped set S defined by $0 \le l \le g(\mathbf{d})$, $\mathbf{d} \in \mathscr{D}$.

(c) It suffices to prove the "if" part of the theorem in the case where $\mathbf{Z}$ is degenerate. This is so because, we can then take convex mixtures and weak limits to get the required result for all random vectors $\mathbf{Z}$.

Let $l_0 > 0$ and $d_0 \in \mathscr{D}$ be arbitrary. Choose a sequence g_m of smooth functions on $\mathscr{D}$ such that $g_m(\mathbf{d}) \to 0$ if $\mathbf{d} \neq \mathbf{d}_0$ and $g_m(\mathbf{d}_0) \to l_0$. Write $\mathbf{z}_0 = l_0\mathbf{d}_0$. From the first paragraph of (b), we know that the distribution of each $U^{1/n}\mathbf{Z}(g_m)$ is star unimodal. Letting $m \to \infty$, we see that $U^{1/n}\mathbf{z}_0$ is star unimodal. This proves the "if" part and completes the proof of the theorem. ■

Remark. Definition 2.1 is based on the fact that intervals in R can be generalized to star-shaped sets in R^n. But intervals can also be generalized to convex sets in R^n. So, it is legitimate to ask how the set of all star unimodal distributions (about $\mathbf{0}$) is related to the closed convex hull $\mathscr{C}$ of the set of all uniform distributions on convex sets containing zero. The answer is that the two classes of distributions are the same. To justify this, suppose $\mathbf{z}_0 \in R^n$ and $\mathbf{z}_0 \neq \mathbf{0}$. Let U be uniform on $(0, 1)$. We show that the distribution of $U^{1/n}\mathbf{z}_0$ is in $\mathscr{C}$. By rotating the axes and changing the scale, we can arrange to have $\mathbf{z}_0 = (1, 0, \ldots, 0)$. If $\mathbf{x} \in R^n$, we write $\mathbf{x} = (x_1, \mathbf{x}')$, where $\mathbf{x}' = (x_2, \ldots, x_n)$. Let

$$C_\delta = \{\mathbf{x} \in R^n : 0 \le x_1 \le 1 \quad \text{and} \quad \|\mathbf{x}'\| < x_1\delta\}.$$

Then C_δ is convex. Let P_δ be the uniform distribution on C_δ. Then $P_\delta \in \mathscr{C}$. Further, as $\delta \to 0$, P_δ converges weakly to some P_0 which must concentrate all its mass on the line segment $[0, \mathbf{z}_0]$. For $0 < t < 1$, let $B_t = \{\mathbf{x} = (x_1, \mathbf{x}') \in R^n : 0 \le x_1 \le t\}$. Then $B_t \cap C_\delta = tC_\delta$ and so

$$\lambda_n(\mathrm{B}_t \cap C_\delta) = t^n \lambda_n(C_\delta)$$

or

$$P_\delta(B_t) = t^n.$$

It follows that P_0 is the distribution of $U^{1/n}\mathbf{z}_0$. Now $P_0 \in \mathscr{C}$. So, the distribution of $U^{1/n}\mathbf{z}_0$ is in $\mathscr{C}$. But then so is the distribution of $U^{1/n}\mathbf{Z}$ where U, Z are independent. Theorem 2.1 now shows that every star unimodal distribution is in $\mathscr{C}$. Since every distribution in $\mathscr{C}$ is clearly star unimodal, the comment in question is justified.

Another definition based on Theorem 1.2 has been given by Shepp (1962). He calls a bivariate distribution with density $g(x, y)$ "unimodal" if for a rectangle R with lower-left corner (a, b) and upper-right corner (c, d) and sides parallel to the axes, the expression

$$g(a, b) - g(a, d) - g(b, c) + g(c, d)$$

is nonnegative if R is contained in quadrants I and III and nonpositive if R is contained in quadrants II and IV. If such a "unimodal" distribution corresponds to the *uniform* distribution of unit mass over a compact set S in the first quadrant, then it is easy to show that S must be a rectangle with corners $(0, 0)$, $(0, D)$, $(C, 0)$ and (C, D) for some positive C and D. Since such a rectangle can be described as a "block," we get the following definition applicable to distributions on R^n.

Definition 2.2. A distribution on R^n is said to be *block unimodal* about $\mathbf{0}$ if it belongs to the closed convex hull of the set of all uniform distributions on rectangles containing $\mathbf{0}$ and having edges parallel to the coordinate axes.

A characterization of block unimodal distributions, due to Shepp (1962), is given in Theorem 2.2 below. Suppose $U_1, \ldots, U_n$ are independently and uniformly distributed on $(0, 1)$ and suppose $\mathbf{z} \in R^n$ is fixed. Then the distribution of the vector $(U_1 z_1, \ldots, U_n z_n)$ is just the uniform distribution on the rectangle with edges parallel to the coordinate axes and with opposite vertices $\mathbf{0}$ and $\mathbf{z}$. Letting $\mathbf{z}$ be the value of a random vector $\mathbf{Z}$, one immediately gets the following theorem.

Theorem 2.2. *A random vector* $\mathbf{X}$ *has a block unimodal distribution if, and only if,* $\mathbf{X}$ *is distributed as* $(U_1 Z_1, \ldots, U_n Z_n)$ *where the vectors* $\mathbf{U}$ *and* $\mathbf{Z}$ *are independent and the* U_i*'s are independent and uniform on* $(0, 1)$.

Since the uniform distributions used in definining block unimodal distributions are all star unimodal, it is seen that the set of all block unimodal distributions (about $\mathbf{0}$) is a proper subset of the set of all star unimodal distributions about $\mathbf{0}$.

Yet another definition of unimodality in R^n can be given through the unimodality of all linear functions. Such a definition was given by Ghosh (1974).

Definition 2.3. A random vector $\mathbf{X} = (X_1, \ldots, X_n)$ is said to be *linear unimodal* about $\mathbf{0}$ if for every $\mathbf{a} \in R^n$, the linear combination $\sum_{i=1}^n a_i X_i$ is univariate unimodal about 0.

While the method of reducing a multivariate problem to a univariate problem by taking linear functions works well in many contexts, Definition 2.3 has a serious geometrical drawback. A linear unimodal distribution may have a density with a local minimum at the "vertex". This is exhibited by the following example, which is taken from Dharmadhikari and Jogdeo (1976a). Another example is given by Kanter (1977).

Example 2.1. Let $0 < a < 1$ and $b < 1$. Consider the bivariate density

$$f(x, y) = K[e^{-a^2(x^2+y^2)/2} - be^{-(x^2+y^2)/2}],$$

where $x \in R$ and $y \in R$. The constant K can be chosen to make f a density. Let

$$h(t) = e^{-a^2 t} - be^{-t}.$$

Then $h'(0) = b - a^2$. Therefore

$$\begin{aligned} b > a^2 &\Rightarrow h \text{ is increasing at } 0 \\ &\Rightarrow f(tx, tx) \text{ is increasing in } t \in [0, \delta) \text{ for some } \delta > 0 \\ &\Rightarrow f \text{ is not star unimodal.} \end{aligned}$$

Now the marginal density f_1 is given by

$$f_1(x) = K\sqrt{2\pi}\left[\frac{e^{-a^2x^2/2}}{a} - be^{-x^2/2}\right].$$

Therefore

$$f_1'(x) = Kx\sqrt{2\pi}\,[be^{-x^2/2} - ae^{-a^2x^2/2}].$$

If $b < a$, then $f_1'(x) < 0$ for $x > 0$ and $f_1'(x) > 0$ for $x < 0$. Therefore f_1 is unimodal whenever $b < a$. Using the circular symmetry of f we see that, if $a^2 < b < a$ then f is linear unimodal but not star unimodal.

Remark. If, in the above example, we take $b > a$, then we get a spherically symmetric distribution on R^2 which is not linear unimodal. Recently, Berk and Hwang (1985) have shown that *every* spherically symmetric distribution on R^n, $n \geq 3$, is linear unimodal.

Yet another type of "unimodality" applicable to multivariate distributions is that of *logconcavity*. This is discussed in detail in Section 2.3.

We now consider some definitions of unimodality applicable to centrally symmetric distributions. First of all we can look at the set of all centrally symmetric star unimodal distributions. This set will clearly be the convex hull of the set of all uniform distributions on centrally symmetric star-shaped sets. The random vector $\mathbf{Z}$ given by Theorem 2.1 will be centrally symmetric. Next, in the definition of linear unimodality one would require all linear functions $\sum_{i=1}^{n} a_i X_i$ to be symmetric and unimodal about $\mathbf{0}$. We now present three more definitions based on Anderson (1955) and Sherman (1955).

Definition 2.4. A distribution on R^n is said to be *convex unimodal* if it has a density f such that, for every $c > 0$, the set $\{\mathbf{x} : f(\mathbf{x}) \geq c\}$ is a centrally symmetric convex set.

This was the definition given by Anderson (1955). From the geometric point of view the definition is quite natural. The graph of a convex unimodal density (which is a subset of R^{n+1}) is a "hyperhill" with convex contours. Also the restriction of such a density to every line in R^n is unimodal in the ordinary sense. Unfortunately, the set of convex unimodal distributions is neither convex nor closed. The set is not closed because we can clearly have limits which are not absolutely continuous. A simple example shows that the set is not convex.

Example 2.2. Suppose C and D are centrally symmetric convex bodies such that $C \cup D$ is not convex. Then the uniform distributions on C and D are convex unimodal but any proper mixture of these two distributions is not convex unimodal.

The "smoothing out" of Anderson's definition is achieved by the following definition based on Sherman (1955).

Definition 2.5. A distribution on R^n is said to be *central convex unimodal* if it is in the closed convex hull of the set of all uniform distributions on centrally symmetric convex bodies in R^n.

Theorem 2.3. *Every convex unimodal distribution is central convex unimodal.*

Proof. Let f be the density of a convex unimodal distribution P on R^n. For $s > 0$, let $C_s = \{\mathbf{x} \in R^n : f(\mathbf{x}) \geq s\}$. Then each C_s is convex and centrally symmetric. If I_A denotes the indicator function of A, then

$$\int_0^{\infty} I_{C_s}(\mathbf{x})\, ds = \int_0^{f(\mathbf{x})} ds = f(\mathbf{x}).$$

Therefore, for a Borel set B in R^n,

$$P(B) = \int_B f(\mathbf{x})\, d\mathbf{x} = \int_B \int_0^\infty I_{C_s}(\mathbf{x})\, ds\, d\mathbf{x}$$

$$= \int_0^\infty \lambda_n(B \cap C_s)\, \mathrm{ds}. \tag{2.4}$$

Suppose W_C denotes the uniform distribution on the set C. Then (2.4) shows that

$$P(B) = \int_0^\infty W_{C_s}(B) \cdot \lambda_n(C_s)\, ds = \int_0^\infty W_{C_s}(B) g(s)\, ds, \tag{2.5}$$

where $g(s) = \lambda_n(C_s)$. Putting $B = R^n$ in (2.5), we see that g is a density. Since the distribution with density g can be approximated by discrete distributions, (2.5) shows that P is central convex unimodal. ■

Remark. Formula (2.5) shows that every convex unimodal distribution is a generalized mixture (that is, an integral) of uniform distributions on centrally symmetric convex sets. Indeed every central convex unimodal distribution is obtainable as a generalized mixture of the type mentioned. This result will be proved in Section 2.5.

Our third definition of unimodality for centrally symmetric distributions is as follows.

Definition 2.6. A distribution P on R^n is said to be *monotone unimodal* if for every $\mathbf{y} \in R^n$ and every centrally symmetric convex set $C \subset R^n$, $P[C + k\mathbf{y}]$ is nonincreasing in $k \in [0, \infty)$.

This definition is based on the results of Anderson (1955) and Sherman (1955). Anderson proved that every convex unimodal distribution is monotone unimodal. Sherman extended the result by showing that every central convex unimodal distribution is monotone unimodal.

We conclude this section by examining whether the various types of unimodality introduced in this section are preserved under the operation of taking cartesian products.

Theorem 2.4. *Products of star (respectively, block, central convex) unimodal distributions are star (respectively, block, central convex) unimodal.*

Proof. We give the proof for star unimodal distributions. The proofs for the other two cases are similar.

Let $\mathscr{U}_m$ denote the set of all distributions on R^m which are star unimodal about $\mathbf{0}$. For $P_2 \in \mathscr{U}_n$, let $D_1(P_2) = \{P_1 \in \mathscr{U}_m : P_1 \times P_2 \in \mathscr{U}_{m+n}\}$. Then $D_1(P_2)$ is convex and closed. Similarly, for $P_1 \in \mathscr{U}_m$, the set $D_2(P_1) = \{P_2 \in \mathscr{U}_n : P_1 \times P_2 \in \mathscr{U}_{m+n}\}$ is closed and convex.

Now, let Q_1 and Q_2 be uniform distributions on star-shaped sets S_1 and S_2 in R^m and R^n, respectively. Then $Q_1 \times Q_2$ is the uniform distribution on $S_1 \times S_2$, which is star-shaped. That is, $Q_1 \times Q_2 \in \mathscr{U}_{m+n}$. In the notation of the preceding paragraph, $D_1(Q_2)$ contains all uniform distributions in $\mathscr{U}_m$. But $D_1(Q_2)$ is closed and convex. Therefore $D_1(Q_2)$ includes $\mathscr{U}_m$. In other words, $P_1 \times Q_2 \in \mathscr{U}_{m+n}$ whenever $P_1 \in \mathscr{U}_m$ and Q_2 is uniform and in $\mathscr{U}_n$. Repeating the above argument, we see that $D_2(P_1)$ includes $\mathscr{U}_n$ whenever $P_1 \in \mathscr{U}_m$. That is, $P_1 \times P_2 \in \mathscr{U}_{m+n}$ whenever $P_1 \in \mathscr{U}_m$ and $P_2 \in \mathscr{U}_n$. This proves the theorem. ■

The next two examples show that analogs of Theorem 2.4 do not hold for linear unimodal and convex unimodal distributions.

Example 2.3. Let F and G be distribution functions on R such that F, G are unimodal about 0, but $F * G$ is not unimodal. If X, Y are independent random variables with distribution functions, F, G, then the joint distribution of X, Y is not linear unimodal, because the distribution of $X + Y$ is not unimodal.

Example 2.4. Suppose a density g is defined on R as follows:

$$g(x) = \begin{cases} \frac{3}{8} & \text{if } |x| \le 1 \\ \frac{1}{8} & \text{if } 1 < |x| \le 2 \\ 0 & \text{elsewhere.} \end{cases}$$

Let X, Y be independent random variables with common density g. Then X and Y have convex unimodal distributions. But the joint distribution of X, Y is not convex unimodal, because $\{(x, y) : g(x)g(y) \ge (\frac{3}{64})\}$ is not convex.

Kanter (1977) studied a class of "symmetric unimodal" distributions. We show in Section 2.5 that his class coincides with the class of central convex unimodal distributions.

2.3. Logconcave Distributions

It was seen in Chapter 1 that the concept of strong unimodality led to the class of logconcave densities. In this section we study logconcavity of distributions in higher dimensions.

We will denote by $\mathscr{B}_n$ the σ-algebra of Borel sets in R^n. The Lebesgue measure on $\mathscr{B}_n$ will be denoted by λ_n. If P is a probability on $\mathscr{B}_n$, then supp P denotes the support of P, namely, the smallest closed set of probability 1.

A nonnegative function f on R^n is called *logconcave* if, for all $\mathbf{x}$, $\mathbf{y}$ in R^n and for all $\theta \in (0, 1)$, we have

$$f[\theta\mathbf{x} + (1-\theta)\mathbf{y}] \geq [f(\mathbf{x})]^\theta [f(\mathbf{y})]^{1-\theta}.$$

Exactly in the same way, a probability P on $\mathscr{B}_n$ is called *logconcave* if, for all nonempty A, B in $\mathscr{B}_n$ and for all $\theta \in (0, 1)$, we have

$$P[\theta A + (1-\theta)B] \geq [P(A)]^\theta [P(B)]^{1-\theta}. \tag{2.6}$$

Measures satisfying (2.6) were studied by Prékopa (1971). More generally, the right side of (2.6) can be replaced by some mean other than the geometric mean. Properties of these more general classes were considered by Borell (1975). Our treatment here is based on Borell's paper and on Rinott (1976). The main result states that a probability P is logconcave if and only if P is absolutely continuous w.r.t. Lebesgue measure on the affine hull of supp P and the density of P is logconcave.

Lemma 2.1. *Let P be a logconcave probability on $\mathscr{B}_n$. Let L be a non-zero real matrix of order $m \times n$. Define a probability Q on $\mathscr{B}_m$ as follows:*

$$Q(A) = P\{\mathbf{x} \in R^n : L\mathbf{x} \in A\}.$$

Then Q is a logconcave probability on $\mathscr{B}_m$.

Proof. For $A \in \mathscr{B}_m$, write $\tilde{A} = \{\mathbf{x} \in R^n : L\mathbf{x} \in A\}$. Let A, B be in $\mathscr{B}_m$, $\theta \in (0, 1)$ and $C = \theta A + (1-\theta)B$. Then $\tilde{C} \supset \theta\tilde{A} + (1-\theta)\tilde{B}$. Therefore

$$\begin{aligned} Q(C) = P(\tilde{C}) &\geq P[\theta\tilde{A} + (1-\theta)\tilde{B}] \\ &\geq [P(\tilde{A})]^\theta [P(\tilde{B})]^{1-\theta} \\ &= [Q(A)]^\theta [Q(B)]^{1-\theta}. \end{aligned}$$

This proves the lemma. ∎

Theorem 2.5. *Let P be a nondegenerate logconcave probability on $\mathscr{B}_n$. Then* supp P *is convex. If H is the affine hull of* supp P, *then P is absolutely continuous w.r.t. Lebesgue measure on H.*

Proof. Let $\mathbf{x}$, $\mathbf{y}$ be in R^n, $\delta > 0$ and $\theta \in (0, 1)$. Then

$$N_\delta[\theta\mathbf{x} + (1-\theta)\mathbf{y}] \supset \theta N_\delta(\mathbf{x}) + (1-\theta)N_\delta(\mathbf{y}).$$

Therefore

$$P[N_\delta(\theta\mathbf{x} + (1-\theta)\mathbf{y})] \geq \{P[N_\delta(\mathbf{x})]\}^\theta \{P[N_\delta(\mathbf{y})]\}^{1-\theta} \tag{2.7}$$

If **x**, **y** are in supp P, then the right side of (2.7) is positive for every $\delta > 0$. Therefore the left side is also positive for every $\delta > 0$. This shows that supp P is convex.

Let H have dimension m. By rotating the axes we may take $H = R^m$ and assume that P is defined on $\mathscr{B}_m$. We note that $m \geq 1$ because P is not degenerate. We want to show that $P \ll \lambda_m$. Write $P = P_1 + P_2$ where $P_1 \ll \lambda_m$ and $P_2 \perp \lambda_m$. Assume that $P_2 \neq 0$. If possible, suppose that supp P_2 has dimension m. Then we can find $(m+1)$ affine independent points $\mathbf{x}_1, \mathbf{x}_2, \ldots, \mathbf{x}_{m+1}$ in supp P_2 (see Section A.1 of the Appendix). Choose $\delta > 0$ so small that $\mathbf{y}_1, \ldots, \mathbf{y}_{m+1}$ are affine independent whenever $\mathbf{y}_i \in N_\delta(\mathbf{x}_i)$, $i = 1, \ldots, m+1$. Now, the singularity of P_2 w.r.t. λ_m shows that [see Rudin (1987), Theorem 7.15] we can choose the $\mathbf{y}_i$ and a sequence $\delta_j \to 0$ in such a way that

$$\delta_j^{-m} P[N_{\delta_j}[\mathbf{y}_i)] \to \infty \quad \text{as} \quad j \to \infty, \qquad \text{for} \quad i = 1, 2, \ldots, m+1.$$

Now (2.7) shows that, if $\theta_i > 0$ for all i and $\sum \theta_i = 1$, then

$$\delta_j^{-m} P[N_{\delta_j}(\sum \theta_i \mathbf{y}_i)] \geq \delta_j^{-m} \prod_{i=1}^{m+1} \{P[N_{\delta_j}(\mathbf{y}_i)]\}^{\theta_i} \to \infty.$$

This means [see Rudin (1987), Theorem 7.14] that $(dP_1/d\lambda_m) = \infty$ on a set of positive λ_m-measure, namely, the convex hull of $\{y_1, \ldots, y_{m+1}\}$. But this is a contradiction because P_1 is a finite measure. This contradiction shows that supp P_2 must have dimension less than m. That is, supp P_2 is contained in a hyperplane H_1, which may be written as $\mathbf{u} \cdot \mathbf{x} = 0$. Define a measure Q on $\mathscr{B}_1$ as follows.

$$Q(B) = P\left\{\mathbf{x} \in R^m : \mathbf{u} \cdot \mathbf{x} \in B\right\}.$$

We note that $Q(\{0\}) \geq P_2(H_1) > 0$. Further Q is nonzero and absolutely continuous w.r.t. λ_1 on the set $R - \{0\}$. Therefore, we can find an interval $A = [a, b]$ such that $0 \notin A$ and $Q(A) > 0$. Take $B = \{0\}$. Then, by the logconcavity of Q, assured by Lemma 2.1,

$$Q([\theta a, \theta b]) = Q[\theta A + (1-\theta)B] \geq \min\{Q(A), Q(B)\}. \tag{2.8}$$

If $\theta \to 0$, the left side of (2.8) goes to zero, whereas the right side remains positive. This contradiction shows that $P_2 = 0$. Thus $P \ll \lambda_m$ and the theorem is proved. ■

Corollary. *Let P be a logconcave probability on $\mathscr{B}_n$. Let H be an affine set in R^n such that $P(H) > 0$. Then $P(H) = 1$.*

Proof. If P is degenerate, the corollary is trivial. So suppose P is not degenerate. Let H_1 be the affine hull of $H \cap (\text{supp } P)$ and let m be the dimension of H_1. If possible, suppose $P(H_1) < 1$. Then supp P must have dimension $v \geq (m+1)$. Theorem 2.5 now shows that P is absolutely continuous w.r.t. Lebesgue measure on a v-dimensional space. Therefore, since H_1 has dimension m, $P(H_1)$ must be zero. But $P(H_1) \geq P(H) > 0$. This contradiction shows that $P(H) = 1$. ∎

The proof of the next theorem uses the argument that was used by Hadwiger and Ohman (1956) to simplify the proof of the Brunn–Minkowski inequality.

Theorem 2.6. *Let P be a probability on $\mathscr{B}_n$ and suppose that P assigns zero mass to every hyperplane in R^n. In order that P is logconcave, it is sufficient that* (2.6) *holds for all rectangles A, B with sides parallel to the coordinate axes.*

Proof. (a) Fix $\theta \in (0, 1)$ and for A, B in $\mathscr{B}_n$, write

$$q(A, B) = P[\theta A + (1-\theta)B] - \{P(A)\}^\theta \{P(B)\}^{1-\theta}.$$

We want to show that $q(A, B) \geq 0$ for all nonempty A and B. We assume that this result holds whenever A, B are rectangles of the type mentioned.

(b) Let $A_1 \in \mathscr{B}_n$ and let H_1 be a hyperplane defined by the equation $\mathbf{u} \cdot \mathbf{x} = c_1$. Let

$$A_1' = A_1 \cap \{\mathbf{x} : \mathbf{u} \cdot \mathbf{x} \geq c_1\} \quad \text{and} \quad A_1'' = A_1 \cap \{\mathbf{x} : \mathbf{u} \cdot \mathbf{x} \leq c_1\}.$$

Let $A_2 \in \mathscr{B}_n$ and let A_2', A_2'' be defined as above with reference to another hyperplane H_2 which is parallel to H_1 and which is such that

$$\frac{P(A_1')}{P(A_1)} = \frac{P(A_2')}{P(A_2)}. \tag{2.9}$$

Write $B = \theta A_1 + (1-\theta)A_2$, $B' = \theta A_1' + (1-\theta)A_2'$ and $B'' = \theta A_1'' + (1-\theta)A_2''$ Then $B \supset B' \cup B''$ and $P(B' \cap B'') = 0$, because P assigns zero mass to hyperplanes. Therefore

$$P(B) \geq P(B') + P(B''). \tag{2.10}$$

Further, condition (2.9) easily shows that

$$\begin{aligned} &\{P(A_1)\}^\theta \{P(A_2)\}^{1-\theta} \\ &\quad = \{P(A_1')\}^\theta \{P(A_2')\}^{1-\theta} + \{P(A_1'')\}^\theta \{P(A_2'')\}^{1-\theta}. \end{aligned} \tag{2.11}$$

Subtracting (2.11) from (2.10), we see that

$$q(A_1, A_2) \geq q(A_1', A_2') + qA_1'', A_2''). \tag{2.12}$$

(c) Let $A = \bigcup_{i=1}^{m} A_i$ and $B = \bigcup_{j=1}^{k} B_j$ be finite disjoint unions of rectangles with sides parallel to the axes. Let $m + k = v$. We use induction on v to show that $q(A, B) \geq 0$. By hypothesis, the result holds when $v = 2$. Suppose the result holds when $v \leq t$ and suppose A, B are as above with $m + k = t + 1$, where $t \geq 2$. We may clearly suppose that $m \geq k$ and so $m \geq 2$. Now A_1 and A_2 are convex. Therefore, they can be (weakly) separated by a hyperplane H_1 which is parallel to a coordinate hyperplane. Now define A', A'' from A and H_1 as in (b) above. Let $\alpha = P(A')/P(A)$. Then we can find a hyperplane H_2 parallel to H_1 in such a way that $P(B')/P(B)$ equals α. Now (2.12) shows that

$$q(A, B) \geq q(A', B') + q(A'', B''). \tag{2.13}$$

But A', B' are unions of rectangles of the required type with $m + k \leq t$ and so are A'' and B''. Therefore $q(A', B') \geq 0$ and $q(A'', B'') \geq 0$, by the induction hypothesis. Therefore, we see from (2.13) that $q(A, B) \geq 0$ whenever A, B are finite disjoint unions of rectangles. Now we can take monotone limits to show that $q(A, B) \geq 0$ whenever A, B are nonempty compact sets. The theorem now follows by the regularity of P. ■

Theorem 2.6 enables us to give a simple proof of the fact that products of logconcave probabilities are again logconcave.

Theorem 2.7. *Let P_1 and P_2 be logconcave probabilities on $\mathscr{B}_m$ and $\mathscr{B}_n$ respectively. Then $P_1 \times P_2$ is a logconcave probability on $\mathscr{B}_m \times \mathscr{B}_n$.*

Proof. Suppose first that supp P_1 has dimension m and that supp P_2 has dimension n. Then $(P_1 \times P_2)$ is absolutely continuous w.r.t. $\lambda_m \times \lambda_n$ and so $P_1 \times P_2$ assigns zero mass to all hyperplanes in $R^m \times R^n$. By Theorem 2.6, we need only show that $P_1 \times P_2$ satisfies the required logconcavity condition on rectangles with sides parallel to the axes. Let A_1, B_1 be in $\mathscr{B}_m$ and let A_2, B_2 be in $\mathscr{B}_n$. Write $P = P_1 \times P_2$, $A = A_1 \times A_2$ and $B = B_1 \times B_2$. For $\theta \in (0, 1)$,

$$\theta A + (1 - \theta)B \supset [\theta A_1 + (1 - \theta)B_1] \times [\theta A_2 + (1 - \theta)B_2].$$

Therefore

$$\begin{aligned} P[\theta A + (1 - \theta)B] &\geq P_1[\theta A_1 + (1 - \theta)B_1] \cdot P_2[\theta A_2 + (1 - \theta)B_2] \\ &\geq \{P_1(A_1)\}^{\theta}\{P_1(B_1)\}^{1-\theta}\{P_2(A_2)\}^{\theta}\{P_2(B_2)\}^{1-\theta} \\ &= [P(A)]^{\theta}[P(B)]^{1-\theta}. \end{aligned}$$

Thus P is logconcave when the supports of P_1 and P_2 have dimensions m and n, respectively. If supp P_1 has dimension $< m$, then P_1 is absolutely continuous w.r.t. Lebesgue measure on some affine set H_1 in R^m. Therefore we can write $P = P_1' \times P_1''$ where P_1'' is a degenerate probability. Similarly, if the support P_2 has dimension $< n$, then we can write $P_2 = P_2' \times P_2''$ where P_2'' is degenerate. Now $P_1 \times P_2 = (P_1' \times P_2') \times (P_1'' \times P_2'')$. By the first part of the proof $P_1' \times P_2'$ is logconcave. Since $P_1'' \times P_2''$ is degenerate, it is very easy to check now that $P_1 \times P_2$ is logconcave. This proves the theorem. ∎

We are now in a position to prove that logconcave probabilities are precisely the ones whose densities w.r.t. Lebesgue measure (on suitable affine sets) are logconcave functions. This result is due to Prékopa (1971, 1973). The simpler proof given here is taken from Rinott (1976). We note that Karlin (1968) calls logconcave densities on R *Polya frequency functions* of type 2. It is easy to see that a density f on R is logconcave if, and only if, the function g defined on R^2 by $g(x, y) = f(x - y)$ has monotone likelihood ratio.

Theorem 2.8. *Suppose P is a probability on $\mathscr{B}_n$ such that the affine hull of* supp P *has dimension n. Then P is logconcave if, and only if, there is a logconcave function g on R^n such that*

$$P(B) = \int_B g(\mathbf{x})\, d\mathbf{x}, \qquad B \in \mathscr{B}_n. \tag{2.14}$$

Proof. (a) Suppose P is logconcave. By Theorem 2.5, $P \ll \lambda_n$. Let $g(\mathbf{x}) = \lim\inf_{\delta \to 0} \{P[N_\delta(\mathbf{x})]/V_\delta\}$, where V_δ is the volume of $N_\delta(\mathbf{x})$. For $0 < \theta < 1$,

$$P\{N_\delta[\theta\mathbf{x} + (1-\theta)\mathbf{y}]\} \geq \{P[N_\delta(\mathbf{x})]\}^\theta \{P[N_\delta(\mathbf{y})]\}^{1-\theta}.$$

Dividing by V_δ and taking lim inf as $\delta \to 0$, we get

$$g(\theta\mathbf{x} + (1-\theta)\mathbf{y}) \geq [g(\mathbf{x})]^\theta [g(\mathbf{y})]^{1-\theta}.$$

But $g(\mathbf{x}) = dP/d\lambda_n$ for almost all $\mathbf{x}$ w.r.t. λ_n. This proves the "only if" part.

(b) For use, in the proof of the "if" part, we define a measure μ on $\mathscr{B}_{n+1}$ by

$$\frac{d\mu}{d\lambda_{n+1}} = e^{-x_{n+1}}.$$

Observe that μ is a product measure. The component measures, namely, the Lebesgue measure λ_1 and the measure with density e^{-x} are easily checked to be logconcave on intervals and hence they are logconcave by Theorem 2.6. Theorem 2.7 then shows that μ is logconcave on $\mathscr{B}_{n+1}$.

(c) To prove the "if" part, suppose (2.14) holds with a logconcave g. If $B \in \mathscr{B}_n$, we write

$$B^* = \{(\mathbf{x}, \alpha): \mathbf{x} \in B, \alpha \in R, \alpha + \log g(\mathbf{x}) \geq 0\}.$$

Then

$$\begin{aligned}\mu(B^*) &= \int_B \left\{\int_{-\log g(\mathbf{x})}^{\infty} e^{-\alpha}\, d\alpha\right\} d\mathbf{x} \\ &= \int_B g(\mathbf{x})\, d\mathbf{x} \doteq P(B).\end{aligned}$$

Further, the concavity of $\log g$ shows that

$$[\theta A + (1-\theta)B]^* \supset \theta A^* + (1-\theta)B^*.$$

Therefore

$$\begin{aligned}P[\theta A + (1-\theta)B] &= \mu\{[\theta A + (1-\theta)B]^*\} \\ &\geq \mu[\theta A^* + (1-\theta)B^*] \\ &\geq [\mu(A^*)]^{\theta}[\mu(B^*)]^{1-\theta} \\ &= [P(A)]^{\theta}[P(B)]^{1-\theta}.\end{aligned}$$

The theorem is thus proved. ■

Theorem 2.8 can be used to prove that certain standard distributions are logconcave. The multivariate normal distribution on R^n with mean vector $\mathbf{m}$ and a (nonsingular) covariance matrix $\boldsymbol{\Sigma}$ has the density

$$f(\mathbf{x}) = K \exp[-\tfrac{1}{2}(\mathbf{x}-\mathbf{m})' \boldsymbol{\Sigma}^{-1}(\mathbf{x}-\mathbf{m})],$$

where K is a suitable constant. Since the matrix $\boldsymbol{\Sigma}$ is positive definite, the function $\mathbf{x}'\boldsymbol{\Sigma}^{-1}\mathbf{x}$ is convex in $\mathbf{x}$. Thus f is logconcave.

Consider now the Wishart distribution with the density

$$g(\mathbf{A}) = K \cdot (\det \mathbf{A})^{(n-p-1)/2} \exp[-\tfrac{1}{2}\operatorname{tr}(\mathbf{A}\boldsymbol{\Sigma}^{-1})],$$

when $\mathbf{A}$ is positive definite and $g(\mathbf{A}) = 0$, otherwise. Here $\mathbf{A}$ is a symmetric random matrix of order $p \times p$ and $\boldsymbol{\Sigma}$ is a fixed positive definite, symmetric matrix of order $p \times p$. Again K is a suitable constant. We assume that $n \geq (p+1)$. If we write $\boldsymbol{\Sigma}^{-1} = (\sigma^{ij})$, then $\operatorname{tr} \mathbf{A}\boldsymbol{\Sigma}^{-1} = \sum\sum a_{ij}\sigma^{ij}$, which is linear in the a_{ij}'s. Therefore, to show that g is logconcave, it is sufficient to show that $\det \mathbf{A}$ is logconcave in the a_{ij}'s. That is, we need to show that

$$\det[\theta\mathbf{A}_1 + (1-\theta)\mathbf{A}_2] \geq (\det \mathbf{A}_1)^{\theta}(\det \mathbf{A}_2)^{1-\theta}, \tag{2.15}$$

whenever $\mathbf{A}_1, \mathbf{A}_2$ are positive definite and $0 < \theta < 1$. But this result is well known; see, for example, Marshall and Olkin (1979, p. 476) or Beckenbach and Bellman (1965, p. 63). Thus, for $n \geq (p+1)$, the Wishart distribution is logconcave.

A third example of a logconcave distribution is the Dirichlet distribution. This has the density

$$h(\mathbf{x}) = K\left(1 - \sum_{i=1}^{n} x_i\right)^{p_{n+1}-1} \prod_{i=1}^{n} x_i^{p_i - 1},$$

whenever $x_i > 0$ for $i = 1, \ldots, n$ and $\sum_{i=1}^{n} x_i < 1$ and $h(\mathbf{x}) = 0$, otherwise. If we assume that $p_i \geq 1$ for $i = 1, \ldots, (n+1)$, then the logconcavity of h follows immediately from the concavity of the logarithmic function.

If f is a logconcave function on R^n, it is easily seen that for every $c > 0$, the set $\{x : f(\mathbf{x}) \geq c\}$ is convex. Therefore, Theorem 2.8 immediately implies the following theorem.

Theorem 2.9. *Every centrally symmetric, absolutely continuous, logconcave distribution is convex unimodal.*

2.4. Preservation of Unimodality Properties under Weak Limits

Several notions of unimodality were introduced in Section 2.2 and the concept of logconcavity of distributions was discussed in Section 2.3. In this section, we show that almost all these properties of unimodality are preserved under weak limits. The classes of distributions which are star, block or central convex unimodal are closed under weak limits by definition. We have already noted that the class of convex unimodal distributions is not closed under weak limits because the limits need not be absolutely continuous. For the remaining types of unimodality, the situation is summarized in the next theorem.

Theorem 2.10. *The properties of linear unimodality, monotone unimodality and logconcavity are preserved under weak limits.*

Proof. (a) For linear unimodality, the result follows easily from Theorem 1.1.

(b) Let $\{P_m\}$ be a sequence of monotone unimodal distributions on R^n and let $P_m \to P$ weakly. We show that P is monotone unimodal. Let C be a compact, centrally symmetric convex body in R^n and let $\mathbf{y} \in R^n$ be nonzero. For $k > 0$, let A_k be the set of all $\alpha > 0$ such that the boundary of $\alpha C + k\mathbf{y}$

has positive P-measure. For fixed k and varying α, the boundaries of $\alpha C + k\mathbf{y}$ are disjoint. Therefore the set A_k is countable. Let $0 < k < t$ and $\alpha \in (A_k \cup A_t)^c$. Then the boundaries of $\alpha C + k\mathbf{y}$ and $\alpha C + t\mathbf{y}$ have P-measure zero. Therefore

$$P_m(\alpha C + k\mathbf{y}) \to P(\alpha C + k\mathbf{y}) \quad \text{and} \quad P_m(\alpha C + t\mathbf{y}) \to P(\alpha C + t\mathbf{y}).$$

But $P_m(\alpha C + k\mathbf{y}) \geq P_m(\alpha C + t\mathbf{y})$. Therefore,

$$P[\alpha C + k\mathbf{y}] \geq P[\alpha C + t\mathbf{y}],$$

whenever $\alpha \in [A_k \cup A_t]^c$. Letting α tend to 1 downward through the dense set $[A_k \cup A_t]^c$, we see that $P[C + k\mathbf{y}] \geq P[C + t\mathbf{y}]$. This proves that P is monotone unimodal.

(c) Let $\{P_\nu\}$ be a sequence of logconcave probabilities on R^n and let $P_\nu \to P$ weakly. If A and B are nonempty closed sets in $\mathscr{B}_n$ such that the boundaries ∂A and ∂B have P-measure zero, then, for $\theta \in (0, 1)$, we have

$$\begin{aligned} P[\theta A + (1-\theta)B] &\geq \lim_{\nu\to\infty} P_\nu[\theta A + (1-\theta)B] \\ &\geq \lim_{\nu\to\infty} [P_\nu(A)]^\theta [P_\nu(B)]^{1-\theta} \qquad (2.15) \\ &= [P(A)]^\theta [P(B)^{1-\theta}. \end{aligned}$$

Now, if G and H are nonempty open sets in R^n then we can approach G and H from below by closed sets A and B whose boundaries have zero P-measure. Therefore, (2.15) shows that

$$P[\theta G + (1-\theta)H] \geq [P(G)]^\theta [P(H)]^{1-\theta}.$$

Now the logconcavity of P follows by its regularity. The theorem is thus proved. ■

2.5. The Choquet Version of Central Convex Unimodality

Kanter (1977) has given, for centrally symmetric distributions on R^n, a definition of unimodality which can be considered to be the Choquet version of central convex unmimodality. For a centrally symmetric, compact, convex set $K \subset R^n$, let W_K denote the uniform distribution on K. Generalized mixtures (that is, integrals) of such uniform distributions are called "symmetric unimodal" by Kanter. We show, in this section, that Kanter's class coincides with the class of central convex unimodal distributions. This is Theorem 2.11 below. For ease of reference, we denote by $\mathscr{C}_n^*$ the class of all generalized mixtures of the distributions W_K defined above and we use $\mathscr{C}_n$ to denote the set of all central convex unimodal distributions on R^n. First we need a lemma which is due to Kanter. The proof given here is only a sketch of a complete proof in the sense that we do not prove the measurability of the relevant maps. For a complete proof, we refer the reader to Kanter (1977).

Lemma 2.2. *The set $\mathscr{C}_n^*$ is closed under weak convergence.*

Sketch of the Proof. Let $\mathscr{L}$ denote the set of all centrally symmetric logconcave probability distributions on R^n. We first observe that $\mathscr{C}_n^*$ coincides with the class of all generalized mixtures of distributions in $\mathscr{L}$. To see this, note that by Theorem 2.9, every absolutely continuous distribution in $\mathscr{L}$ is convex unimodal and by the remark following Theorem 2.3, every convex unimodal distribution is in $\mathscr{C}_n^*$. Therefore $\mathscr{L} \subset \mathscr{C}_n^*$ and consequently, every generalized mixture of distributions in $\mathscr{L}$ is again in $\mathscr{C}_n^*$. On the other hand, the uniform distributions W_K are all in $\mathscr{L}$ by Theorem 2.8. Therefore every distribution in $\mathscr{C}_n^*$ is a generalized mixture of distributions in $\mathscr{L}$. The advantage of this representation is that $\mathscr{L}$ is closed under weak convergence (see Theorem 2.10).

Suppose $P_m \in \mathscr{C}_n^*$ and let $P_m \to P_0$ weakly. By the preceding paragraph, we can write

$$P_m = \int_{\mathscr{L}} Q \; d\sigma_m(Q),$$

where σ_m is a probability measure on $\mathscr{L}$. We show that the sequence $\{\sigma_m\}$ is tight. Let ε, δ be positive numbers less than 1. Since $\{P_m\}$ converges weakly, there exists a compact set $K \subset R^n$ such that $P_m(K) > 1 - \varepsilon\delta$ for all m. This means that

$$\int_{\mathscr{L}} Q(K) \; d\sigma_m(Q) > 1 - \varepsilon\delta,$$

for all m. Thus $Q(K)$, as a function of Q, is bounded above by 1 and its σ_m-expectation exceeds $1 - \varepsilon\delta$. But if a real random variable X is bounded above by A and $k < A$, then, by elementary arguments,

$$P(X \geq k) \geq [E(X) - k]/(A - k).$$

Therefore, we conclude that

$$\sigma_m\{Q : Q(K) \geq 1 - \delta\} > 1 - \varepsilon,$$

for all m. Now a double application of Prohorov's theorem (see Theorem A.4 in the Appendix) shows that $\{\sigma_m\}$ is tight. By going to a subsequence, we may assme that σ_m converges weakly to some probability σ_0 on $\mathscr{L}$. But then it easily follows that

$$P_m \to \int_{\mathscr{L}} Q \, d\sigma_0(Q).$$

Since P_m also converges to P_0, we see that

$$P_0 = \int_{\mathscr{L}} Q \, d\sigma_0(Q).$$

Thus $P_0 \in \mathscr{C}_n^*$. The lemma is thus proved. ■

Theorem 2.11. *A distribution P on R^n is central convex unimodal if, and only if, it is a generalized mixture of uniform distributions on centrally symmetric convex sets in R^n.*

Proof. Recall that $\mathscr{C}_n^*$ denotes the set of all generalized mixtures of uniform distributions on symmetric convex sets and that $\mathscr{C}_n$ denotes the set of all central convex unimodal distributions. Suppose $K \subset R^n$ is a centrally symmetric convex body. Then W_K, the uniform distribution on K, is in $\mathscr{C}_n^*$. Since $\mathscr{C}_n^*$ is clearly convex and is also closed by Lemma 2.2, $\mathscr{C}_n^*$ must include the closed convex hull of the set of all such distributions W_K. But the latter set is just $\mathscr{C}_n$ because of Definition 2.5. Thus $\mathscr{C}_n \subset \mathscr{C}_n^*$.

Now let $K \subset R^n$ be a centrally symmetric convex set. We show that $W_K \in \mathscr{C}_n$. If K has dimension n, then K is a convex body and $W_K \in \mathscr{C}_n$, by definition. If K has dimension $m < n$, assume, without loss of generality, that $\mathbf{x} \in K \Rightarrow x_i = 0$ for $i > m$. Write $R^n = R^m \times R^{n-m}$. Let N_δ be the ball in R^{n-m} with center $\mathbf{0}$ and radius δ. Since $K \times N_\delta$ is a centrally symmetric convex body, the uniform distribution on $K \times N_\delta$ is in $\mathscr{C}_n$. Letting $\delta \to 0$, we see that $W_K \in \mathscr{C}_n$.

Let $P \in \mathscr{C}_n^*$. Then P has the representation

$$P = \int_{\Theta_n} W_K \, d\nu(W_K),$$

where Θ_n is the set of all uniform distributions on centrally symmetric convex sets in R^n. The preceding paragraph shows that $P \in \mathscr{C}_n$ if ν is degenerate. The same results holds if ν has finite support because $\mathscr{C}_n$ is convex. Now suppose that ν is arbitrary. There is a sequence ν_m of measures with finite support such that $\nu_m \to \nu$ weakly. Define

$$P_m = \int_{\Theta_n} W_K \, d\nu_m(W_K).$$

Then $P_m \to P$ weakly. But each $P_m \in \mathscr{C}_n$ and $\mathscr{C}_n$ is closed. Therefore $P \in \mathscr{C}_n$. Thus $\mathscr{C}_n^* \subset \mathscr{C}_n$ and the theorem is proved. ■

2.6. Interrelationships among the Definitions

In this section we study how the various defnitions in Sections 2.2 and 2.3 are related to one another. Note that we have four definitions applicable to all distributions. These are: star unimodality, block unimodality, linear unimodality and logconcavity.

Theorem 2.12. *Block unimodality ⇒ star unimodality, and logconcavity ⇒ star unimodality. There are no other implications among star unimodality, block unimodality, linear unimodality and logconcavity.*

Proof. For ease of reference we first list four distributions in R^2. Let A be the triangle with vertices $(0, 0)$, $(1, 1)$ and $(1, -1)$. Then the uniform distribution on $A \cup (-A)$ will be denoted by P_1. The symbol P_2 will denote the uniform distribution on $B \cup (-B)$, where B is the unit square. Next, P_3 will be the uniform distribution on an ellipse and P_4 will be the uniform distribution on a triangle.

(a) *Star vs block.* Since a rectangle containing $\mathbf{0}$ is also star-shaped about $\mathbf{0}$, the first implication of the theorem follows from the definitions of star and block unimodality. The reverse implication does not hold because the distribution P_1 is star unimodal but not block unimodal.
(b) *Star vs linear.* Let (X_1, X_2) have the distribution P_2. Then (X_1, X_2) is star unimodal but not linear unimodal because $X_1 + X_2$ has a bimodal distribution. Example 2.1 gives a linear unimodal distribution which is not star unimodal.
(c) *Star vs logconcave.* Again the distribution P_2 is star unimodal but not logconcave. Now suppose that f is a logconcave density. Then, for every $c > 0$, the set $\{\mathbf{x} : f(\mathbf{x}) \geq c\}$ is convex and a fortiori, star-shaped. Consequently every logconcave distribution is star unimodal and the second implication of the theorem is verified.
(d) *Block vs linear.* The density of Example 2.1 is linear unimodal but not

star unimodal. Therefore, the density cannot be block unimodal. Again, the distribution P_2 is block unimodal but not linear unimodal.

(e) *Block vs logconcave.* The distribution P_2 is block unimodal but it is not logconcave because its support is not convex. The distribution P_3 is logconcave but not block unimodal.

(f) *Linear vs logconcave.* A linear unimodal distribution which is not star unimodal cannot be logconcave. The distribution P_4 is logconcave but not linear unimodal because, even though each linear function has a unimodal distribution, there is no "vertex" of linear unimodality.

All the assertions of the theorem are now verified. ∎

Theorem 2.13. *For centrally symmetric distributions, block unimodality ⇒ star unimodality, logconcavity ⇒ star unimodality, logconcavity ⇒ linear unimodality and there are no other implications among star unimodality, block unimodality, linear unimodality and logconcavity.*

Proof. With the exception of the distribution P_4, all the other distributions used in the proof of Theorem 2.12 are centrally symmetric. Therefore we only need to prove the third implication of the present theorem. Let $\mathbf{X}$ have a centrally symmetric logconcave distribution P. By Lemma 2.1, every linear function $\Sigma\, a_i x_i$ has a unimodal and centrally symmetric distribution. Thus each $\Sigma\, a_i x_i$ is unimodal about 0 and hence P is linear unimodal. The theorem follows. ∎

For centrally symmetric distributions, Theorem 2.15 below gives some additional interrelationships among the various definitions. For ease of later reference, we isolate one important relation as Theorem 2.14 and refer to it as *Anderson's Theorem.* It is a slightly extended version of the main theorem of Anderson (1955). The proof given here is based on the following lemma which was given by Fáry and Rédéi (1950) and was rediscovered by Sherman (1955).

Lemma 2.3. *Let C and D be convex bodies in R^n and let $\lambda_n(A)$ denote the Lebesgue measure of A in R^n. Then $\psi(\mathbf{x}) = \{\lambda_n[C \cap (D + \mathbf{x})]\}^{1/n}$ is concave on its support.*

Proof. According to the Brunn-Minkowski inequality (see Section A.4 of the Appendix), for two convex bodies A and B in R^n and $0 < \theta < 1$,

$$[\lambda_n(\theta A + (1-\theta)B)]^{1/n} \geq \theta[\lambda_n(A)]^{1/n} + (1-\theta)[\lambda_n(B)]^{1/n}. \qquad (2.16)$$

We use this inequality with $A = C \cap (D + \mathbf{x}_1)$ and $B = C \cap (D + \mathbf{x}_2)$ where

$\mathbf{x}_1$ and $\mathbf{x}_2$ are in the support of ψ. Since $\psi(x_1)$ and $\psi(x_2)$ are positive, A and B are convex bodies. We note that

$$C \cap (D + \theta\mathbf{x}_1 + (1-\theta)\mathbf{x}_2) \supset \theta A + (1-\theta)B. \tag{2.17}$$

To see this, let $\mathbf{y} \in \theta A + (1-\theta)B$. Then there exist $\mathbf{y}_1$, $\mathbf{y}_2$ such that $\mathbf{y}_1 \in C \cap (D + \mathbf{x}_1)$, $\mathbf{y}_2 \in C \cap (D + \mathbf{x}_2)$ and $\mathbf{y} = \theta\mathbf{y}_1 + (1-\theta)\mathbf{y}_2$. The convexity of C shows that $\mathbf{y} \in C$. Further $\mathbf{y}_1 - \mathbf{x}_1$ and $\mathbf{y}_2 - \mathbf{x}_2$ both belong to D, which is convex. Therefore $\theta(\mathbf{y}_1 - \mathbf{x}_1) + (1-\theta)(\mathbf{y}_2 - \mathbf{x}_2)$ belongs to D. This means that $\theta\mathbf{y}_1 + (1-\theta)\mathbf{y}_2 - [\theta\mathbf{x}_1 + (1-\theta)\mathbf{x}_2] \in D$. Thus $\mathbf{y} \in D + \theta\mathbf{x}_1 + (1-\theta)\mathbf{x}_2$ and (2.17) is proved. Now (2.16) and (2.17) show that

$$\begin{aligned}\psi[\theta\mathbf{x}_1 + (1-\theta)\mathbf{x}_2] &\geq \{\lambda_n[\theta A + (1-\theta)B]\}^{1/n} \\ &\geq \theta[\lambda_n(A)]^{1/n} + (1-\theta)[\lambda_n(B)]^{1/n} \\ &= \theta\psi(\mathbf{x}_1) + (1-\theta)\psi(\mathbf{x}_2).\end{aligned}$$

This proves the lemma. ■

Theorem 2.14. (Anderson's Theorem). *Every centrally convex unimodal distribution is monotone unimodal.*

Proof. Let P be the uniform distribution on a centrally symmetric convex body C. For a centrally symmetric convex body D, let

$$g(\mathbf{x}) = [P(D + \mathbf{x})]^{1/n} = \frac{\{\lambda_n[C \cap (D + \mathbf{x})]\}^{1/n}}{[\lambda_n(C)]^{1/n}}.$$

Lemma 2.3 shows that $g(\mathbf{x})$ is concave on its support. But in the present case, $g(\mathbf{x})$ is also centrally symmetric. This easily implies that for every fixed $\mathbf{x}$, $g(k\mathbf{x})$ is symmetric unimodal in $k \in R$. Thus P is monotone unimodal. Now the set of all monotone unimodal distributions is closed (by Theorem 2.10) and trivially convex. The theorem thus follows from Definition 2.5. ■

Theorem 2.15. *For centrally symmetric distributions:*

(a) *Logconcavity ⇒ convex unimodality ⇒ central convex unimodality ⇒ monotone unimodality*
(b) *Montone unimodality ⇒ star unimodality*
(c) *Monotone unimodality ⇒ linear unimodality*
(d) *None of the implications in* (a), (b) *and* (c) *can be reversed, in general.*
(e) *No implication relationship exists between block unimodality on the one hand and monotone unimodality, central convex unimodality or convex unimodality on the other hand.*

Proof. (1) The first implication in (a) is Theorem 2.9, the second implication is Theorem 2.3 and the third implication is Theorem 2.14.

The first implication cannot be reversed even on the line because the Cauchy distribution is convex unimodal but not logconcave. The second implication cannot be reversed because the set of all convex unimodal distributions is neither convex nor closed.

Sherman (1955) had conjectured that monotone unimodality implies central convex unimodality. This conjecture was settled in the negative by Wells (1978) who verified a (possible) counterexample proposed by Dharmadhikari and Jogdeo (1976a). The example is as follows. Let ABC be an equilateral triangle with centroid at the origin and let $A'B'C'$ be its reflexion through the origin. Let Q be the distribution which is the $(\frac{1}{2}, \frac{1}{2})$ mixture of the uniform distributions on ABC and $A'B'C'$. If Q is a mixture of uniform distributions on centrally symmetric convex sets $D_1, D_2, \ldots$, then it is clear that A, B, C must belong to distinct sets D_i. Consequently, the density on the inner hexagon $(ABC) \cap (A'B'C')$ must be at least three times the density on the six outer triangles. But for Q, the density on the inner hexagon is only twice that on the outer triangles. Therefore Q is not central convex unimodal. The verification that Q is monotone unimodal is somewhat long and was carried out by Wells (1978). Thus the third implication in (a) cannot be reversed.

(2) Suppose P is monotone unimodal and let Q_δ denote the uniform distribution on the ball C_δ with center $\mathbf{0}$ and radius δ. The density of $P * Q_\delta$ at $\mathbf{x}$ is then proportional to $P(C_\delta + \mathbf{x})$. The monotone unimodality of P shows that $P(C_\delta + k\mathbf{x})$ is unimodal in k. Thus $P * Q_\delta$ is star unimodal. Letting $\delta \to 0$, we see that P is star unimodal. We now show that P is also linear unimodal. It is sufficient to prove that P_1, *the* x_1*-marginal of* P, is unimodal. This is because any non-singular linear transformation on R^n takes convex bodies into convex bodies. Let $\delta > 0$, $D = \{\mathbf{x} \in R^n : -\delta < x_1 < \delta\}$ and $\mathbf{y} = (1, 0, \ldots, 0)$. Then $P_1[(k - \delta, k + \delta)] = P(D + k\mathbf{y})$, which is unimodal in k. This means that the convolution of P_1 with the uniform distribution on $(-\delta, \delta)$ is unimodal. Once again we let $\delta \to 0$ to see that P_1 is unimodal. Thus (b) and (c) are verified.

Now a star unimodal distribution which is not linear unimodal cannot be monotone unimodal. Similarly a linear unimodal distribution which is not star unimodal cannot be monotone unimodal. Such distributions do exist in view of Theorem 2.13. Consequently the implicatons (b) and (c) cannot be reversed.

(3) Once again, we recall from Theorem 2.13 that block unimodality does not imply linear unimodality. Therefore block unimodality cannot imply

monotone unimodality or central convex unimodality or convex unimodality. The uniform distribution on the unit ball is convex unimodal but not block unimodal. Statement (e) is thus verified and the proof of the theorem is complete. ■

Remark. It is easy to see that, for spherically symmetric distributions, star unimodality, monotone unimodality, central convex unimodality and convex unimodality are all equivalent.

Remark. Suppose $\mathbf{X}$ is uniformly distributed on some set $S \subset R^2$. If every linear function $\Sigma\, a_i X_i$ is known to have a unimodal distribution, the set S may not even be star-shaped. However, if every linear function has a unimodal distribution about 0, then, under mild assumptons, the set S is convex and centrally symmetric. Thus, for uniform distributions on sets in R^2, linear unimodality implies convex unimodality. For details, see Converse (1977) and Dharmadhikari and Jogdeo (1973).

2.7. Marginal Distributions

In applications, it is important to know whether conditions on a multivariate distribution are inherited by its marginals. We examine this question for the various notions of unimodality introduced in this chapter. Marginals are considered to be projections on lower dimensional spaces, which are not necessarily of dimension one.

Lemma 2.1 essentially states that marginals of logconcave distributions are logconcave. A direct and simple proof of this result in the absolutely continuous case was given by Brascamp and Lieb (1975). We present it below.

Theorem 2.16. *Let f be a logconcave density on R^{m+n}* and let

$$g(\mathbf{x}) = \int_{R^n} f(\mathbf{x}, \mathbf{y})\, d\mathbf{y}.$$

Then g is logconcave on R^m.

Proof. Firstly, we may assume that $m = 1$ because we only need to show that the restriction of g to any straight line is logconcave. Secondly, a simple induction argument shows that we may also take $n = 1$. So let $m = n = 1$ and let x_1, x_2 be in R. Write $F(x, y) = e^{ax} f(x, y)$ where a depends on x_1, x_2 and is chosen in such a way that

$$\sup_y F(x_1, y) = \sup_y F(x_2, y). \tag{2.18}$$

We note that the logconcavity of f and F are equivalent. Define

$$G(x) = \int F(x, y)\, dy = e^{ax} g(x).$$

Again the logconcavity of g and G are equivalent. So we show that G is logconcave.

For $x \in R$ and $z \geq 0$, let $C(x, z) = \{y : F(x, y) \geq z\}$. Let $\alpha(x, z)$ be the Lebesgue measure of $C(x, z)$. The set $C(x, z)$ is convex because F is logconcave. Further (2.18) ensures that $\alpha(x_1, z) > 0$ if, and only if, $\alpha(x_2, z) > 0$. Now let $0 < \theta < 1$. Then

$$C[\theta x_1 + (1-\theta)x_2, z] \supset \theta C(x_1, z) + (1-\theta)C(x_2, z).$$

Therefore, the Brunn–Minkowski inequality in one dimension shows that

$$\alpha(\theta x_1 + (1-\theta)x_2, z) \geq \theta\alpha(x_1, z) + (1-\theta)\alpha(x_2, z). \tag{2.19}$$

By a standard argument

$$G(x) = \int_R F(x, y)\, dy = \int_R \left\{ \int_0^\infty I_{C(x,z)}\, dz \right\} dy = \int_0^\infty \alpha(x, z)\, dz.$$

Therefore, (2.19) shows that

$$\begin{aligned} G(\theta x_1 + (1-\theta)x_2] &\geq \theta G(x_1) + (1-\theta)G(x_2) \\ &\geq [G(x_1)]^\theta [G(x_2)]^{1-\theta}, \end{aligned}$$

where the last step follows by the arithmetic mean-geometric mean inequality. Thus G is logconcave. The proof of the theorem is complete. ■

We now present a theorem about the marginals of the various other types of unimodal distributions.

Theorem 2.17.

(a) *Marginals of star unimodal distributions need not be star unimodal.*
(b) *If* $(X_1, \ldots, X_n)$ *has a block unimodal distribution, then for* $1 \leq m < n$, *the distribution of* $(X_1, \ldots, X_m)$ *is block unimodal.*
(c) *Marginals of linear unimodal distributions are linear unimodal.*
(d) *Marginals of convex unimodal distributions need not be convex unimodal.*

However, uniform distributions on symmetric convex bodies have convex unimodal marginals.

(e) *Marginals of central convex unimodal distributions are central convex unimodal.*

(f) *Marginals of monotone unimodal distributions are monotone unimodal.*

Proof. (1) We have seen in Theorem 2.12 that a star unimodal distribution need not be linear unimodal. Consequently, the marginals of star unimodal distributions need not be star unimodal. This verifies (a). Verification of (b) is immediate and (c) follows trivially from Definition 2.3.

(2) Suppose f is a convex unimodal density on R^2 and let g be the uniform density on a centrally symmetric convex set $C \subset R^2$. If we write $h(\mathbf{x}, \mathbf{y}) = f(\mathbf{x} - \mathbf{y})g(\mathbf{y})$, then h is a convex unimodal density on R^4. Further the $\mathbf{x}$-marginal of h is $f * g$. Now Sherman (1955) showed [see part (c) of the proof of Theorem 2.19] that the density f and the set C can be chosen in such a way that $f * g$ is not convex unimodal. Thus the marginals of the convex unimodal density h need not be convex unimodal. This verifies the first part of (d). To see the second part, let P be the uniform distribution on a centrally symmetric convex body. Then P has a logconcave density. Therefore, by Theorem 2.16, any marginal of P has a logconcave density. Since this marginal is also centrally symmetric, it is convex unimodal. Statement (d) is thus completely verified.

(3) Again let P be the uniform distribution on a centrally symmetric convex body. Any marginal of P is convex unimodal and so it is also central convex unimodal. By taking mixtures and weak limits we find that central convex unimodal distributions have central convex unimodal marginals. This verifies (e).

(4) Let $\mathbf{X} = (\mathbf{X}_1, \ldots, X_n)$ have a monotone unimodal distribution P and let $\mathbf{Y} = (X_1, \ldots, X_m)$, $m < n$. Let $C \subset R^m$ be centraly symmetric and convex and let $\mathbf{y} \in R^m$ be nonzero. Write $D = C \times R^{n-m}$ and $\mathbf{x} = (\mathbf{y}, \mathbf{0}) \in R^n$. Let Q denote the distribution of $\mathbf{Y}$. Then $Q(C + k\mathbf{y}) = P(D + k\mathbf{x})$, which is unimodal in k. Therefore Q is monotone unimodal. This proves (f) and completes the proof of the theorem. ■

2.8. Convolutions

In probability and statistics, convolution is one of the common and useful operations. Therefore it is important to know whether a unimodality property is preserved under convolutions. From the univariate case we know that without some conditions on the shape of the distributions, like symmetry or logconcavity, convolutions of unimodal distributions are, in general, not unimodal. The notion of symmetry which is important for multivariate applications is central symmetry. Other notions of symmetry can also be used and will be considered in Chapter 3.

Theorem 2.18. *Let P and Q be logconcave distributions on R^n. Then the convolution $P * Q$ is logconcave.*

Proof. It suffices to prove the theorem in the case where P and Q are absolutely continuous. So let f, g be the densities of P and Q, respectively. By Theorem 2.8, f and g can be chosen to be logconcave. Now the function $h(\mathbf{x}, \mathbf{y}) = f(\mathbf{x} - \mathbf{y})g(\mathbf{y})$ is logconcave on R^{2n}. Therefore, by Theorem 2.16,

$$(f * g)(\mathbf{x}) = \int_{R^n} h(\mathbf{x}, \mathbf{y})\, dy$$

is logconcave. Again, Theorem 2.8 shows that $P * Q$ is logconcave. ∎

For notions of unimodality other than that of logconcavity we consider centrally symmetric distributions only and separate the results on convolutions into positive and negative ones.

Theorem 2.19. *Consider centrally symmetric distributions on R^n. The convolution of two star (respectively, block, convex) unimodal distributions need not be star (respectively, block, convex) unimodal.*

Proof. (a) Let X_1, X_2 be independent random variables with density $p(x) = |x|$, $-1 < x < 1$. Then X_1 is distributed as $\sqrt{U_1} \cdot Z_1$ where U_1 is uniform on $(0, 1)$, Z_1 is independent of U_1 and $P(Z_1 = \pm 1) = \frac{1}{2}$. Thus $(X_1, 0)$ is distributed as $\sqrt{U_1} \cdot (Z_1, 0)$ and is therefore star unimodal by Theorem 2.1. Similarly $(0, X_2)$ is also star unimodal. But their sum (X_1, X_2) is not star unimodal because its density, namely,

$$f(x, y) = |xy|, \qquad |x| < 1, \qquad |y| < 1$$

does not decrease along rays. Thus the convolution of two star unimodal distributions need not be star unimodal.

(b) Consider two densities f and g on R^2, where f corresponds to the uniform distribution on $A \cup (-A)$, A is the unit square and g corresponds to the uniform distribution on the square with vertices $(\pm\frac{1}{2}, \pm\frac{1}{2})$. Then $(f * g)(-\frac{1}{2}, -\frac{1}{2}) = \frac{1}{2}$ and $(f * g)(\frac{1}{2}, \frac{1}{2}) = \frac{1}{2}$ but $(f * g)(0, 0) = \frac{1}{4}$. Thus $(f * g)$ is not star unimodal and so cannot be block unimodal. But f and g are clearly block unimodal.

(c) Sherman (1955) gave an example of the following type to show that convolutions of convex unimodal distributions need not be convex unimodal. Let

$$f(x, y) = \begin{cases} \alpha & \text{if } |x| \leq 1, |y| \leq 1, \\ \beta & \text{if } |x| \leq 1, 1 < |y| \leq 5 \\ 0 & \text{elsewhere.} \end{cases}$$

We require $4\alpha + 16\beta = 1$ and $\beta \leq \alpha$ so that f is a convex unimodal density. Let

$$g(x, y) = \begin{cases} \frac{1}{4} & \text{if } |x| \leq 1, |y| \leq 1, \\ 0 & \text{otherwise.} \end{cases}$$

Let $h = f * g$. Then it is easy to see geometrically that

$$h(0, 4) = \beta, \qquad h(\tfrac{1}{2}, 2) = \tfrac{3}{4}\beta \text{ and } h(1, 0) = \frac{\alpha}{2}.$$

Therefore, if we choose $\alpha > \frac{3}{2}\beta$, then h is not convex unimodal. We can choose, for instance, $\alpha = 1/(12)$ and $\beta = 1/(24)$. The proof of the theorem is now complete. ■

Theorem 2.20. *Consider centrally symmetric distributions P and Q on R^n.*

(a) *P and Q linear unimodal $\Rightarrow P * Q$ is linear unimodal.*
(b) *P and Q central convex unimodal $\Rightarrow P * Q$ is central convex unimodal.*
(c) *$\left.\begin{array}{l}P \text{ central convex unimodal} \\ \text{and } Q \text{ monotone unimodal}\end{array}\right\} \Rightarrow P * Q$ is monotone unimodal.*

Proof. (1) The result (a) follows immediately from the univariate result of Wintner (Theorem 1.6).

(2) Theorem 2.14 shows that the convolution of uniform distributions on centrally symmetric convex bodies is convex unimodal and hence central convex unimodal. Now Defnition 2.5 implies that the set of central convex unimodal distributions is closed under convolutions. Thus (b) is proved.

(3) To prove (c), we first note that a distribution P on R^n is monotone unimodal if, and only if, the convolution $P * Q$ is star unimodal for every central convex unimodal distribution Q on R^n. This is because, if Q is the uniform distribution on a centrally symmetric convex set C, then $P(C + k\mathbf{y})$ is (apart from a multiplicative constant) just the density of $P * Q$ at the point $k\mathbf{y}$.

Suppose now that f is a central convex unimodal density and g is a monotone unimodal density. Let h be an arbitrary central convex unimodal density. Then

$$(g * f) * h = g * (f * h) = g * f_1,$$

where $f_1 = f * h$. But f_1 is central convex unimodal by part (b). Therefore $g * f_1$ is star unimodal. It follows that $(g * f)$ is monotone unimodal. This simple proof of part (c) of the theorem was shown to us by Professors S. Das Gupta (1976c) and M. Perlman (1976).

The theorem is thus proved. ∎

Remark. It is not known whether the convolution of two monotone unimodal distributions is monotone unimodal.

3 Some More Notions of Unimodality

3.0. Summary

In Chapter 2, we studied several different notions of unimodality for distributions in higher dimensions. The current chapter supplements and generalizes some of the results of Chapter 2. We first consider the concept of Schur concavity of distributions and indicate how this is just another concept of unimodality. Next, we discuss the results of Olshen and Savage (1970) on the concept of generalized unimodality. These results are extensions and completions of the results on star unimodality proved in Chapter 2. The final section considers a notion of concavity for measures which generalizes the notion of logconcavity discussed in Chapter 2. We continue to use the notation introduced in Chapter 2.

3.1. Schur Concavity and Related Concepts of Unimodality

In Chapter 2, we defined central convex unimodal distributions as weak limits of mixtures of uniform distributions on centrally symmetric convex bodies. One of the useful results of Chapter 2 is that, if P is a central convex unimodal distribution on R^n, then $P(C + k\mathbf{y})$ is nonincreasng in $k \in [0, \infty)$, for every centrally symmetric convex set C in R^n and for every $\mathbf{y} \in R^n$. We thus have a notion of convexity and a notion of symmetry. We also have a partial order on a family of sets whereby $C + k\mathbf{y}$ dominates $C + k'\mathbf{y}$ whenever $k < k'$. The end result is a probability inequality which states that $P(A) \geq P(B)$ whenever A dominates B. It is possible to give some more examples of this phenomenon by using some other notions of convexity and symmetry. This is done in the present section.

We begin with a discussion of the concept of Schur convexity. In view of the excellent monograph on majorization by Marshall and Olkin (1979), we restrict ourselves here to emphasizing those aspects of the theory which have direct interpretations in terms of unimodality. We first consider the bivariate case.

Let L be the equiangular line $x_1 = x_2$. Consider two vectors $\mathbf{a}$ and $\mathbf{b}$ such that $a_1 + a_2 = b_1 + b_2$. We say that the vector $\mathbf{b}$ *majorizes* the vector $\mathbf{a}$ and we write $\mathbf{a} < \mathbf{b}$ if $\mathbf{a}$ is closer to L than $\mathbf{b}$. Thus $\mathbf{a} < \mathbf{b}$ if, and only if $\mathbf{a}$ is in the line segment joining (b_1, b_2) to (b_2, b_1). Note that the points (a_1, a_2) and (a_2, a_1) are considered equivalent with regard to the partial ordering of majorization.

Definition 3.1.

(i) A measurable set B is said to be *Schur convex* if

$$\mathbf{b} \in B \quad \text{and} \quad \mathbf{a} < \mathbf{b} \Rightarrow \mathbf{a} \in B.$$

(ii) A real-valued measurable function f is said to be *Schur concave* if

$$\mathbf{a} < \mathbf{b} \Rightarrow f(\mathbf{a}) \geq f(\mathbf{b}).$$

In Chapter 1, it was seen (see Theorem 1.5) that the class of all symmetric unimodal distributions on R is just the closed convex hull of the set of all uniform distributions on symmetric intervals. Now Definition 3.1(i) says that a Schur convex set is one whose linear sections perpendicular to L are symmetric intervals centered on L. If we use the uniform distributions on Schur convex sets as building blocks, we get the following definition.

Definition 3.2. A distribution P is called *Schur unimodal* if P is in the closed convex hull of the set of all uniform distributions on Schur convex sets.

If a distribution P has a continuous density f, then it is easy to show that P is Schur unimodal if, and only if, f is Schur concave.

We recall another useful result from Chapter 1, namely, the corollary following Theorem 1.6. This states that, under a symmetric unimodal distribution, an interval gets decreasing amounts of mass if its center is moved away from the mode. Exactly in the same way, a Schur concave density assigns a decreasing amount of mass to a Scur convex set B as the central line of B is moved away from L in a direction perpendicular to L. This last result can be restated in an equivalent form as follows. The convolution of a Schur concave density f and the uniform density on a Schur convex set B is again Schur concave. Once again, we can take convex mixtures and weak limits to show that the convolution of two Schur unimodal distributions is Schur unimodal.

We now generalize the above concepts to R^n. If $\mathbf{a}$ and $\mathbf{b}$ are in R^n, we say that $\mathbf{b}$ *majorizes* $\mathbf{a}$ if $\mathbf{a}$ is in the convex hull of the set of $n!$ points obtained by permuting the components of $\mathbf{b}$. The $n!$ vectors obtained by permuting the components of a given vector $\mathbf{a}$ are all considered to be equivalent under the partial order of majorization. An alternative equivalent way of defining majorization is as follows. Rearrange the coordinates of $\mathbf{a}$ and $\mathbf{b}$ so that $a_1 \geq a_2 \geq \cdots \geq a_n$ and $b_1 \geq b_2 \geq \cdots \geq b_n$. Then $\mathbf{a} < \mathbf{b}$ if

$$\sum_{i=1}^{k} a_i \leq \sum_{i=1}^{k} b_i, \qquad k = 0, 1, \ldots, n-1$$

and

$$\sum_{i=1}^{n} a_i = \sum_{i=1}^{n} b_i.$$

These conditions express the fact that $\mathbf{a}$ is closer to "equality" than $\mathbf{b}$ is. Suppose a_i and b_i represent the wealths of individuals in two groups of size n. Then $\mathbf{a} < \mathbf{b}$ means that the wealthiest k individuals in the first group have a smaller combined wealth than the wealthiest k individuals in the second group.

Once majorization has been defined, Schur convex sets and Schur concave functions can be defined as in Definition 3.1 and Schur unimodal distributions are defined as in Definition 3.2.

The elegance and richness of the concepts of majorization and Schur convexity result from the fact that these are essentially bivariate concepts. Thus, it can be shown that, if $\mathbf{a} < \mathbf{b}$, then we can find "intermediate" vectors $\mathbf{c}_1, \ldots, \mathbf{c}_m$ such that

$$\mathbf{a} < \mathbf{c}_1 < \mathbf{c}_2 < \cdots < \mathbf{c}_m < \mathbf{b},$$

where adjacent vectors differ in at most two components. Therefore, to check whether a function is Schur concave, one only needs to look at its bivariate restrictions. Similarly, the Schur–Ostrowski criterion for checking the Schur concavity of a differentiable function φ is also bivariate in nature. The criterion requires that

$$(x_i - x_j)\left[\frac{\partial\varphi(\mathbf{x})}{\partial x_i} - \frac{\partial\varphi(\mathbf{x})}{\partial x_j}\right] \le 0, \tag{3.1}$$

for every $\mathbf{x}$ and for every i, j. That (3.1) is a simple consequence of unimodality considerations can be seen as follows. Without loss of generality, assume that φ is defined on R^2. Suppose $\mathbf{a}$ is such that $a_1 > a_2$. Schur concavity requires unimodality along the line $x_1 + x_2 = a_1 + a_2$. Now for all $\Delta > 0$, $(a_1 + \Delta, a_2 - \Delta) > \mathbf{a}$. Therefore,

$$\varphi(a_1 + \Delta, a_2 - \Delta) - \varphi(\mathbf{a}) \le 0.$$

We can rewrite the last inequality as

$$\varphi(a_1 + \Delta, a_2) - \varphi(\mathbf{a}) - [\varphi(a_1 + \Delta, a_2) - \varphi(a_1 + \Delta, a_2 - \Delta)] \le 0.$$

This last inequality is equivalent to

$$\frac{\partial\varphi(\mathbf{a})}{\partial a_1} - \frac{\partial\varphi(\mathbf{a})}{\partial a_2} \le 0,$$

which verifies (3.1) because $a_1 > a_2$. A similar result holds when $a_2 > a_1$.

The following important result was given by Marshall and Olkin (1974). It generalizes the bivariate version discussed earlier in this section. It has its roots in Wintner's Theorem (see Theorem 1.6) and Marshall and Olkin show several applications of it in statistics.

Theorem 3.1. *Suppose $\mathbf{X} = (X_1, \ldots, X_n)$ possesses a Schur concave density on R^n and suppose $A \subset R^n$ is a Schur convex set. Then the function h defined by*

$$h(\mathbf{z}) = P[\mathbf{X} \in A + \mathbf{z}], \qquad \mathbf{z} \in R^n,$$

is Schur concave in $\mathbf{z}$. More generally, the class of all Schur concave densities is closed under convolutions.

Another set of reslts can be obtained by replacing permutation symmetry by sign invariance. A measurable set $D \subset R^n$ is said to be *sign invariant* if

$$(x_1, \ldots, x_n) \in D \Rightarrow (\varepsilon_1 x_1, \ldots, \varepsilon_n x_n) \in D,$$

where each ε_i is ± 1. A measurable set D is said to be *axially convex* if D is sign invariant and

$$(\mathbf{x} \in D \text{ and } |y_i| \le |x_i| \text{ for all } i) \Rightarrow \mathbf{y} \in D.$$

Thus the concept of "majorization" relevant in this case is simply the partial ordering by absolute values. A measurable function f on R^n is called *axially unimodal* if f is sign invariant and $|\mathbf{y}| \leq |\mathbf{x}| \Rightarrow f(\mathbf{y}) \geq f(\mathbf{x})$. Such a function can also be described as "decreasing in absolute value." Again axial unimodality is a "spliced version" of univariate unimodality (with symmetry) where splicing is done along the axes. The following result, due to Jogdeo (1977) is easy to verify.

Theorem 3.2.

(a) *Let* $\mathbf{X}$ *have an axially unimodal density on* R^n *and let* D *be an axially convex set. Then the function* g *defined by*

$$g(\mathbf{z}) = P[X \in D + \mathbf{z}], \qquad \mathbf{z} \in R^n,$$

is axially unimodal in $\mathbf{z}$.

(b) *Convolutions of axially unimodal functions are axially unimodal.*

It is natural to investigate reflection groups other than the group of permutations and the group of sign changes. The first general result in this direction was given by Mudholkar (1966). His motivation was to generalize Anderson's theorem (Theorem 2.14) by replacing central symmetry by invariance under a group of transformations. His theorem follows.

Theorem 3.3. *Let* G *be a subgroup of the group of orthogonal transformations on* R^n. *Let* f *be a convex unimodal* G*-invariant density on* R^n *and let* $E \subset R^n$ *be a* G*-invariant convex set. Then, for every* $\mathbf{y} \in R^n$ *and for every* $\mathbf{z}$ *in the convex hull of the* G*-orbit of* $\mathbf{y}$, *we have*

$$\int_E f(\mathbf{x} + \mathbf{z})\, d\mathbf{x} \geq \int_E f(\mathbf{x} + \mathbf{y})\, d\mathbf{x}.$$

Proof. Suppose λ_n denotes Lebesgue measure in R^n. For $s > 0$, let $C_s = \{\mathbf{x}: f(\mathbf{x}) \geq s\}$. For $\mathbf{x} \in R^n$, let $H_{\mathbf{x}}(s) = \lambda_n[(E + \mathbf{x}) \cap C_s]$. Then

$$\begin{aligned}
\int_E f(\mathbf{x} + \mathbf{y})\, d\mathbf{x} = \int_{E+\mathbf{y}} f(\mathbf{x})\, d\mathbf{x} &= \int_{E+\mathbf{y}} \int_0^{f(x)} ds\, d\mathbf{x} \\
&= \int_{E+\mathbf{y}} \int_0^{\infty} I_{C_s}(\mathbf{x})\, ds\, d\mathbf{x} \\
&= \int_0^{\infty} \lambda_n[(E + \mathbf{y}) \cap C_s]\, ds \\
&= \int_0^{\infty} H_{\mathbf{y}}(s)\, ds.
\end{aligned}$$

Similarly

$$\int_E f(\mathbf{x}+\mathbf{z})\,d\mathbf{x} = \int_0^\infty H_{\mathbf{z}}(s)\,ds.$$

Therefore, the theorem will follow if we prove that $H_{\mathbf{z}}(s) \geq H_{\mathbf{y}}(s)$ for all $s > 0$. This is done by using the Brunn-Minkowski inequality.

Since $\mathbf{z}$ is in the convex hull of the G-orbit of $\mathbf{y}$, we can find a finite subset $\{g_1, \ldots, g_N\}$ of G and nonnegative numbers α_i with $\sum \alpha_i = 1$ such that $\mathbf{z} = \sum \alpha_i g_i \mathbf{y}$. Now the convexity and G-invariance of E and C_s show that

$$(E+\mathbf{z}) \cap C_s \supset \sum \alpha_i [(E + g_i\mathbf{y}) \cap C_s]$$

and

$$\lambda_n[(E + g_i\mathbf{y}) \cap C_s] = \lambda_n[(E+\mathbf{y}) \cap C_s].$$

Therefore, the Brunn-Minkowski inequality shows that

$$\begin{aligned}
[H_{\mathbf{z}}(s)]^{1/n} &= \lambda_n^{1/n}[(E+\mathbf{z}) \cap C_s] \\
&\geq \sum_{i=1}^{n} \alpha_i \lambda_n^{1/n}[(E + g_i\mathbf{y}) \cap C_s] \\
&= \lambda_n^{1/n}[(E+\mathbf{y}) \cap C_s] = [H_{\mathbf{y}}(s)]^{1/n}.
\end{aligned}$$

Thus $H_{\mathbf{z}}(s) \geq H_{\mathbf{y}}(s)$ and the theorem is proved. ■

We can state the above theorem in terms of convolutions. Call a function h on R^n *G-monotone* if $h(\mathbf{z}) \geq h(\mathbf{y})$ whenever $\mathbf{z}$ is in convex hull of the G-orbit of $\mathbf{y}$. The theorem is then equivalent to the statement that the convolution $f_1 * f_2$ of two G-invariant convex unimodal functions is G-monotone. Here the condition of G-invariance on f_1 and f_2 is natural. But the condition of convex unimodality is somewhat strong. For instance, convex unimodality is stronger than Schur unimodality. Consequently, the theorem can be generalized. One such generalization has been given by Das Gupta (1976a). In a detailed study, Eaton and Perlman (1977) give a variety of conditions which lead to the G-monotonicity of convolutions. We omit the details and refer the reader to the papers by Das Gupta and by Eaton and Perlman. See also Conlon, Leon, Proschan and Sethuraman (1977).

3.2. Generalized Unimodality Indexed by a Positive Parameter

In an important paper, Olshen and Savage (1970) gave a definition of generalized unimodality indexed by a positive parameter α and applicable to distributions on R^n. Their results generalize and illuminate the original work of Khintchine on unimodality. This section presents the basic results on α-unimodal distributions. It will turn out that star unimodality of distributions on R^n is equivalent to their n-unimodality. A basic result on convolutions will also explain why the convolution of two unimodal distributions on R may fail to be unimodal. In addition, we will show how the concept of α-unimodality can be used to obtain an improvement of a bivariate Chebyshev-type inequality.

The definition of α-unimodality that follows generalizes a property of unimodal distributions stated in Theorem 1.4. We assume that $\alpha > 0$.

Definition 3.3. A random n-vector $\mathbf{X}$ is said to have an *α-unimodal* distribution about $\mathbf{0}$ if, for every bounded, nonnegative, Borel measurable function g on R^n, the quantity $t^\alpha E[g(t\mathbf{X})]$ is nondecreasing in $t \in (0, \infty)$.

Throughout this section, we use the term α-unimodal to mean α-unimodal about $\mathbf{0}$.

Remark. It is trivial to see from the above definition that if $\mathbf{X}$ is α-unimodal and $\beta > \alpha$, then $\mathbf{X}$ is β-unimodal. Therefore, the best index α of unimodality is the smallest possible value of α.

Let $\mathcal{U}(n, \alpha)$ denote the class of all α-unimodal distributions on R^n. If we write $t^\alpha E[g(t, \mathbf{X})]$ in terms of the distribution P of $\mathbf{X}$, it becomes clear that $\mathcal{U}(n, \alpha)$ is convex under mixtures. It can also be easily shown to be closed under weak convergence. It is not immediately clear what the extreme points of $\mathcal{U}(n, \alpha)$ are. One would also like to know whether there is a Choquet-type representation theorem for α-unimodal distributions. These questions are addressed after the proof of Theorem 3.5 below.

In view of Theorem 1.4, we see that the usual unimodality of a distribution on R is equivalent to its 1-unimodality. Olshen and Savage (1970) have given several characterizations of α-unimodality. To motivate one of these, we present a simple lemma.

Lemma 3.1. *Let X be a positive real random variable. Then X is α-unimodal if, and only if, X^α is unimodal.*

Proof. Let $Y = X^\alpha$. For a nonnegative, bounded and Borel measurable function g, let $h(x) = g(x^\alpha)$. Also write $u = t^\alpha$. Then

$$uE[g(uY] = t^\alpha E(g(uX^\alpha)] = t^\alpha E[h(tX)]$$

The first expression above is nondecreasing in u if, and only if, the last expression is nondecreasing in t. The lemma now follows from Definition 3.3 and Theorem 1.4. ■

Remark. Lemma 3.1 and Theorem 1.3 show that a positive real random variable X is α-unimodal if, and only if, X is distributed as $U^{1/\alpha}Z$, where U is uniform on $(0, 1)$ and Z is a (positive) random variable independent of U.

The result stated in the preceding remark can be proved for general α-unimodal random vectors. For this purpose we again use the polar transformation which writes $\mathbf{x} \in R^n$ as $\mathbf{x} = l \cdot \mathbf{d}$ where $l = \|x\|$ is the length and $\mathbf{d}$ the direction vector of $\mathbf{x}$. Similarly we write $\mathbf{X} = L \cdot \mathbf{D}$.

Theorem 3.4. *$\mathbf{X}$ is α-unimodal if, and only if, L given $\mathbf{D}$ is one-dimensional α-unimodal with probability one.*

Proof. (a) Suppose L given $\mathbf{D}$ is α-unimodal with probability one. Let g be bounded, nonnegative and measurable on R^n. Then for $0 < s < t$

$$t^\alpha E[g(t\mathbf{X})] - s^\alpha E[g(s\mathbf{X})] = E\{E[t^\alpha g(tL\mathbf{D}) - s^\alpha g(sL\mathbf{D}) \mid \mathbf{D}]\}$$

which is nonnegative because the expression in the bracket is nonnegative. Thus $\mathbf{X}$ is α-unimodal.

(b) Suppose L given $\mathbf{D}$ is not α-unimodal. Then we can find points $0 < s < t$ and a nonnegative, bounded, Borel measurable function g on $[0, \infty)$ such that

$$s^\alpha E[g(sL) \mid \mathbf{D}] > t^\alpha E[g(tL) \mid \mathbf{D}],$$

on a set of positive probability. Let $h(\mathbf{D})$ be the indicator of this last set. Then, taking $v(\mathbf{x}) = h(\mathbf{d})g(l)$, we see that

$$E[s^\alpha v(s\mathbf{X}) - t^\alpha v(t\mathbf{X})] = E\{h(\mathbf{D})E[s^\alpha g(sL) - t^\alpha g(tL) \mid \mathbf{D}]\},$$

which is positive. Thus $\mathbf{X}$ is not α-unimodal. The theorem is thus proved. ■

Theorem 3.5. *A random n-vector $\mathbf{X}$ is α-unimodal if, and only if, $\mathbf{X}$ is distributed as $U^{1/\alpha}\mathbf{Z}$ where U is uniform on $(0, 1)$ and $\mathbf{Z}$ is independent of U.*

Proof. (a) Suppose $\mathbf{X}$ is α-unimodal. Then by Theorem 3.4, L given $\mathbf{D}$ is a one-dimensional, positive, α-unimodal random variable with probability one. Therefore, by the remark above, L given $\mathbf{D}$ is distributed as $U^{1/\alpha}Z_{\mathbf{D}}$ where U is uniform on $(0, 1)$ and $Z_{\mathbf{D}}$ is a positive random variable (independent of

U) whose distribution depends on $\mathbf{D}$. Now $\mathbf{X}$ is distributed as $U^{1/\alpha}\mathbf{Z}$ where $\mathbf{Z}/\|\mathbf{Z}\|$ is distributed like $\mathbf{D}$ and $\|\mathbf{Z}\|$ given $\mathbf{D}$ is distributed like $Z_{\mathbf{D}}$.

(b) Suppose $\mathbf{X}$ is distributed as $U^{1/\alpha}\mathbf{Z}$ with the stated conditions on U and $\mathbf{Z}$. Then

$$\begin{aligned} t^{\alpha}E[g(t\mathbf{X})] &= t^{\alpha}E[g(tU^{1/\alpha}\mathbf{Z})] \\ &= t^{\alpha}\int_0^1 E[g(tu^{1/\alpha}\mathbf{Z})\, du \\ &= \alpha\int_0^t w^{\alpha-1}E[g(w\mathbf{Z})]\, dw, \end{aligned}$$

which is nondecreasing in $t > 0$ whenever g is bounded, nonnegative and measurable. Thus $\mathbf{X}$ is α-unimodal and the theorem is proved. ∎

Theorem 3.5 enables us to obtain a Choquet-type representation for α-unimodal distributions. For $\mathbf{z} \in R^n$, let $W(\alpha, \mathbf{z})$ be the distribution of $U^{1/\alpha}\mathbf{z}$, where U is uniform on $(0, 1)$. If P is the distribution of an α-unimodal random vector $\mathbf{X}$ and Q is the distribution of the random vector $\mathbf{Z}$ given by Theorem 3.5, then we have

$$P = \int_{R^n} W(\alpha, \mathbf{z})\, dQ(z). \tag{3.2}$$

Thus every α-unimodal distribution has a representation of the form (3.2). It is trivial to see that Q determines P. But it is also true that P determines Q. In other words, the representation (3.2) is unique. To prove this, let us write the representation $\mathbf{X} = U^{1/\alpha}\mathbf{Z}$ in terms of characteristic functions. If φ and ψ are the characteristic functions of $\mathbf{X}$ and $\mathbf{Z}$ respectively, then, for $\mathbf{v} \in R^n$,

$$\varphi(\mathbf{v}) = E[\exp(iU^{1/\alpha}\mathbf{v}'Z)] = \int_0^1 \psi(u^{1/\alpha}\mathbf{v})\, du.$$

Therefore, for $t > 0$, we have

$$\varphi(t\mathbf{v}) = \int_0^1 \psi(tu^{1/\alpha}\mathbf{v})\, du = \alpha t^{-\alpha}\int_0^t w^{\alpha-1}\psi(w\mathbf{v})\, dw.$$

Consequently,

$$\frac{d}{dt}\{t^{\alpha}\varphi(t\mathbf{v})\} = \alpha t^{\alpha-1}\psi(t\mathbf{v}).$$

Setting $t = 1$, we see that φ determines ψ, which is, of course, the same as saying that P determines Q.

A by-product of the uniqueness of the representation (3.2) is that the distributions $W(\alpha, \mathbf{z})$ are now seen to be the extreme points of $\mathscr{U}(n, \alpha)$. We thus have a highly satisfactory convex structure for the set of α-unimodal distributions.

Theorem 3.5 and Theorem 2.1 show that star unimodal distributions are just n-dimensional, n-unimodal distributions. We have seen in Chapter 2 that an absolutely continuous star unimodal distribution has a density which is nonincreasing along rays emanating from the mode. The next theorem shows that for an α-unimodal distribution, the density may increase along rays but that the rate of increase can be controlled.

Theorem 3.6. *Let* $\mathbf{X}$ *be an n-dimensional absolutely continuous random vector. Then* $\mathbf{X}$ *is* α*-unimodal about* $\mathbf{0}$ *if, and only if,* $\mathbf{X}$ *has a density f such that for every* $\mathbf{x} \neq 0$, $t^{n-\alpha} f(t\mathbf{x})$ *is nonincreasing in* $t \in (0, \infty)$.

Proof. Write $\mathbf{X} = L\mathbf{D}$ as before. We recall that the Jacobian of the polar transformation is $l^{n-1}h(\mathbf{d})$ where h is an easily calculated function. By Theorem 3.4, the α-unimodality of $\mathbf{X}$ is equivalent to the α-unimodality of L given $\mathbf{D}$. This means, by Lemma 3.1, that L^{α} given $\mathbf{D}$ should have a density which is nonincreasing on $(0, \infty)$. So, if $q(l \mid \mathbf{d})$ denotes the conditional density of L given $\mathbf{D} = \mathbf{d}$, then $l^{1-\alpha} q(l \mid \mathbf{d})$ should be nonincreasng in l for every fixed $\mathbf{d}$. Now let ξ denote the marginal density of $\mathbf{D}$. Then

$$\begin{aligned} l^{n-\alpha} f(l\mathbf{d}) &= l^{n-\alpha} q(l \mid \mathbf{d}) \xi(\mathbf{d}) l^{1-n} / \mathrm{h}(\mathbf{d}) \\ &= l^{1-\alpha} q(l \mid \mathbf{d}) \xi(\mathbf{d}) / h(\mathbf{d}). \end{aligned}$$

Therefore $l^{n-\alpha} f(l\mathbf{d})$ should be nonincreasing in l for every fixed $\mathbf{d}$. This essentially proves the theorem. A rigorous proof would take account of the fact that densities are not unique and have to be suitably chosen so that the required nonincreasing property holds everywhere. For details, see the paper by Olshen and Savage (1970). ■

We now turn to an important property of α-unimodal distributions. For use in the proof of the property, we need a couple of definitions. Suppose $S \subset R^n$ is a set which is star-shaped about $\mathbf{0}$. The *Minkowski functional* π_S of S is defined by

$$\pi_S(\mathbf{x}) = \inf\{c > 0 : \mathbf{x} \in cS\}, \qquad \mathbf{x} \in R^n.$$

If there is no $c > 0$ for which $\mathbf{x} \in cS$, then we set $\pi_S(\mathbf{x}) = \infty$. Some facts follow immediately from this definition.

(i) The functional π_S is positively homogeneous. That is, $\pi_S(k\mathbf{x}) = k\pi_S(\mathbf{x})$ for all $\mathbf{x} \in R^n$ and $k > 0$.

(ii) For $k > 0$,

$$\pi_S(\mathbf{x}) < k \Rightarrow \mathbf{x} \in kS \Rightarrow \pi_S(\mathbf{x}) \leq k.$$

Therefore, if $\mathbf{X}$ is a random n-vector, then

$$P[\pi_S(\mathbf{X}) < k] \leq P[\mathbf{X} \in kS] \leq P[\pi_S(\mathbf{X}) \leq k].$$

The second definition we need is that of a function which decreases along rays emanating from the origin. We call a nonnegative, bounded, measurable function g on R^n *stardown* if, for every $\mathbf{x} \neq \mathbf{0}$, $g(t\mathbf{x})$ is nonincreasing on $[0, \infty)$.

Theorem 3.7. *Let* $\mathbf{X}$ *be an* α*-unimodal random n-vector. For a stardown g on* R^n *and* $t > 0$, *let* $B(t, g) = E[g(t^{-1/\alpha}\mathbf{X})]$. *Then* $B(\cdot, g)$ *is concave on* $(0, \infty)$.

Proof. (a) First suppose that $g = I_S$, where S is star shaped about $\mathbf{0}$. Then

$$B(t, g) = P[t^{-1/\alpha}\mathbf{X} \in S] = P[\mathbf{X} \in t^{1/\alpha}S].$$

By property (ii) above, with $k = t^{1/\alpha}$, we have

$$P[\pi_S(\mathbf{X}) < t^{1/\alpha}] \leq B(t, g) \leq P[\pi_S(\mathbf{X}) \leq t^{1/\alpha}].$$

Now write $Y_S = [\pi_S(\mathbf{X})]^\alpha$. Then

$$P(Y_S < t) \leq B(t, g) \leq P(Y_S \leq t). \tag{3.3}$$

By Theorem 3.5, $\mathbf{X} = U^{1/\alpha}\mathbf{Z}$ where U is uniform on $(0, 1)$ and U, $\mathbf{Z}$ are independent. Property (i) above shows that

$$\pi_S(\mathbf{X}) = U^{1/\alpha}\pi_S(\mathbf{Z})$$

and so

$$Y_S = [\pi_S(\mathbf{X})]^\alpha = U \cdot [\pi_S(\mathbf{Z})]^\alpha. \tag{3.4}$$

Since U and $\mathbf{Z}$ are independent, (3.4) and Theorem 1.2 show that Y_S has a unimodal distribution on R with mode 0. Therefore $P[Y_S \leq t]$ is concave and thus continuous on $(0, \infty)$. Therefore, the extreme two quantities in (3.3) are equal and also concave in t. Thus $B(\cdot, g)$ is concave on $(0, \infty)$.

(b) Suppose now that g is a general star down function. Let $C_s = \{\mathbf{x}: g(\mathbf{x}) \geq s\}$ and $g_s = I_{C_s}$. Then by an argument which has been repeatedly used before,

$$g = \int_0^\infty g_s \, ds.$$

Therefore,

$$B(t, g) = \int_0^\infty B(t, g_s)\, ds. \tag{3.5}$$

By the first part of the proof, each $B(\cdot, g_s)$ is concave on $(0, \infty)$. Therefore (3.5) shows that $B(\cdot, g)$ is also concave on $(0, \infty)$. The theorem is thus proved. ■

Corollary. *If* $\mathbf{X}$ *is* α*-unimodal, then for every fixed* $h > 0$*, the probability*

$$P[t^{1/\alpha} < \| \mathbf{X} \| \leqq (t+h)^{1/\alpha}] \tag{3.6}$$

is nonincreasing in $t > 0$.

Proof. This follows from Theorem 3.7 by setting $g = I_S$, where S is the unit ball. ■

Remark. The result (3.6) is similar in spirit to Anderson's theorem. In Anderson's case,the probability carried by a centrally symmetric convex set decreases when the set is translated. In the present case the probability carried by a shell decreases when the shell is dilated in a specific way.

We now turn to a discussion of products, marginals and convolutions of α-unimodal distributions. The situation is summarized by the following theorem.

Theorem 3.8.

(a) *The marginals of an* α*-unimodal distribution are* α*-unimodal.*
(b) *The cartesian product of an* α*-unimodal distribution with a* β*-unimodal distribution is* $(\alpha + \beta)$*-unimodal.*
(c) *The convolution of an* α *unimodal distribution with a* β*-unimodal distribution is* $(\alpha + \beta)$*-unimodal.*

Proof. Assertion (a) follows easily from Definition 3.3. To prove (b), let $\mathbf{X}$ be an α-unimodal random m-vector and let $\mathbf{Y}$ be a β-unimodal random n-vector. Suppose $\mathbf{X}$ and $\mathbf{Y}$ are independent. Write $\mathbf{Z} = (\mathbf{X}, \mathbf{Y})$. We want to show that $\mathbf{Z}$ is $(\alpha + \beta)$-unimodal. Since the set of all absolutely continuous α-unimodal distributions is dense in the set of all α-unimodal distributions, we may assme that $\mathbf{X}$ and $\mathbf{Y}$ are absolutely continuous with densities f and g, respectively. If we write h for the density of Z and write $\mathbf{z} = (\mathbf{x}, \mathbf{y})$, then, for $t > 0$,

$$t^{m+n-\alpha-\beta}h(t\mathbf{z}) = t^{m+n-\alpha-\beta}f(t\mathbf{x})g(t\mathbf{y})$$
$$= t^{m-\alpha}f(t\mathbf{x}) \cdot t^{n-\alpha}g(t\mathbf{y}).$$

By Theorem 3.6, the two factors on the right side of the last expression are nonincreasing on $[0, \infty)$. Therefore, the left side is also nonincreasing and the same theorem shows that $\mathbf{Z}$ is $(\alpha+\beta)$-unimodal. This proves (b). Now (c) follows trivially from (a) and (b). ∎

Example 3.1 below shows that the index $(\alpha+\beta)$ given by part (c) of Theorem 3.8 is, in general, the best possible. Therefore, if P and Q are two unimodal distributions on R, then the convolution $P * Q$ will, in general, be 2-unimodal rather than 1-unimodal.This explains why $P * Q$ may not be unimodal.

Example 3.1. Let X, Y be independent random variables such that, X has the density $\frac{1}{2}\alpha|x|^{\alpha-1}$, $0<|x|<1$ and Y has the density $\frac{1}{2}\beta|y|^{\beta-1}$, $0<|y|<1$. Then $(X,0)$ and $(Y,0)$ are independent α-unimodal and β-unimodal random vectors, respectively. Write $Z=(X, Y)$. Then $Z=(X,0)+(0, Y)$ and, therefore, the distribution of Z is the convolution of the distributions of $(X,0)$ and $(0, Y)$. The density h of Z is given by

$$4h(x, y) = \alpha\beta|x|^{\alpha-1}|y|^{\beta-1}, \qquad 0<|x|<1, \qquad 0<|y|<1.$$

Since Z is 2-dimensional, Theorem 3.6 shows that Z is $(\alpha+\beta)$-unimodal about 0 and not γ-unimodal about 0 for any $\gamma<(\alpha+\beta)$.

Suppose now that Z is γ-unimodal about some nonzero vertex (a, b). Then, for every fixed (x, y), the function g defined by

$$g(t) = t^{2-\gamma}h[a+t(x-a),\ b+t(y-b)]$$

must be nonincreasing in $t \in (0, \infty)$. Since the density of Z has mirror symmetry about the coordinate axes and vanishes outside the square with vertices $(\pm 1, \pm 1)$, we may assume that $0 \le a \le 1$ and $0 \le b \le 1$. Two cases arise

Case 1. Suppose that both a and b are positive. Then, by choosing $x=-a$ and $y=-b$, we get

$$g(t) = \alpha\beta|a|^{\alpha-1}\cdot|b|^{\beta-1}\cdot|1-2t|^{\alpha+\beta-2}\cdot t^{2-\gamma}$$

valid at least for $0<t<1$. Now, if $(\alpha+\beta) \neq 2$, then $g(\frac{1}{2})$ is either zero or infinite and g cannot be nonincreasng on $(0, 1)$. On the otherhand, if $\alpha+\beta=2$, then $g(t)$ is a multiple of $t^{2-\gamma}$ and so g can be nonincreasing on $(0, 1)$ only if

$\gamma \geq 2$. So, in the present case, the index γ of unimodality of Z about (a, b) must satisfy $\gamma \geq (\alpha + \beta)$.

Case 2. Suppose now that $b = 0$. Then $a > 0$ and we choose $x = 0$ and $0 < y < 1$ to get

$$g(t) = \alpha\beta\,|a|^{\alpha-1} \cdot |y|^{\beta-1} \cdot |1-t|^{\alpha-1} \cdot t^{1+\beta-\gamma},$$

valid for $0 < t < 1 + \delta$, where δ is a suitable small positive number. Now, if $\alpha \neq 1$, then $g(1)$ is either 0 or ∞ and g cannot be nonincreasing on $(0, 1 + \delta)$. So, let $\alpha = 1$. Then $g(t)$ is a multiple of $t^{1+\beta-\gamma}$ and so g cannot be nonincreasing on $(0, 1 + \delta)$ unless $\gamma \geq 1 + \beta = \alpha + \beta$.

We thus see that, even with a changed vertex, the best index of unimodality for the distribution of Z is $(\alpha + \beta)$.

Despite the conclusion of Example 3.1, one can show that the index $(\alpha + \beta)$ given by Theorem 3.8(c) can be improved if attention is restricted to distributions on the line and if one of the distributions is symmetric and unimodal. This is done in Theorem 3.9 below, which is taken from Dharmadhikari and Jogdeo (1974). We first prove a lemma.

Lemma 3.2. *Let $\alpha \geq 2$ and $g(x) = x^{1-\alpha}[(x+1)^{\alpha} - (x-1)^{\alpha}]$. Then g is nonincreasing on $[1, \infty)$.*

Proof. Clearly, for $x > 1$,

$$\alpha^{-1} g(x) = x^{1-\alpha} \int_{-1}^{1} (x+y)^{\alpha-1}\, dy = \int_{-1}^{1} \left(1 + \frac{y}{x}\right)^{\alpha-1} dy.$$

Therefore

$$\alpha^{-1} g'(x) = (1-\alpha) x^{-2} \int_{-1}^{1} h(y)\, dy, \tag{3.7}$$

where $h(y) = y[1 + (y/x)]^{\alpha-2}$. Now, it is easy to see that $h(y) + h(-y) \geq 0$ for all $y \in (0, 1)$. Therefore, the right side of (3.7) is nonpositive. The Lemma is thus proved. ■

Theorem 3.9. *Suppose P and Q are distributions on R such that P is symmetric and unimodal about 0 and Q is α-unimodal about 0.*

(a) *If $1 \leq \alpha \leq 2$, then $P * Q$ is $\frac{1}{2}(2 + \alpha)$-unimodal about 0.*
(b) *If $\alpha \geq 2$, then $P * Q$ is α-unimodal about 0.*

Proof. Recall from Theorem 3.5 that α-unimodal random variables are precisely those distributed as $U^{1/\alpha}Z$, where U is uniform on $(0,1)$ and U, Z are independent. Therefore, in proving the theorem, we may asusme that Q is the distribution of $c \cdot U^{1/\alpha}$, where c is a real constant. Similarly, by Theorem 1.5, we may assume that P is the uniform distribution on $(-\delta, \delta)$, where δ is a positive real constant. Further, we can make a common scale change and assume that $c = 1$. Thus Q has the density

$$q(x) = \alpha x^{\alpha-1}, \qquad 0 < x < 1.$$

Let $k(x)$ denote the density of the convolution $P * Q$. To show that $P * Q$ is β-unimodal about 0, we need to show (see Theorem 3.6) that $x^{1-\beta}k(x)$ and $x^{1-\beta}k(-x)$ are nonincreasing in $x \in (0, \infty)$.

We observe that $k(x) = 0$ for $x < -\delta$ and for $x > 1 + \delta$. Consider two cases.

Case 1. Let $\delta \le \frac{1}{2}$. Then

$$2\delta k(x) = \begin{cases} (x+\delta)^\alpha, & -\delta \le x \le \delta \\ (x+\delta)^\alpha - (x-\delta)^\alpha, & \delta \le x \le 1-\delta \\ 1-(x-\delta)^\alpha, & 1-\delta \le x \le 1+\delta. \end{cases}$$

Write $g(x) = x^{1-\beta}(x+\delta)^\alpha$. Then

$$x^\beta g'(x) = (x+\delta)^{\alpha-1}[(\alpha+1-\beta)x + (1-\beta)\delta],$$

which is negative for $x \in (0, \delta)$ whenever $2(\beta - 1) \ge \alpha$. Thus $x^{1-\beta}k(x)$ is nonincreasing on $[0, \delta]$ if $\beta \ge (2+\alpha)/2$.

To get the same result for $x \in [\delta, 1-\delta]$, consider

$$h(x) = x^{-\alpha/2}[(x+\delta)^\alpha - (x-\delta)^\alpha].$$

By easy algebra,

$$2x^{(2+\alpha)/2}h'(x) = \alpha(x+\delta)(x-\delta)[(x+\delta)^{\alpha-2} - (x-\delta)^{\alpha-2}].$$

Therefore, if $1 \le \alpha \le 2$, then h is nonincreasing on $[\delta, \infty)$ and hence $x^{-\alpha/2}k(x)$ is nonincreasng on $[\delta, 1-\delta]$. On the other hand, if $\alpha \ge 2$, then Lemma 3.2 shows that $x^{1-\alpha}k(x)$ is nonincreasing on $[\delta, 1-\delta]$.

Now $k(x)$ is nonincreasing on $[1-\delta, \infty)$ and nondecreasing on $(-\infty, 0]$. Therefore, putting everything together, we see that k is $(2+\alpha)/2$-unimodal for $1 \le \alpha \le 2$ and α-unimodal for $\alpha \ge 2$.

Case 2. Let $\delta \geq \frac{1}{2}$. Then

$$2\delta k(x) = \begin{cases} (x+\delta)^\alpha, & -\delta \leq x \leq 1-\delta \\ 1, & 1-\delta \leq x \leq \delta \\ 1-(x-\delta)^\alpha, & \delta \leq x \leq 1+\delta. \end{cases}$$

Since $(1-\delta) \leq \delta$, the argument involving the function g in the first case shows that $x^{-\alpha/2}k(x)$ is nonincreasing on $[0, 1-\delta]$. Since k is nonincreasing on $[1-\delta, \infty)$ and nondecreasing on $(-\infty, 0]$, we see that k is $(2+\alpha)/2$-unimodal for every $\alpha \geq 1$.

The proof of the theorem is now complete. ■

Remark. It is clear that the result of Part (b) of Theorem 3.9 is the best possible even when Q is symmetric. This is because P could be degenerate. One might therefore ask whether the result of Part (a) can be improved if Q is also assumed to be symmetric. If $\alpha = 1$, then Wintner's result (Theorem 1.5) shows that the index $(\frac{3}{2})$ can be reduced to 1. However, if $1 < \alpha < 2$, then no improvement is possible and the best index of unimodality of $P * Q$ is still $(2+\alpha)/2$. In this sense Wintner's result is somewhat exceptional. For details, see Dharmadhikari and Jogdeo (1974).

To conclude the section, we show that the concept of α-unimodality can be used to obtain an improvement of a bivariate Chebyshev-type inequality. This result is just a continuation of the discussion of Section 1.5 concerning the Gauss inequality.

First we present a generalization of Theorem 1.11.

Theorem 3.10. *Let X be an α-unimodal real random variable. Then for every $k > 0$ and $s > 0$,*

$$P(|X| \geq k) \leq \left[\frac{s}{s+\alpha}\right]^{s/\alpha} \frac{E(|X|^s)}{k^s}.$$

Proof. We may assume that $X = U^{1/\alpha}Z$, where U, Z are independent and U is uniform on $(0, 1)$. Write $Y = |X|^\alpha$. Then $Y = U|Z|^\alpha$ and so Y is unimodal. Therefore, by Theorem 1.11,

$$P(Y \geq t) \leq \left[\frac{r}{r+1}\right]^r \frac{E(Y^r)}{t^r},$$

for all $t > 0$ and $r > 0$. The theorem now follows if we set $t = k^\alpha$ and $r = s/\alpha$.

The bivariate Chebyshev-type inequality that we want to improve was given by Berge (1937) and is stated in the next theorem.

Theorem 3.11. *Let X_1, X_2 be random variables with zero means, unit variances and correlation coefficient ρ. Then, for all $k > 0$,*

$$P[|X_1| \geq k \quad \text{or} \quad |X_2| \geq k] \leq [1 + \sqrt{(1-\rho^2)}]/k^2. \tag{3.8}$$

Proof. Write points $\mathbf{x}$ in R^2 as row vectors. Let $B_k = \{\mathbf{x} \in R^2 : |x_1| \geq k$ or $|x_2| \geq k\}$. For $|t| < 1$, define the matrix

$$A_t = (1-t^2)^{-1}\begin{pmatrix} 1 & -t \\ -t & 1 \end{pmatrix}.$$

Then

$$\mathbf{x}A_t\mathbf{x}' = \frac{x_1^2 + x_2^2 - 2tx_1x_2}{(1-t^2)} = x_1^2 + \frac{(tx_1 - x_2)^2}{1-t^2}.$$

Therefore

$$\mathbf{x} \in B_k \Rightarrow \mathbf{x}A_t\mathbf{x}' \geqq k^2.$$

Consequently,

$$\begin{aligned} P(\mathbf{X} \in B_k) &\leq E(\mathbf{X}A_t\mathbf{X}')/k^2 \\ &= 2(1-\rho t)/[k^2(1-t^2)]. \end{aligned} \tag{3.9}$$

The minimum of (3.9) over $t \in (-1, 1)$ occurs when

$$t = [1 - \sqrt{(1-\rho^2)}]/\rho.$$

Substituting this value of t in (3.9) and simplifying the resulting expression, one gets (3.8). The theorem is thus proved. ■

An improvement of (3.8) in the presence of α-unimodality was given by Dharmadhikari and Joag-dev (1986) and is presented next.

Theorem 3.12. *Let $\mathbf{X} = (X_1, X_2)$ be an α-unimodal random vector whose components have zero means, unit variances and correlation coefficient ρ. Then, for all $k > 0$,*

$$P[|X_1| \geq k \quad \text{or} \quad |X_2| \geq k] \leq \left[\frac{2}{2+\alpha}\right]^{2/\alpha}\left[\frac{1+\sqrt{(1-\rho^2)}}{k^2}\right].$$

Proof. We use the notation given in the proof of Theorem 3.11. We may assume that $\mathbf{X} = U^{1/\alpha}\mathbf{Z}$ where U, $\mathbf{Z}$ are independent and U is uniform on $(0, 1)$. Then

$$\mathbf{X}A_t\mathbf{X}' = U^{2/\alpha}\mathbf{Z}A_t\mathbf{Z}'.$$

Therefore $\mathbf{X}A_t\mathbf{X}'$ is $(\alpha/2)$-unimodal. Now Theorem 3.10 shows that

$$P(\mathbf{X}A_t\mathbf{X}' \geqq k^2) \leq \left[\frac{2}{2+\alpha}\right]^{2/\alpha} \frac{E(\mathbf{X}A_t\mathbf{X}')}{k^2}.$$

In other words, (3.9) is improved by the factor $[2/(2+\alpha)]^{2/\alpha}$. But we can again take minimum over $t \in (-1, 1)$ to show that (3.8) is also improved by the same factor. The theorem follows. ■

When one applies Theorem 3.12, the case of the most practical interest occurs when (X_1, X_2) is star unimodal, that is, when $\alpha = 2$. In this case, (3.8) can be improved by a factor of $(\frac{1}{2})$. We write this result as a corollary.

Corollary. *Let (X_1, X_2) have a star unimodal distribution about* $\mathbf{0}$. *Suppose that X_1, X_2 have zero means, unit variances and correlation coefficient ρ. Then for all $k > 0$,*

$$P[\,|X_1| \geq k \quad \text{or} \quad |X_2| \geq k] \leq [1 + \sqrt{(1-\rho^2)}]/(2k^2).$$

3.3. Classes of Concave Densities and Measures

This section generalizes the results of Chapter 2 on logconcave distributions. The generalization starts by defining an index s of concavity for densities and measures in such a way that logconcavity corresponds to the case $s = 0$. The main results investigate the relationship between s-concave densities and t-concave measures. For $t > 0$, the results closely follow the development of the Brunn-Minkowski inequality and can be found in Dinghas (1957). For $t < 0$, the results were given by Borell (1975). Simplified proofs of Borell's results were given by Rinott (1976) and Das Gupta (1980). Our proofs follow the approach given by Das Gupta (1976b and 1980). For many interesting remarks on the topic of this section, see Uhrin (1984).

Call f *quasi-concave* if, for all $\mathbf{x}$, $\mathbf{y}$ in R^n and θ in $(0, 1)$, we have

$$f[(1-\theta)\mathbf{x} + \theta\mathbf{y}] \geq \min\{f(\mathbf{x}), f(\mathbf{y})\}. \tag{3.10}$$

The property of quasi-concavity of f is equivalent to the property that, for every $c>0$, the set $\{\mathbf{x}: f(\mathbf{x}) \geq c\}$ is convex. For $n=1$, *quasi-concavity of f* is equivalent to the unimodality of f. If $n>1$ and f is centrally symmetric, then f is quasi-concave if and only if, f is convex unimodal. In any case, it is easy to check that

$$f \text{ logconcave} \Rightarrow f \text{ quasi-concave.}$$

The reverse implication is not true even for $n=1$; the Cauchy density is a counterexample. So one might ask whether there are properties of f which are intermediate between quasi-concavity and logconcavity. One might also look for properties stronger than logconcavity. These questions are discussed in the present section.

Let us recall that a logconcave function f is one which satisfies

$$f[(1-\theta)\mathbf{x}+\theta\mathbf{y}] \geq [f(\mathbf{x})]^{1-\theta}[f(\mathbf{y})]^{\theta} \tag{3.11}$$

for all $\mathbf{x}, \mathbf{y}$ in R^n and θ in $(0,1)$. Now the right sides of (3.10) and (3.11) are seen to be special types of means of $f(\mathbf{x})$ and $f(\mathbf{y})$. We can therefore, replace them by some generalized mean. This observation forms the starting point of the results of this section.

Let $a \geq 0$, $b \geq 0$ and $\theta \in (0,1)$. The sth *generalized mean* $M_s(a,b;\theta)$ is defined by

$$M_s(a,b;\theta) = [(1-\theta)a^s + \theta b^s]^{1/s}. \tag{3.12}$$

The right side of (3.12) is clearly well defined if $0<s<\infty$. It is also well defined if $s<0$ and a, b are both positive. If $s<0$ and $ab=0$, we set $M_s(a,b;\theta)=0$. The cases $s=0, \infty, -\infty$ are handled through continuity. That is,

$$M_0(a,b;\theta)=a^{1-\theta}b^{\theta}, \qquad M_{-\infty}(a,b;\theta)=\min(a,b)$$

and

$$M_{\infty}(a,b;\theta)=\max(a,b).$$

It is well known that $M_s(a,b;\theta)$ is nondecreasing in s when a, b and θ are fixed. Now (3.10) is the same as

$$f[(1-\theta)\mathbf{x}+\theta\mathbf{y}] \geq M_{-\infty}[f(\mathbf{x}),\ f(\mathbf{y});\ \theta]$$

and (3.11) is the same as

$$f[(1-\theta)\mathbf{x}+\theta\mathbf{y}] \geq M_0[f(\mathbf{x}),\ f(\mathbf{y});\theta].$$

Motivated by this, one can give the following definition.

Definition 3.4. A nonnegative real-valued function f on an open convex set $C \subset R^n$ is said to be *s-concave* on C if, for every choice of $\mathbf{x}_0$, $\mathbf{x}_1$ in C and $\theta \in (0, 1)$, we have

$$f[(1-\theta)\mathbf{x}_0 + \theta\mathbf{x}_1] \geq M_s[f(\mathbf{x}_0), f(\mathbf{x}_1); \theta].$$

Remark. The term "s-concave" was used by Uhrin (1984). Earlier, Das Gupta (1976b) used the term "s-unimodal" to define the same concept.

Let $\mathscr{F}_s(C)$ denote the class of all s-concave functions on C. We write $\mathscr{F}_{s,n}$ for $\mathscr{F}_s(R^n)$. We note that $\mathscr{F}_{0,n}$ is just the class of all logconcave functions on R^n.

Remark. Let $s \leq 0$ and $f \in \mathscr{F}_s(C)$. Define f_0 on the whole of R^n by setting $f_0 = f$ on C and $f_0 = 0$ on $R^n - C$. Then $f_0 \in \mathscr{F}_{s,n}$. On the other hand, if $g \in \mathscr{F}_{s,n}$ then $gI_C \in \mathscr{F}_s(C)$. Therefore, for $s \leq 0$, one only needs to look at the class $\mathscr{F}_{s,n}$. However, if $s > 0$ and $f \in \mathscr{F}_s(C)$, then Lemma 3.3 below shows that either $f(\mathbf{x}) = 0$ for all $\mathbf{x} \in C$ or $f(\mathbf{x}) > 0$ for all $\mathbf{x} \in C$.

The following facts are obvious from Definition 3.4 and the above remark.

(a) $f \in \mathscr{F}_{0,n} \Leftrightarrow f(\mathbf{x}) = \exp[Q(\mathbf{x})]$, where Q is concave on R^n into $[-\infty, \infty)$.
(b) For $-\infty < s < 0$, $f \in \mathscr{F}_{s,n} \Leftrightarrow f(\mathbf{x}) = [Q(\mathbf{x})]^{1/s}$, where Q is convex on R^n into $[0, \infty)$.
(c) For $0 < s < \infty$, $f \in \mathscr{F}_s(C) \Leftrightarrow f(\mathbf{x}) = [Q(\mathbf{x})]^{1/s}$, where *either* $Q(\mathbf{x}) = 0$ for all $\mathbf{x} \in C$ *or* Q is concave on C into $(0, \infty)$.

We noted earlier that $M_s(a, b; \theta)$ is nondecreasing in s for fixed a, b and θ. Therefore,

$$s < t \Rightarrow \mathscr{F}_s(C) \supset \mathscr{F}_t(C).$$

In particular, if $f \in \mathscr{F}_s(C)$ for some s, then $f \in \mathscr{F}_{-\infty}(C)$ and consequently, f is quasi-concave and the support of f is a convex set.

Suppose $\mathscr{B}(C)$ denotes the σ-algebra of Borel sets in C. We write $\mathscr{B}_n$ for $\mathscr{B}(R^n)$. In this section, whenever a reference is made to a measure μ on $\mathscr{B}(C)$, it will be understood that μ is nonnegative and assigns finite values to bounded sets in $\mathscr{B}(C)$. We note that if μ is a measure on $\mathscr{B}(C)$, then supp μ can be equal to C even though C is open in R^n. The reason is that supp μ is only required to be closed in C and not necessarily in R^n.

In analogy with Definition 3.4, we define s-concavity of measures as follows.

Definition 3.5. A measure μ on $\mathscr{B}(C)$ is called *s-concave* if, for every choice of nonempty sets A_0, A_1 in $\mathscr{B}(C)$ and $\theta \in (0, 1)$, we have

$$\mu[(1-\theta)A_0 + \theta A_1] \geq M_s[\mu(A_0), \mu(A_1); \theta]. \tag{3.13}$$

The class of all s-concave measures on $\mathscr{B}(C)$ will be denoted by $\mathscr{M}_s(C)$. We write $\mathscr{M}_{s,n}$ for $\mathscr{M}_s(R^n)$. We note that $\mathscr{M}_{0,n}$ is the class of all logconcave measures on $\mathscr{B}_n$.

The main result of this section states that t-concave measures and s-concave densities correspond to each other in a natural way. Here s and t are related by a simple formula. Precise statements are given in Theorems 3.16 and 3.17.

We note again that $s < t \Rightarrow \mathscr{M}_s(C) \supset \mathscr{M}_t(C)$. Therefore if $\mu \in \mathscr{M}_s(C)$ for some s and C, then $\mu \in \mathscr{M}_{-\infty}(C)$ and hence, for every pair of nonempty sets A_0, A_1 in $\mathscr{B}(C)$ and for every $\theta \in (0, 1)$,

$$\mu[(1-\theta)A_0 + \theta A_1] \geq \min\{\mu(A_0), \mu(A_1)\}.$$

This last result shows that the support of μ is convex. Let H be the affine hull of supp μ and write $C_\mu = C \cap H$. Then $\mu \in \mathscr{M}_s(C_\mu)$. Lemma 3.3 below shows that $C_\mu = C$ if $s > 0$ and μ is nonzero. However, for $s \leqq 0$, the dimension of H may be strictly smaller than the dimension of C.

Lemma 3.3. *Let C be an open convex set in R^n and let $s > 0$. Suppose $f \in \mathscr{F}_s(C)$ and $\mu \in \mathscr{M}_s(C)$.*

(a) *If μ is nonzero, then* supp$(\mu) = C$.
(b) *If f is nonzero, then* supp $(f) = C$.

Proof. We prove (a). Assertion (b) can be proved in a similar way. Suppose μ is nonzero. Then supp (μ) is nonempty. Let $\mathbf{x} \in \text{supp}(\mu)$ and let $\delta > 0$. Write $N_\delta(\mathbf{x})$ for the ball in R^n with center $\mathbf{x}$ and radius δ. We know that $\mu[N_\delta(\mathbf{x})] > 0$ for all δ. Now let $\mathbf{y} \in C$. Since C is open, we can find $\mathbf{z} \in C$ and $\theta \in (0, 1)$ such that $\mathbf{y} = (1-\theta)\mathbf{x} + \theta\mathbf{z}$. Now the s-concavity of μ on C shows that

$$\mu[N_\delta(\mathbf{y})] \geqq M_s[\mu(N_\delta(\mathbf{x})), \mu(N_\delta(\mathbf{z})); \theta].$$

But $\mu[N_\delta(\mathbf{z})] \geq 0$. Therefore,

$$\mu[N_\delta(\mathbf{y})] \geq (1-\theta)^{1/s}\mu(N_\delta(\mathbf{x})) > 0,$$

for all δ. Thus $\mathbf{y} \in \text{supp } \mu$ and supp $\mu = C$ as claimed.

It was noted earlier that $\mathscr{M}_0(C)$ is just the class of all logconcave measures on C. These measures were discussed in Chapter 2 and some of their properties carry over to s-concave measures. We summarize these in the next theorem.

Theorem 3.13

(a) *Let μ be a nondegenerate s-concave measure on $\mathscr{B}(C)$, where C is an open convex set in R^n. Let H be the affine hull of supp μ. Then μ is absolutely continuous w.r.t. Lebesgue measure on $C \cap H$.*

(b) *Let μ be a measure on $\mathscr{B}(C)$ such that $\mu(C \cap L) = 0$ for every hyperplane L in R^n. In order that μ be s-concave on $\mathscr{B}(C)$ it is sufficient that (3.13) holds for all rectangles A_0, A_1 in C, with sides parallel to the coordinate axes.*

Proof. The proofs of both assertions are almost identical with the proofs of Theorems 2.5 and 2.6. The only changes needed are:

(i) For the case $s > 0$, we have to restrict attention to the convex set C.

(ii) The 0-mean has to be replaced by the s-mean. ■

Theorem 3.14. *Let μ be an s-concave measure on an open convex set C in R^n. Then either μ is the zero measure or $s \leq 1/n$.*

Proof. Suppose μ is nonzero. If possible, let $s > 1/n$. By Lemma 3.3, supp $\mu = C$ and so, by Theorem 3.13(a), μ is absolutely continuous w.r.t. Lebesgue measure λ_n on C. Let N_δ denote the ball $\{\mathbf{x} \in R^n : \| \mathbf{x} \| < \delta\}$. By the Lebesgue–Vitali theorem [see Theorem 7.10 in Rudin (1987)], the limit

$$f(\mathbf{x}) = \lim_{\delta \to 0} \frac{\mu(\mathbf{x} + N_\delta)}{\lambda_n(\mathbf{x} + N_\delta)}$$

exists and is finite for every $\mathbf{x} \in B$, where $\lambda_n[C - B] = 0$. Indeed, $f(\mathbf{x})$ coincides with the density of μ w.r.t. λ_n. Since μ is non-zero, there is an $\mathbf{x}_0 \in B$ such that $f(\mathbf{x}_0) > 0$. Without loss of generality, we can take $\mathbf{x}_0 = 0$. Now use (3.13) with $A_0 = \{0\}$ and $A_1 = N_\delta$. We have $(1 - \theta)A_0 + \theta A_1 = N_{\theta\delta}$ and $\mu(A_0) = \mu(\{0\}) = 0$. Therefore,

$$\mu(N_{\theta\delta}) \geq \theta^{1/s} \mu(N_\delta).$$

Now

$$\frac{\mu(N_{\theta\delta})}{\lambda_n(N_{\theta\delta})} \geq \frac{\theta^{1/s}\mu(N_\delta)}{\theta^n \lambda_n(N_\delta)} = \frac{\theta^{(1/s)-n}\mu(N_\delta)}{\lambda_n(N_\delta)} \tag{3.14}$$

Let $\theta \to 0$. Then the extreme left side of (3.14) goes to the finite positive limit $f(\mathbf{x}_0)$. But the extreme right side becomes unbounded because $s > 1/n$. This contradiction proves that $s \leq 1/n$ and completes the proof of the theorem. ■

Remark. Theorem 3.14 shows that, for $s > (1/n)$, the only measure in $\mathscr{M}_{s,n}$ is the zero measure. For $s \leq (1/n)$, the class $\mathscr{M}_{s,n}$ does contain nonzero measures. It turns out that the only measures which are $(1/n)$-concave are multiples of the Lebesgue measure λ_n. When $\mu = \lambda_n$, the requirement (3.13) reduces to the Brunn-Minkowski inequality which has already been used in this chapter and in Chapter 2. (See Theorems 2.14 and 3.3.)

We now proceed to study the relationship between s-concave functions and s-concave measures. We saw in Chapter 2 that logconcave densities lead to logconcave measures and conversely logconcave measures have logconcave densities. We will show in this section (see Theorems 3.16 and 3.17) that s-concave functions and t-concave measures are in natural correspondence when s and t are related by a simple formula. First we present a lemma which is an elaboration of Hölder's inequality.

Lemma 3.4. *Let a_0, a_1, b_0, b_1 be nonnegative quantities and let $\theta \in (0, 1)$. For $0 < u < \infty$ and $-u \le s \le \infty$, write $t = (s^{-1} + u^{-1})^{-1}$ where we take $t = -\infty$ when $u = -s$. Then*

$$M_t(a_0 b_0, a_1 b_1; \theta) \le M_s(a_0, a_1; \theta) \cdot M_u(b_0, b_1; \theta). \tag{3.15}$$

Proof. (i) If $0 < s \le \infty$, then the lemma follows from Hölder's inequality.

(ii) Let $-u < s < 0$. Then $-\infty < t < 0$. If $a_0 a_1 b_0 b_1 = 0$, then the left side of (3.15) is zero while the right side is nonnegative. So, let $a_0 a_1 b_0 b_1 > 0$. Write $s' = -s$ and $t' = -t$. Then (3.15) is equivalent to

$$M_{s'}(a_0^{-1}, a_1^{-1}; \theta) \le M_u(b_0, b_1; \theta) \cdot M_{t'}[(a_0 b_0)^{-1}, (a_1 b_1)^{-1}; \theta].$$

Since $t' > 0$ and $s' = [u^{-1} + (t')^{-1}]^{-1}$, the last inequality again follows from Hölder's inequality.

(iii) Let $s = 0$. Then $t = 0$. Now if $a_0 a_1 = 0$, then both sides of (3.15) vanish. So, let $a_0 a_1 > 0$. Then (3.15) reduces to $M_0(b_0, b_1; \theta) \le M_u(b_0, b_1; \theta)$, which is true because $u > 0$.

(iv) Let $s = -u$. Then, as stated, we take $t = -\infty$. Therefore (3.15) reduces to

$$\min\{a_0 b_0, a_1 b_1\} \le M_{-u}(a_0, a_1; \theta) \cdot M_u(b_0, b_1; \theta). \tag{3.16}$$

Again, if $a_0 a_1 = 0$, then both sides of (3.16) vanish. So, let $a_0 a_1 > 0$. Then (3.16) is equivalent to

$$\min\{a_0 b_0, a_1 b_1\} M_u(a_0^{-1}, a_1^{-1}; \theta) \le M_u(b_0, b_1; \theta),$$

which follows trivially from the definition of M_u.

The lemma is now proved. ■

Theorem 3.15. *For $i=0$, 1, let f_i be a nonnegative measurable function on R^n, let A_i be a nonempty subset of the support of f_i and assume that f_i is integrable w.r.t. Lebesgue measure on A_i. Fix $\theta \in (0, 1)$ and let f be a nonnegative measurable function on R^n such that*

$$f(\mathbf{x}) \geq M_s[f_0(\mathbf{x}_0), f_1(\mathbf{x}_1); \theta], \tag{3.17}$$

whenever $\mathbf{x} = (1-\theta)\mathbf{x}_0 + \theta\mathbf{x}_1$ and $\mathbf{x}_i \in A_i$, $i = 0, 1$. Assume that $-(1/n) \leq s \leq \infty$. Then

$$\int_{(1-\theta)A_0+\theta A_1} f(\mathbf{x})\, d\mathbf{x} \geq M_t\left[\int_{A_0} f_0(\mathbf{x})\, d\mathbf{x}, \int_{A_1} f_1(\mathbf{x})\, d\mathbf{x}; \theta\right], \tag{3.18}$$

where

$$t = \begin{cases} \dfrac{s}{(1+ns)}, & \text{if} \quad -\dfrac{1}{n} < s < \infty \\ -\infty, & \text{if} \quad s = -\dfrac{1}{n} \\ \dfrac{1}{n}, & \text{if} \quad s = \infty. \end{cases}$$

Proof. (i) In view of the monotone convergence theorem, we may assume that the functons f_0, f_1 and the sets A_0, A_1 are bounded. Write

$$a_i = \int_{A_i} f_i(\mathbf{x})\, d\mathbf{x} \quad \text{and} \quad a = \int_{(1-\theta)A_0+A_1} f(\mathbf{x})\, d\mathbf{x}.$$

Suppose first that $a_0 = 0$. Choose $\mathbf{x}_0 \in A_0$. Then

$$\int_{(1-\theta)A_0+\theta A_1} f(\mathbf{x})\, d\mathbf{x} \geq \int_{(1-\theta)\mathbf{x}_0+\theta A_1} f(\mathbf{x})\, d\mathbf{x} = \int_{A_1} f[(1-\theta)\mathbf{x}_0 + \theta\mathbf{x}_1]\theta^n\, d\mathbf{x}_1$$
$$\geq \int_{A_1} \theta^{1/s} f_1(\mathbf{x}_1)\theta^n\, d\mathbf{x}_1 = M_t[0, a_1; \theta],$$

which is the required result. Therefore, in the rest of the proof, we assume that $a_i > 0$, $i = 0, 1$.

(ii) Assume now that $n = 1$. Let c_i be the supremum of f_i on A_i. Write $A = (1-\theta)A_0 + \theta A_1$ and $b = M_s(c_0, c_1; \theta)$. Define

$$B_i = \{(x, y) \in R^2 : x \in A_i,\ y > 0 \text{ and } f_i(x) > c_i y\},$$

and

$$B = \{(x, y) \in R^2 : x \in A,\ y > 0 \quad \text{and} \quad f(x) > by\}.$$

Let $B_i(y)$, $B(y)$ denote the y-sections of B_i and B. Then (3.17) shows that $B(y) \supset (1-\theta)B_0(y) + \theta B_1(y)$. Observe that $B_i(y)$ is empty if $y > 1$ and nonempty if $y < 1$. However, $B(y)$ may be nonempty for some $y > 1$. Now a simple application of Fubini's theorem shows that

$$a_i = \int_{A_i} f_i(x)\,dx = c_i \int_0^1 \lambda_1[B_i(y)]\,dy, \tag{3.19}$$

and

$$a = \int_{(1-\theta)A_0+\theta A_1} f(x)\,dx \geq b \int_0^1 \lambda_1[B(y)]\,dy. \tag{3.20}$$

The Brunn–Minkowski inequality for R^1 shows that

$$\lambda_1[B(y)] \geq (1-\theta)\lambda_1[B_0(y)] + \theta\lambda_1[B_1(y)]. \tag{3.21}$$

Now (3.19), (3.20) and (3.21) easily give:

$$\begin{aligned} a &\geq b[(1-\theta)c_0^{-1}a_0 + \theta c_1^{-1}a_1] \\ &= M_s(c_0, c_1; \theta) \cdot M_1(c_0^{-1}a_0, c_1^{-1}a_1; \theta). \end{aligned} \tag{3.22}$$

Lemma 3.4 with $u = 1$ now shows that the right side of (3.22) is $\geq M_t(a_0, a_1; \theta)$, with $t = s/(1+s)$. The theorem thus holds for $n = 1$.

(iii) The proof of the theorem can now be completed by induction on n. Suppose $n \geq 2$ and that the theorem holds in R^m, $m \leq n-1$. Write $R^n = \mathscr{X} \times \mathscr{Y}$, where $\mathscr{X} = R$ and $\mathscr{Y} = R^{n-1}$. Let $A_i(x) = \{\mathbf{y} \in \mathscr{Y} : (x, \mathbf{y}) \in A_i\}$. Similarly, $A(x)$ will denote the x-section of $A = (1-\theta)A_0 + \theta A_1$. Let A_i^* and A^* be the projections of A_i and A on $\mathscr{X}$. That is, $A_i^* = \{x \in \mathscr{X} : (x, \mathbf{y}) \in A_i \text{ for some } \mathbf{y} \in \mathscr{Y}\}$. Define the functions h_0, h_1, h by

$$h_i(x) = \int_{A_i(x)} f_i(x, \mathbf{y})\,d\mathbf{y} \quad \text{and} \quad h(x) = \int_{A(x)} f(x, \mathbf{y})\,d\mathbf{y}.$$

Suppose that $x_0 \in A_0^*$, $x_1 \in A_1^*$ and $x = (1-\theta)x_0 + \theta x_1$. Then $A_0(x_0)$ and $A_1(x_1)$ are nonempty and $A(x) \supset (1-\theta)A_0(x_0) + \theta A_1(x_1)$. Therefore, the induction hypothesis shows that

$$\begin{aligned} h(x) &\geq \int_{(1-\theta)A_0(x_0)+\theta A_1(x_1)} f(x, \mathbf{y})\,d\mathbf{y} \\ &\geq M_u\left[\int_{A_0(x_0)} f_0(x_0, \mathbf{y})\,d\mathbf{y}, \int_{A_1(x_1)} f_1(x_1, \mathbf{y})\,d\mathbf{y}; \theta\right] \\ &= M_u[h_0(x_0), h_1(x_1); \theta], \end{aligned}$$

where $u = s/[1 + (n-1)s]$. Thus h, h_0 and h_1 satisfy the conditions of the theorem with s relaced by u. Again, $A^* \supset (1-\theta)A_0^* + \theta A_1^*$. Therefore,

$$\int_{A^*} h(x)\,dx \geq \int_{(1-\theta)A_0^* + \theta A_1^*} h(x)\,dx$$

$$\geq M_t\left[\int_{A_0^*} h_0(x)\,dx, \int_{A_1^*} h_1(x)\,dx; \theta\right],$$

where $t = u/(1+u) = s/(1+ns)$. The equivalence of this last result to (3.18) is a simple consequence of Fubini's theorem. Our theorem thus holds in R^n if it holds in R^{n-1}. This completes the proof. ■

When the three functions f_0, f_1 and f in the statement of Theorem 3.15 coincide, we get the following important theorem.

Theorem 3.16. *Let f be an s-concave function on an open convex set C in R^n. Let d be the dimension of the support S of f and let λ be the Lebesgue measure on S. Suppose f is integrable over S with respect to λ. Define a measure μ on $\mathscr{B}(C)$ by $\mu(A) = \int_{A \cap S} f(\mathbf{x})\,d\lambda(\mathbf{x})$. If $-(1/d) \leq s \leq \infty$, then μ is a t-concave measure on $\mathscr{B}(C)$, where*

$$t = \begin{cases} \dfrac{s}{(1+sd)}, & \text{if } -\dfrac{1}{d} < s < \infty \\ -\infty, & \text{if } s = -\dfrac{1}{d} \\ \dfrac{1}{d}, & \text{if } s = \infty. \end{cases}$$

(*Note*: $d = n$ *if* $s > 0$).

Proof. The theorem follows trivially from 3.12 because, with $f = f_0 = f_1$ and $n = d$, condition (3.17) is equivalent to the s-concavity of f and the consequence (3.18) is equivalent to the t-concavity of μ. ■

It is interesting to note that Theorem 3.16 has a converse.

Theorem 3.17. *Let C be an open convex set in R^n and let μ be a t-concave measure on $\mathscr{B}(C)$. If d is the dimension of the support S of μ, write $f = d\mu/d\lambda$, where λ is the Lebesgue measure on S. If $-\infty \leq t \leq 1/d$, then f can be chosen* *to be s-concave on $C \cap H$, where*

$$s = \begin{cases} \dfrac{t}{(1-td)}, & \text{if } -\infty < t < \dfrac{1}{d} \\ -\dfrac{1}{d}, & \text{if } t = -\infty \\ \infty, & \text{if } t = \dfrac{1}{d}. \end{cases}$$

(*Note*: $d = n$ *if* $t > 0$).

Proof. In this proof we use x_k (respectively, x_{ik}) to denote the k-th component of a vector $\mathbf{x}$ (respectively, $\mathbf{x}_i$). In proving the theorem, we may assume that $C \subset R^d$ and that supp μ has dimension d. Fix $\theta \in (0,1)$. Let $\mathbf{x}_i \in C$, $i = 0, 1$ and write $\mathbf{x} = (1-\theta)\mathbf{x}_0 + \theta\mathbf{x}_1$. Let $\mathbf{a}_i \in R^d$, $(i = 0, 1)$, be a vector of positive components and write $\mathbf{a} = (1-\theta)\mathbf{a}_0 + \theta\mathbf{a}_1$. Let A_i be the rectangular block in R^d with sides $[x_{ik}, x_{ik} + a_{ik}]$ and let A be the block with sides $[x_k, x_k + a_k]$. Then $A = (1-\theta)A_0 + \theta A_1$ and the t-concavity of μ shows that

$$\mu(A) \geq M_t[\mu(A_0), \mu(A_1); \theta]. \tag{3.23}$$

Now (3.23) easily implies that a density f of μ can be chosen in such a way that the inequality

$$f(\mathbf{x}) \prod_k a_k \geq M_t\left[f(\mathbf{x}_0) \prod_k a_{0k}, f(\mathbf{x}_1) \prod_k a_{1k}; \theta\right] \tag{3.24}$$

holds for small values of the a_{ik}'s. But since both sides of (3.24) are homogeneous of degree d in the a_{ik}'s, the inequality holds for all positive values of a_{ik}.

Suppose $t \in (-\infty, 0) \cup (0, 1/d)$. Let $s = t/(1 - td)$. Write $c_i = f(\mathbf{x}_i)$ and put $a_{ik} = c_i^s$ in (3.24). If we note that $a_k = (1-\theta)a_{0k} + \theta a_{1k}$, we get

$$f(\mathbf{x})[(1-\theta)c_0^s + \theta c_1^s]^d \geq M_t[c_0^{1+sd}, c_1^{1+sd}; \theta]. \tag{3.25}$$

Now the right side of (3.25) equals

$$[(1-\theta)c_0^{t(1+sd)} + \theta c_1^{t(1+sd)}]^{1/t} = [(1-\theta)c_0^s + \theta c_1^s]^{1/t}.$$

Therefore (3.25) reduces to

$$f(\mathbf{x}) \geq [(1-\theta)c_0^s + \theta c_1^s]^{(1/t)-d} = M_s(c_0, c_1; \theta).$$

This shows that f is s-concave on C.

Suppose $t = -\infty$. Then we take $s = -1/d$. That is, $1 + sd = 0$ and so (3.25) gives

$$f(\mathbf{x})[(1-\theta)c_0^s + \theta c_1^s]^{-1/s} \geq 1,$$

which again shows that f is s-concave.

Suppose $t = 1/d$. Then we put $a_{0k} = 1$ and $a_{1k} = 0$ for all k in (3.24) to obtain $f(\mathbf{x}) \geq f(\mathbf{x}_0)$. Similarly, $f(\mathbf{x}) \geq f(\mathbf{x}_1)$ and hence f is s-concave with $s = \infty$.

Finally, if $t = 0$, then we put $a_{ik} = 1$ for all i and k to show that f is 0-concave on C.

The proof of the theorem is now complete. ■

Remark. Theorems 3.16 and 3.17 show that s-concave functions and t-concave measures with d-dimensional supports correspond to each other if s and t are related by the formulas

$$t = \frac{s}{(1+sd)}, \qquad s = \frac{t}{(1-td)},$$

where $-\infty \leq t \leq (1/d)$ and $(-1/d) \leq s \leq \infty$. Two comments are now in order.

(i) In view of Theorem 3.14, we see that nonzero t-concave measures for all possible indices t get linked with appropriate s-concave densities. On the other hand, s-concave densities with $-\infty \leq s < (-1/d)$ escape the above correspondence.

(ii) Suppose μ is nonzero and $(1/n)$-concave on $\mathscr{B}_n$. Then the density f of μ is ∞-concave, which means that f must be a constant. Therefore, μ is a multiple of the Lebesgue measure λ_n on $\mathscr{B}_n$.

Theorems 3.16 and 3.17 enable us to identify t-concave measures by looking at their densities. In Chapter 2, we saw that several standard densities are logconcave (that is 0-concave) and so they determine 0-concave measures. However, there are standard densities appearing in statistical work which are not 0-concave but which can be shown to be s-concave for suitable negative values of s. Similarly, a density which is known to be 0-concave may, in fact, have the stronger property of being s-concave for some positive value of s. We give some examples.

(i) Let f_m denote the density of the chi-square distribution with m degrees of freedom. If $m \geq 2$, then f_m is 0-concave and not s-concave for any $s > 0$. So, let $m < 2$. Then f_m is not 0-concave. Now, given $c > 0$, the function $x^u e^{cx}$ is convex on $(0, \infty)$ if, and only if, $u \geq 1$. So, by choosing c and u suitably, we can show that f_m is s-concave with $s \leq [2/(m-2)]$.

We note that as m varies over the interval $(0, 2)$, the values of $2/(m-2)$ vary over the interval $(-\infty, -1)$. Therefore the probability measures determined on R by the densities f_m, $0<m<2$, are not t-concave for any t.

(ii) Let g_m denote the density of the t-distribution with m degrees of freedom. Since g_m does not have a finite moment of order m, g_m cannot be logconcave. Therefore, if g_m is s-concave, then s has to be negative. Now, for $a \geq 0$, the function $(a+x^2)^u$ is easily shown to be convex if $u \geq (1/2)$. Therefore, using suitable values of a and u, we can show that g_m is s-concave for $s \leq [-1/(m+1)]$. As m varies over $(0, \infty)$, the values of $[-1/(m+1)]$ vary over $(-1, 0)$. Therefore, Theorem 3.13 shows that the probability measure determined on R by g_m is $(-1/m)$-concave. In particular, the Cauchy density is $(-1/2)$-concave and the Cauchy measure is (-1)-concave.

(iii) Let $h_{m,n}$ be the density of the F-distribution with degrees of freedom m, n. Once again $h_{m,n}$ does not have a finite moment of order $(n/2)$ and so $h_{m,n}$ cannot be logconcave. Now the function $x^{\alpha}(1+x)^{\beta}$ can be shown to be convex on $(0, \infty)$ if $\alpha \leq 0$ and $\beta \geq 1-\alpha$. Therefore, the density $h_{m,n}$ is s-concave with $s \leq [-2/(n+2)]$ when $m \geq 2$. As n varies over $(0, \infty)$, $[-2/(n+2)]$ varies over $(-1, 0)$. Therefore, for $m \geq 2$, the measure determined on R by $h_{m,n}$ is $(-2/n)$-concave.

(iv) Let $\alpha > 1$ and consider the density

$$p_{\alpha}(x) = \alpha x^{\alpha-1}, \qquad 0<x<1.$$

Then p_{α} is s-concave with $s \leq [1/(\alpha-1)]$ and the measure determined by p_{α} is t-concave with $t \leq (1/\alpha)$. Thus the index of concavity of a density or measure can be positive.

We now turn to products, convolutions and marginals of concave measures and functions.

Theorem 3.18. *For $i=1,2$, let μ_i be an s_i-concave measure on $\mathscr{B}(C_i)$ where C_i is an open convex set in R^{n_i}. Suppose $s_1 > 0$, $-s_1 \leq s_2 \leq \infty$ and $s = (s_1^{-1} + s_2^{-1})^{-1}$, where we take $s = -\infty$ if $s_2 = -s_1$. Then $\mu_1 \times \mu_2$ is an s-concave measure on $\mathscr{B}(C_1 \times C_2)$.*

Proof. If we write $\mu = \mu_1 \times \mu_2$ and $C = C_1 \times C_2$, then we want to show that

$$\mu[(1-\theta)A_0 + \theta A_1] \geq M_s[\mu(A_0), \mu(A_1); \theta], \tag{3.26}$$

where $\theta \in (0, 1)$ and A_0, A_1 are nonempty subsets of $\mathscr{B}(C)$. By Theorem 3.13(b), it is sufficient to prove (3.26) for the case where A_0, A_1 are rectangles in C

with sides parallel to the axes. But in this special case, (3.26) easily follows from Lemma 3.4. ■

Theorem 3.19. *For $i = 1, 2$, let f_i be an $s_i$concave function on an open convex set C_i in R^{n_i}. Let $s_1 > 0$, $-s_1 \leq s_2$ and $s = (s_1^{-1} + s_2^{-1})^{-1}$, where we take $s = -\infty$ if $s_1 = -s_2$. Let $g(x, y) = f_1(x)f_2(y)$. Then g is s-concave on $C_1 \times C_2$.*

Proof. This theorem is a direct consequence of Lemma 3.4. ■

In Theorems 3.18 and 3.19, the conditions on the indices s_1, s_2 are somewhat stringent. But a simple example shows that, if $s_1 + s_2 < 0$, then $\mu_1 \times \mu_2$ or $f_1(\mathbf{x})f_2(\mathbf{y})$ may not even be $(-\infty)$-concave.

Example 3.2. Suppose $0 < \beta < \alpha < 1$ and $C = (0, \infty)$. For $x \in C$, let $f(x) = (1 + x)^{-1/\alpha}$ and $g(x) = x^{1/\beta}$. Then f is $(-\alpha)$-concave and g is β-concave on C. Let $h(x, y) = f(x)g(y)$ and

$$C_1 = \{(x, y) : h(x, y) \geq 1\} = \{(x, y) : y \geq (1 + x)^{\beta/\alpha}\}.$$

Since $\beta < \alpha$, the function $(1 + x)^{\beta/\alpha}$ is strictly concave on C. Therefore C_1 is not convex. Thus h is not $(-\infty)$-concave.

Suppose now that μ and ν are the measures determined on $\mathscr{B}(C)$ by f and g, respectively. Let $t = \alpha/(\alpha - 1)$ and $u = \beta/(1 + \beta)$. Since $-1 < -\alpha$ and $-1 < \beta$, Theorem 3.16 shows that μ is t-concave and ν is u-concave. We know that the density h of $\mu \times \nu$ is not $(-\infty)$-concave. Therefore, h cannot be $(-\frac{1}{2})$-concave and so $\mu \times \nu$ is not $(-\infty)$-concave.

The next two theorems give results on marginals of s-concave measures and functions.

Theorem 3.20. *Let μ be an s-concave measure on $\mathscr{B}(C)$, where C is an open convex set in R^n. Let L be a real matrix of order $m \times n$ and rank m. Write $D = \{L\mathbf{x} : \mathbf{x} \in D\}$. For $A \in \mathscr{B}(D)$, let $\mu^*(A) = \mu\{\mathbf{x} \in C : L\mathbf{x} \in A\}$. Then the measure μ^* is s-concave on $\mathscr{B}(D)$.*

Proof. Straightforward. ■

Theorem 3.21. *Let f be s-concave on an open convex set in R^{m+n}. Let C^* be the projection of C on R^m and for $\mathbf{x} \in C^*$, let $C(\mathbf{x})$ be the $\mathbf{x}$-section of C. Define*

$$f^*(\mathbf{x}) = \int_{C(\mathbf{x})} f(\mathbf{x}, \mathbf{y})\, d\mathbf{y}, \qquad \mathbf{x} \in C^*.$$

If $-1/n \leq s \leq \infty$, *then* f^* *is* s^**-concave on* C^*, *where* $s^* = s/(1+ns)$ *with the usual conventions when* $s = -1/n$ *or* $s = \infty$.

Proof. Let $\mathbf{x}_i \in C^*$, $i = 0, 1$ and $0 < \theta < 1$ be fixed. Let $\mathbf{x} = (1-\theta)\mathbf{x}_0 + \theta\mathbf{x}_1$. Write $g_i(\mathbf{y}) = f(\mathbf{x}_i, \mathbf{y})$, $i = 0, 1$ and $g(\mathbf{y}) = f(\mathbf{x}, \mathbf{y})$. Then the s-concavity of f shows that

$$g[(1-\theta)\mathbf{y}_0 + \theta\mathbf{y}_1] \geq M_s[g_0(\mathbf{y}_0), g_1(\mathbf{y}_1); \theta],$$

whenever $\mathbf{y}_i \in C(\mathbf{x}_i)$, $i = 0, 1$. Further $C(\mathbf{x}) \supset (1-\theta)C(\mathbf{x}_0) + \theta C(\mathbf{x}_1)$. Therefore, Theorem 3.15 shows that

$$\int_{C(\mathbf{x})} g(\mathbf{y})\, d\mathbf{y} \geq M_{s^*}\left[\int_{C(\mathbf{x}_0)} g_0(\mathbf{y})\, d\mathbf{y}, \int_{C(\mathbf{x}_1)} g_1(\mathbf{y})\, d\mathbf{y}; \theta\right].$$

This shows that f^* *is* s^*-concave on C^*. ■

The following is a simple example to show that the conclusions of the last two theorems are the best possible under the given conditions.

Example 3.3. First suppose $s > -1$ and let $C = \{(x, y) \in R^2 : 0 < x < y < \infty\}$. Define f on C by $f(x, y) = x^{1/s}$. Then f is s-concave on C. Further, in the notation of Theorem 3.21,

$$f^*(x) = \int_0^x f(x, y)\, dy = x^{(s+1)/s}, \qquad 0 < s < \infty.$$

Therefore, f^* is s^*-concave with $s^* = s/(s+1)$ and f^* is not u-concave for any $u > s^*$. The index s^* given by Theorem 3.21 is thus the best possible.

Now suppose that $s > -\frac{1}{2}$. Then we can write $s = t/(1-2t)$, where $-\infty \leq t < \frac{1}{2}$. The measure μ determined on $\mathscr{B}(C)$ by f is therefore t-concave. Let L denote the projection from R^2 onto the x-axis. Then μ and L determine a measure μ^* on $(0, \infty)$ whose density is f^* (given above). Now the index of concavity of f^* is

$$s^* = \frac{s}{1+s} = \frac{t/(1-2t)}{1 + [t/(1-2t)]} = \frac{t}{1-t}.$$

Therefore, μ^* is t^*-concave, where $t^* = s^*/(1+s^*) = t$. The index t given by Theorem 3.20 is thus also the best possible.

A theorem on product densities (respectively, measures) can be obviously combined with a theorem on marginal distributions to obtain a theorem on convolutions of densities (respectively, measures). Therefore, we can state the following two theorems.

Theorem 3.22. *Let $0 < s_1 \leq (1/n)$, $-s_1 \leq s_2 \leq (1/n)$. For $i = 1, 2$, let μ_i be s_i-concave on $\mathscr{B}(C_i)$ where C_i is an open convex set in R^n. Write $s = (s_1^{-1} + s_2^{-1})^{-1}$,* where we take $s = -\infty$ if $s_2 = -s_1$. *Then the convolution $\mu_1 * \mu_2$ is s-concave on $\mathscr{B}(C_1 + C_2)$.*

Theorem 3.23. *For $i = 1, 2$, let f_i be an s_i-concve function on an open convex set C_i in R^n. Let $s_1 > 0$ and $-s_1 \leq s_2$. Write $s = (s_1^{-1} + s_2^{-1})^{-1}$. If $(-1/n) \leq s$, then the convolution $f_1 * f_2$ is s^*-concave on $C_1 + C_2$, where $s^* = s/(1 + ns)$.*

Remark. While the conditions of Theorem 3.23 are rather detailed, we note that, in many situations, the index s^* will be greater than $-\infty$. In any case, one does need some condition to even assert that the convolution $f_1 * f_2$ is $(-\infty)$-concave. To justify this statement we refer to Sherman's example [used in part (c) of the proof of Theorem 2.19] in which we have two bivariate densities f_1, f_2 such that f_1 is 0-concave, f_2 is $(-\infty)$-concave and $f_1 * f_2$ is not $(-\infty)$-concave. On the other hand, in the univariate case $f_1 * f_2$ is $(-\infty)$-concave as soon as f_1 is 0-concave and f_2 is $(-\infty)$-concave.

To conclude this section, we present a theorem which is obviously motivated by Anderson's theorem.

Theorem 3.24. *Let f be an s-concave probability density on R^n with $s \geq (-1/n)$. Write $t = s/(1 + ns)$. Let P be the probability defined on $\mathscr{B}_n$ by f. Then, for every convex set $C \subset R^n$, the function h defined on R^n by $h(\mathbf{x}) = P[C + \mathbf{x}]$ is t-concave. In particular, $P[C + k\mathbf{y}]$ is unimodal in $k \in R$ for every fixed convex set $C \subset R^n$* and every fixed $\mathbf{y} \in R^n$.

Proof. By Theorem 3.16, we know that P is t-concave on $\mathscr{B}_n$. Let $\mathbf{x}$, $\mathbf{y}$ be in R^n and let $\theta \in (0, 1)$. Then for every convex set $C \subset R^n$,

$$C + (1 - \theta)\mathbf{x} + \theta\mathbf{y} \supset (1 - \theta)(C + \mathbf{x}) + \theta(C + \mathbf{y}).$$

Therefore

$$P[C + (1 - \theta)\mathbf{x} + \theta\mathbf{y}] \geq M_t[P(C + \mathbf{x}), P(C + \mathbf{y}); \theta],$$

which means that

$$h[(1 - \theta)\mathbf{x} + \theta\mathbf{y}] \geq M_t[h(\mathbf{x}), h(\mathbf{y}); \theta].$$

Thus h is t-concave and hence quasi-concave. The second assertion is now immediate. ■

There are important differences between Anderson's theorem and Theorem 3.24. In Anderson's theorem, f is required to be centrally symmetric but only $(-\infty)$-concave. The convex set C is also required to be centrally symmetric. In Theorem 3.24, the density f or the set C need not be symmetric. But this strengthening is achieved at the expense of the stronger assumption that f is s-concave with $s \geq (-1/n)$.

4 Unimodality for Discrete Distributions

4.0. Summary

According to the definition of unimodality given in Chapter 1, the only discrete distributions which are unimodal are the degenerate ones. For this reason, an alternative definition is given in this chapter for use in the discrete case. We give the basic theorems on the convex structure of the set of all discrete unimodal distributions. It is found that a discrete uniform distribution can sometimes have more than one representation as a mixture of discrete uniform distributions. Strong unimodality is discussed next and here Ibragimov's characterization carries over nicely to the discrete case. Finally, we present some results on the unimodality of high convolutions.

4.1. The Structure of Unimodal Discrete Distributions

Consider a distribution which concentrates its mass on the set of integers. We denote such a distribution by a sequence $\{p_n, -\infty < n < \infty\}$, where p_n denotes the mass assigned to the integer n. The commonly accepted definition of unimodality of such distributions is as follows: see Keilson and Gerber (1971) or Medgyessy (1972).

Definition 4.1. A distribution $\{p_n, -\infty < n < \infty\}$ is called *unimodal* about a mode M if

$$p_n \geq p_{n-1} \qquad \text{for } n \leq M$$

and

$$p_n \leq p_{n-1} \qquad \text{for } n \geq M+1.$$

According to the above definition, a unimodal distribution $\{p_n\}$ is one for which the sequence $\{p_n - p_{n-1}\}$ has only one change of sign, when the zero terms are ignored. Let $\mathscr{U}_d$ denote the set of all distributions on the set of integers which are unimodal about 0. Then the set $\mathscr{U}_d$ is convex under mixtures. The extreme points of $\mathscr{U}_d$ are the uniform distributions on $\{-j, -j+1, \ldots, k\}$, where j and k are nonnegative integers. This is shown in Theorem 4.1 below. In contrast, in the general (non-discrete) case, one gets only the unilateral uniform distributions as extreme points. The reason is that the uniform distribution on $(-a, b)$ is a mixture of the uniform distributions on $(-a, 0)$ and $(0, b)$ when a and b are positive. Such a result cannot hold in the discrete case because the mass at 0 gets inflated when we take a mixture.

Before we state Theorem 4.1, we fix some notation. For nonnegative integers j and k, the uniform distribution on $\{-j, \ldots, k\}$ will be denoted by $\mathbf{u}_{jk}$. The unilateral uniform distribution $\mathbf{u}_{0k}$ will sometimes be denoted by $\mathbf{v}_k$. Similarly $\mathbf{w}_j$ will sometimes be written for $\mathbf{u}_{j0}$. Of course, $\mathbf{u}_{00} = \mathbf{v}_0 = \mathbf{w}_0$. For a given sequence $\{a_n\}$ we write $\Delta a_n = a_{n+1} - a_n$ and $a_n' = a_{-n}$. The next three theorems are taken from Dharmadhikari and Jogdeo (1976b).

Theorem 4.1. *Let $\mathscr{U}_d$ denote the set of all distributions on the set of integers which are unimodal about 0. A distribution $\mathbf{p} = \{p_n\}$ belongs to $\mathscr{U}_d$ if, and only if, it has a representation*

$$\mathbf{p} = \sum_{j=0}^{\infty} \sum_{k=0}^{\infty} c_{jk} \mathbf{u}_{jk}, \tag{4.1}$$

where $c_{jk} \geq 0$ and $\sum_0^\infty \sum_0^\infty c_{jk} = 1$. Moreover, the distributions $\mathbf{u}_{jk}$ are extreme in $\mathscr{U}_d$.

Proof. The "if" part of the first assertion is trivial. So, let $\mathbf{p}$ be in $\mathscr{U}_d$. To establish (4.1), we have to show that the following two equations have a nonnegative solution.

$$p_n = \sum_{j=0}^{\infty} \sum_{k=n}^{\infty} \left(\frac{c_{jk}}{j+k+1} \right), \qquad n \geq 0, \tag{4.2}$$

and

$$p'_m = \sum_{j=m}^{\infty} \sum_{k=0}^{\infty} \left(\frac{c_{jk}}{j+k+1} \right), \qquad m \geq 0. \tag{4.3}$$

Write $b_{jk} = c_{jk}/(j+k+1)$. Then (4.2) and (4.3) are equivalent to

$$-\Delta p_k = \sum_{j=0}^{\infty} b_{jk}, \qquad k \geq 0 \quad \text{and} \quad -\Delta p'_j = \sum_{k=0}^{\infty} b_{jk}, \qquad j \geq 0. \tag{4.4}$$

The two parts of (4.4) are consistent because

$$\sum_{k=0}^{\infty} \Delta p_k = \sum_{j=0}^{\infty} \Delta p'_j = -p_0. \tag{4.5}$$

Thus (4.4) can be solved. For instance, take $b_{jk} = (\Delta p_k)(\Delta p'_j)/p_0$. Then (4.5) shows that (4.4) holds. The nonnegativity of this choice of the b_{jk} follows because both Δp_k and $\Delta p'_j$ are nonpositive. That the c_{jk} must add to 1 follows by a simple application of Fubini's theorem. The "only if" part of the first assertion is thus proved.

Suppose now, that in the representation (4.1), we replace $\mathbf{p}$ by $\mathbf{u}_{JK}$. Since $\Delta p_k = 0$ for $0 \leq k < K$ and for $k \geq K+1$, (4.4) shows that $b_{jk} = 0$ whenever $k \neq K$. Similarly $b_{jk} = 0$ whenever $j \neq J$. Thus, in the representation (4.1), we must have $c_{JK} = 1$. This proves that $\mathbf{u}_{JK}$ is an extreme point of $\mathscr{U}_d$. The proof of the theorem is now complete. ■

Remark. Theorem 4.1 enables us to bring out two points of contrast between the discrete case being considered here and the general case considered in Chapter 1. These are:

(a) Bilateral uniform distributions also arise as extreme points of $\mathscr{U}_d$, and
(b) The representation (4.1) may not be unique. To verify (b), observe that

$$\tfrac{1}{2}[\mathbf{u}_{k0} + \mathbf{u}_{0k}] = \frac{(2k+1)}{2(k+1)} \mathbf{u}_{kk} + \frac{1}{2(k+1)} \mathbf{u}_{00}.$$

The preceding remark raises a natural question. What are the distributions which are obtainable as mixtures of the unilateral uniform distributions $\mathbf{v}_j$ and $\mathbf{w}_k$? The answer is given in the next theorem.

Theorem 4.2. *A distribution $\mathbf{p} = \{p_n\}$ is unimodal about 0 and satisfies $p_0 \geq p_1 + p'_1$ if, and only if, $\mathbf{p}$ has a representation*

$$\mathbf{p} = a_0 \mathbf{v}_0 + \sum_{n=1}^{\infty} (a_n \mathbf{v}_n + a'_n \mathbf{w}_n), \tag{4.6}$$

where the a_n are nonnegative constants such that $a_0 + \sum_1^\infty (a_n + a_n') = 1$. Moreover, the representation (4.6) *is unique.*

Proof. Suppose (4.6) holds. Then **p** is unimodal about 0. Further

$$p_0 = a_0 + \sum_1^\infty \left[\frac{(a_n + a_n')}{(n+1)}\right],$$

$$p_1 = \sum_1^\infty \left[\frac{a_n}{(n+1)}\right] \quad \text{and} \quad p_1' = \sum_1^\infty \left[\frac{a_n'}{(n+1)}\right].$$

Therefore, $p_0 - p_1 - p_1' = a_0 \geq 0$. This proves the "if" part of the first assertion.

Conversely, suppose that **p** is unimodal about 0 and that $p_0 \geq p_1 + p_1'$. To prove (4.6), we have to show that the following equations have a nonnegative solution.

$$p_0 = a_0 + \sum_1^\infty \left[\frac{(a_n + a_n')}{(n+1)}\right] \tag{4.7}$$

$$p_n = \sum_n^\infty \left[\frac{a_k}{(k+1)}\right] \quad \text{and} \quad p_n' = \sum_n^\infty \left[\frac{a_k'}{(k+1)}\right]. \tag{4.8}$$

Now, for $n \geq 1$, (4.8) easily implies

$$-\Delta p_n = \frac{a_n}{(n+1)} \quad \text{and} \quad -\Delta p_n' = \frac{a_n'}{(n+1)}. \tag{4.9}$$

The unimodality of **p** about 0 shows that, for $n \geq 1$, the values of a_n and a_n' calculated from (4.9) are nonnegative and unique. Further, (4.7) shows that $a_0 = p_0 - p_1 - p_1' \geq 0$. Thus (4.6) holds with nonnegative constants a_n. Again, the a_n must add to 1 because of Fubini's theorem. This completes the proof of the first assertion. The second assertion follows immediately from (4.9) and (4.7). This proves the theorem. ∎

The following theorem gives a representation theorem for symmetric uniform distributions on the set of integers. Its proof follows the same lines as that of Theorem 4.2 and so we omit it.

Theorem 4.3. *A symmetric distribution* **p** *on the set of integers is unimodal if, and only if, it has a representation*

$$\mathbf{p} = \sum_{k=0}^\infty a_k \mathbf{u}_{kk},$$

where a_k, $k \geq 0$ are nonnegative constants such that $\sum_0^\infty a_k = 1$. Such a representation is unique.

Suppose F is a distribution function on R which is twice differentiable. Then F is unimodal about 0 if, and only if, $-xF''(x)$ is a probability density function; see the second remark following Theorem 1.3. A discrete analog of this result is given in the next theorem, which is due to Medgyessy (1972).

Theorem 4.4. *Suppose $\theta \in (0, 1)$ is arbitrary. A sequence $\mathbf{p} = \{p_n\}$ defines a distribution on the set of integers which is unimodal about 0 if, and only if, the sequence $\mathbf{q} = \{q_n\}$ defined by*

$$q_n = (\theta - n)(p_n - p_{n-1}) \tag{4.10}$$

gives a distribution on the set of integers.

Proof. Suppose that $\mathbf{p}$ is a distribution which is unimodal about 0. Then $p_n - p_{n-1} \geq 0$ or ≤ 0 according as $n \leq 0$ or $n \geq 1$. Also $(\theta - n) > 0$ or < 0 according as $n \leq 0$ or $n \geq 1$. Therefore, q_n given by (4.10) is nonnegative for all n. Now it is easy to see that

$$\sum_{n=-\infty}^{\infty} q_n = \sum_{n=-\infty}^{\infty} p_n = 1.$$

Thus $\mathbf{q}$ is a distribution on the set of integers.

Conversely, suppose $\mathbf{q}$ is a distribution on the set of integers and define p_n by (4.10). Then, for $n \geq 0$, we have

$$p_n = \sum_{k=n+1}^{\infty} \left[\frac{q_k}{(k - \theta)}\right] \quad \text{and} \quad p_n' = \sum_{k=n}^{\infty} \left[\frac{q_k'}{(k + \theta)}\right].$$

These values of p_n and p_n' are clearly nonnegative. Further $p_n - p_{n-1} \geq 0$ or ≤ 0 according as $n \leq 0$ or $n \geq 1$. Finally, one gets $\sum_{-\infty}^{\infty} p_n = \sum_{-\infty}^{\infty} q_n = 1$. Thus, $\mathbf{p}$ is a distribution on the integers which is unimodal about 0. The theorem is thus proved. ■

Theorem 4.4 can, of course, be written in terms of characteristic functions. Let φ and ψ, respectively, be the characteristic functions of $\mathbf{p}$ and $\mathbf{q}$. Write $\xi(t) = (1 - e^{it})\varphi(t) = \sum (p_n - p_{n-1})e^{int}$. Then (4.10) gives $\psi(t) = \theta\xi(t) + i\xi'(t)$, whose solution is

$$i\xi(t) = e^{-i\theta t} \int_0^t e^{-i\theta u}\psi(u)\, du, \quad t \in R.$$

or

$$\varphi(t) = \frac{e^{i\theta t}}{i(1-e^{it})} \int_0^t e^{-i\theta u}\psi(u)\, du, \qquad t \neq 2n\pi.$$

An important consequence of this formula connecting φ and ψ is that φ is now seen to be differentiable on every interval which does not include any of the points $2n\pi$, $n=0, \pm 1, \ldots$. This fact will be used later.

Again, let F be a distribution function on R which is twice differentiable. Let $g(x) = -xF''(x)$. If g is a probability density, then, as noted above, not only F is unimodal about 0, but F is also the distribution function of UZ where U is uniform on (0, 1), Z has density g and U, Z are independent. In the discrete case, given a distribution $\mathbf{p}$ which is unimodal about 0, the distribution $\mathbf{q}$ given in Theorem 4.4 is the discrete analog of the density g. One might ask whether there is also a discrete analog of the representation of the type UZ. The impossibility of such a representation has been proved by Dharmadhikari and Jogdeo (1976b).

Many of the standard discrete distributions like the binomial, Poisson and negative binomial are easily checked to be unimodal. As will be shown later, these distributions have the stronger property that their convolutions with all unimodal discrete distributions are again unimodal. The next result, due to Holgate (1970), shows that certain compound Poisson distributions are discrete unimodal.

Theorem 4.5. *Let f be a unimodal density* $(0, \infty)$. *Define*

$$p_n = \frac{1}{n!} \int_0^\infty e^{-\lambda}\lambda^n f(\lambda)\, d\lambda, \qquad n \geq 0. \tag{4.11}$$

Then $\mathbf{p}$ *is a unimodal distribution on the set of integers.*

Proof. Write $g(n, \lambda) = e^{-\lambda}\lambda^n/(n!)$ and $G(n, \lambda) = \int_0^\lambda g(n, t)\, dt$. It is well known that

$$G(n, \lambda) = \sum_{k=n+1}^{\infty} g(k, \lambda). \tag{4.12}$$

To prove the theorem, we may clearly assume that f is differentiable on $(0, \infty)$. Integrate the right side of (4.11) by parts to obtain

$$p_n = [G(n, \lambda)f(\lambda)]_0^\infty - \int_0^\infty G(n, \lambda)f'(\lambda)\, d\lambda. \tag{4.13}$$

The first term on the right side of (4.13) vanishes because G is bounded, f

is integrable and $G(n, \lambda) = O(\lambda^{n+1})$ as $\lambda \to 0$. Therefore,

$$p_n = -\int_0^{\infty} G(n, \lambda) f'(\lambda)\, d\lambda.$$

Now (4.12) shows that

$$\Delta p_n = \int_0^{\infty} g(n+1, \lambda) f'(\lambda)\, d\lambda. \tag{4.14}$$

Let λ_0 be a mode of f. Now

$$\frac{g(n+1, \lambda)}{g(n, \lambda)} = \frac{\lambda}{(n+1)} = \frac{\lambda_0}{(n+1)} \cdot \frac{\lambda}{\lambda_0}.$$

Therefore

$$\frac{g(n+1, \lambda)}{g(n, \lambda)} \lesseqqgtr \frac{\lambda_0}{(n+1)} \qquad \text{according as } \lambda \lesseqqgtr \lambda_0.$$

But $f'(\lambda) \leq 0$ for $\lambda \geq \lambda_0$ and $f'(\lambda) \geq 0$ for $\lambda \leq \lambda_0$. It now follows from (4.14) that

$$\Delta p_n \leq \frac{\lambda_0}{(n+1)} \int_0^{\infty} g(n, \lambda) f'(\lambda)\, d\lambda = \frac{\lambda_0}{(n+1)} \Delta p_{n-1}.$$

Thus $\Delta p_{n-1} \leq 0 \Rightarrow \Delta p_n \leq 0$. The distribution $\mathbf{p}$ is therefore, unimodal.

Bertin and Theodorescu (1984) have given a slightly different definition of the mode of discrete unimodal distribution. They call a distribution $\mathbf{p}$ *unimodal with mode M* if

$$p_n \geq p_{n-1} \quad \text{for} \quad n \leq M$$

and (4.15)

$$p_n \leq p_{n-1} \quad \text{for} \quad n \geq M+2$$

The rationale behind their definition is that if (4.15) holds and F is the distribution function defined by $\mathbf{p}$ then F restricted to the integers, is convex on $(-\infty, M]$ and concave on $[M, \infty)$. It is clear that a distribution $\mathbf{p}$ which satisfies (4.15) must have its maximum mass at M or $M+1$. But it may happen that $p_M < p_{M+1}$. Thus the maximum mass may *not* be at the "Mode." While this goes against accepted practice, two points can be stated in favor of the modified definition.

(1) The set of all unimodal distributions (with all possible modes) is the same under both definitions.

(2) Let $\mathscr{U}_d^*$ denote the set of all discrete distributions which are unimodal with mode 0 under the new definition (4.15). The convex structure of $\mathscr{U}_d^*$ is

simpler than the convex structure of the set $\mathcal{U}_d$ of Theorem 4.1. This is because the extreme points of $\mathcal{U}_d^*$ are the (unilateral) uniform distributions on $\{-j, \ldots, 0\}$ and $\{1, \ldots, k\}$, where $j \geq 0$ and $k \geq 1$. As a consequence, every $\mathbf{p}$ in $\mathcal{U}_d^*$ has a unique representation in terms of these extreme points. ■

The proof of the next theorem is omitted because it is straightforward.

Theorem 4.6. *The class of all unimodal distributions on the set of integers is closed under weak limits. The same holds for the set of all distributions on the set of integers which are unimodal with a fixed mode M.*

4.2. Convolutions of Discrete Unimodal Distributions and Strong Unimodality.

In this section we show how some of the results of Chapter 1 on convolutions and strong unimodality can be carried over to the discrete case.

First, we give an example to show that the convolution of two discrete unimodal distributions may not be unimodal. The example is similar to Example 1.1.

Example 4.1. Let $\mathbf{v}_k$ denote the uniform distribution on $\{0, 1, \ldots, k\}$. Let $n \geq 2$ and let $\mathbf{p} = \frac{1}{2}(\mathbf{v}_0 + \mathbf{v}_n)$. Write $\mathbf{q} = \mathbf{p} * \mathbf{p}$. We show that $\mathbf{q}$ is not unimodal. We have

$$4\mathbf{q} = \mathbf{v}_0 + 2\mathbf{v}_n + (\mathbf{v}_n * \mathbf{v}_n).$$

Therefore,

$$4q_0 = 1 + 2(n+1)^{-1} + (n+1)^{-2},$$

$$4q_1 = 2(n+1)^{-1} + 2(n+1)^{-2},$$

and

$$4q_2 = 2(n+1)^{-1} + 3(n+1)^{-2}.$$

Thus

$$4(q_0 - q_1) = 1 - (n+1)^{-2} > 0$$

and

$$4(q_1 - q_2) = -(n+1)^{-2} < 0.$$

Therefore, $q_1 < \min\{q_0, q_2\}$ and q is not unimodal.

Next, we prove the discrete analog of Wintner's theorem (see Theorem 1.6).

Theorem 4.7. *The convolution of two symmetric discrete unimodal distribution is symmetric unimodal.*

Proof. Let k, l be nonnegative integers. Recall that $\mathbf{u}_{kk}$ is the uniform distribution on $\{-k, \ldots, 0, \ldots, k\}$. The convolution of $\mathbf{u}_{kk}$ and $\mathbf{u}_{ll}$ is triangular if $k = l$ and trapezoidal if $k \neq l$. Thus $\mathbf{u}_{kk} * \mathbf{u}_{ll}$ is always symmetric unimodal. Now let $\mathbf{p}$ and $\mathbf{q}$ be arbitrary symmetric unimodal distributions. Using Theorem 4.3, we can express $\mathbf{p} * \mathbf{q}$ as a mixture of the convolutions $\mathbf{u}_{kk} * \mathbf{u}_{ll}$. The same theorem again shows that $\mathbf{p} * \mathbf{q}$ is symmetric and unimodal. ■

Let us now turn to strong unimodality. Following Ibragimov (1956), Keilson and Gerber (1971) gave the following definition of strong unimodality for discrete distributions.

Definition 4.2. A distribution $\mathbf{p}$ on the set of integers is called *strongly unimodal* if the convolution $\mathbf{p} * \mathbf{q}$ is unimodal for every unimodal distribution $\mathbf{q}$ on the set of integers.

The following results follow immediately from the above defnition.

(a) A strongly unimodal distribution is unimodal.
(b) All degenerate distributions are strongly unimodal.
(c) The set of all strongly unimodal distributions is closed under weak limits.

We saw in Section 4.1 that there are some important differences between the discrete case and the general case with regard to the convex structures of the sets of uniform distributions. In contrast, Ibragimov's characterization of strong unimodality carries over nicely to the discrete case. This was established by Keilson and Gerber (1971) and we give their proof below. While the basic ideas behind the proof are similar to those used in Section 1.5, it should be noted that the discrete case is inherently simpler.

Theorem 4.8. *A discrete distribution* $\mathbf{h}$ *is strongly unimodal if, and only if,*

$$h_n^2 \geq h_{n-1} h_{n+1}, \qquad \text{for all } n. \tag{4.16}$$

Proof. Suppose (4.16) holds. Let $\mathbf{p}$ be unimodal about 0 and write $\mathbf{q} = \mathbf{h} * \mathbf{p}$.

We show that **q** is unimodal. Assume first that $h_n > 0$ for all n. Observe that

$$q_n = \sum_{m=-\infty}^{\infty} h_m p_{n-m}.$$

Therefore,

$$\Delta q_n = \sum_{m=-\infty}^{\infty} h_m \Delta p_{n-m} = \sum_{m=-\infty}^{\infty} h_{n-m} \Delta p_m. \tag{4.17}$$

Now (4.16) shows that h_n/h_{n+1} is nondecreasing in m. Therefore

$$\frac{h_{n-m}}{h_{n+1-m}} \le \frac{h_n}{h_{n+1}} \qquad \text{if } m \ge 0$$

and

$$\frac{h_{n-m}}{h_{n+1-m}} \ge \frac{h_n}{h_{n+1}} \qquad \text{if } m \le 0.$$

But $\Delta p_m \ge 0$ for $m \le 0$ and $\Delta p_m \le 0$ for $m \ge 0$. Therefore, (4.17) shows that

$$\Delta q_n \ge \frac{h_n}{h_{n+1}} \sum_{m=-\infty}^{\infty} h_{n+1-m} \Delta p_m = \frac{h_n}{h_{n+1}} \Delta q_{n+1}.$$

It follows that $\Delta q_n < 0 \Rightarrow \Delta q_{n+1} < 0$. Thus **q** is unimodal. We now remove the condition that $h_n > 0$ for all n. Suppose, for instance, that $h_n > 0$ for $n < 0$ and $h_n = 0$ for $n \ge 0$. Define a distribution $\mathbf{h}(\delta)$ as follows.

$$h_n(\delta) = \frac{h_n}{c_\delta}, \qquad n < 0$$

and

$$h_n(\delta) = \frac{\delta}{c_\delta}\left(\frac{\delta}{h_{-1}}\right)^n, \qquad n \ge 0.$$

Here c_δ is a normalizing constant. Then, for sufficiently small but positive δ, $\mathbf{h}(\delta)$ satisfies (4.16). Therefore, $\mathbf{h}(\delta)$ is strongly unimodal for all sufficiently small positive δ. Taking limits as $\delta \to 0$, we see that **h** is strongly unimodal. This proves the "if" part of the theorem.

Conversely, suppose **h** is a distribution for which $h_1^2 < h_0 h_2$. We construct a unimodal distribution **p** for which $\mathbf{q} = \mathbf{h} * \mathbf{p}$ is not unimodal. Both h_0 and h_2 must clearly be positive. If $h_1 = 0$, then **h** is not unimodal and we can take **p** to be any degenerate distribution. Assume, therefore, that $h_1 > 0$. Define a distribution $\mathbf{p}$ as follows.

$$p_n = \begin{cases} \beta/(N+1), & -(N-1) \le n \le 0, \\ (\beta+\gamma)/(N+1), & n = 1, \\ \gamma/(N+1), & 2 \le n \le (N+1), \\ 0, & \text{elsewhere.} \end{cases}$$

Here N is a positive integer and β, γ are positive numbers such that $(\beta + \gamma) = 1$. Formula (4.17) shows that

$$(N+1)\Delta q_1 = \gamma h_1 - \beta h_0 + \beta h_{N+1} - \gamma h_{-N}, \tag{4.18}$$

and

$$(N+1)\Delta q_2 = \gamma h_2 - \beta h_1 + \beta h_{N+2} - \gamma h_{-N+1} \tag{4.19}$$

Since $h_1^2 < h_0 h_2$, we can find a $\beta \in (0, 1)$ such that

$$\frac{h_1}{h_0} < \frac{\beta}{1-\beta} < \frac{h_2}{h_1}.$$

For the chosen value of β, γ is then determined by the condition $(\beta + \gamma) = 1$. Now (4.18) shows that

$$\lim_{N\to\infty} \frac{(N+1)\Delta q_1}{\gamma h_0} = \frac{h_1}{h_0} - \frac{\beta}{(1-\beta)} < 0,$$

and

$$\lim_{N\to\infty} \frac{(N+1)\Delta q_2}{\gamma h_1} = \frac{h_2}{h_1} - \frac{\beta}{(1-\beta)} > 0.$$

Thus, for large N, $\Delta q_1 < 0$ whereas $\Delta q_2 > 0$. So, $\mathbf{q}$ cannot be unimodal for large N. This proves the "only if" part and completes the proof of the theorem. ■

Theorem 4.8 enables us to show that certain standard distributions are strongly unimodal.

(i) The uniform distributions $\mathbf{u}_{jk}$ are all strongly unimodal.
(ii) Every distribution on a 2-point set $\{n, n+1)\}$ is strongly unimodal.
(iii) Since a binomial distribution is the convolution of several 2-point distributions, we see from (ii) that every binomial distribution is strongly unimodal.

(iv) Since limits of strongly unimodal distributions are again strongly unimodal, we see from (iii) that all Poisson distributions are strongly unimodal.

(iv) Consider the negative binomial distribution **p** with parameters $r \in (0, \infty)$ and $\theta \in (0, 1)$. Here

$$p_n = \frac{r(r+1)\ldots(r+n-1)(1-\theta)^r\theta^n}{(n!)}, \qquad n = 0, 1, 2, \ldots.$$

Therefore

$$\left(\frac{p_n}{p_{n-1}}\right) = \left[\frac{(r+n-1)\theta}{n}\right] = \theta\left\{1 + \left[\frac{(r-1)}{n}\right]\right\},$$

which is nonincreasing in n if, and only if $r \geq 1$. Thus a negative binomial distribution with parameters r and θ is strongly unimodal if and only if $r \geq 1$.

Remark. Suppose **p** is strongly unimodal and let $\alpha_n = p_n/p_{n-1}$. Now α_n is nonincreasing and must eventually become less than 1. Thus p_n goes to zero at a geometric rate as $n \to \pm\infty$. Therefore, **p** must have a moment generating function and, in particular, **p** must have finite moments of all orders.

4.3. Unimodality of High Convolutions

Medgyessy (1972) has mentioned a conjecture due to Rényi, which states that, for a given discrete distribution **p**, the m-fold convolution $\mathbf{p}^{*m}$ of **p** with itself is unimodal for all sufficiently large m. The continuous version of the conjecture would be that the m-fold convolution f^{*m} of a continuous density f with itself is unimodal for all sufficiently large m. The rationale behind the conjecture is that $\mathbf{p}^{*m}$ or f^{*m}, suitably shifted and normalized, converges to the normal distribution which is unimodal. If true, the conjecture would immediately imply the unimodality of all stable laws. However, the conjecture is false. Counterexamples have been given recently by Brockett and Kemperman (1982) and by Ushakov (1982). We first present the examples given by Ushakov.

Let $\mathscr{T}_m$ denote the set of all distributions **p** on the integers such that $p_n > 0$ for $n < m$, $p_m = 0$, $p_{m+1} > 0$ and $p_n = 0$ for all $n \geq (m+2)$. Write $\mathscr{T} = \bigcup_{-\infty}^{\infty} \mathscr{T}_m$. We note that $\mathscr{T}$ is dense in the set of all distributions on the integers under any reasonable sense of distance.

Lemma 4.1. *The set $\mathscr{T}$ is closed under convolutions.*

Proof. Let $\mathbf{p} \in \mathscr{T}_m$, $\mathbf{q} \in \mathscr{T}_k$ and $h = \mathbf{p} * \mathbf{q}$. Then $h_n = 0$ for $n > (m+k+2)$, $h_{m+k+2} = p_{m+1}q_{k+1} > 0$, $h_{m+k+1} = p_{m+1}q_k + p_m q_{k+1} = 0$ and, for $n \leq (m+k)$, $h_n \geq p_{m+1}q_{n-m-1} > 0$. Therefore, $\mathbf{h} \in \mathscr{T}_{m+k+1}$. This proves the lemma. ■

Since none of the distributions in $\mathscr{T}$ is unimodal, the following corollary is immediate.

Corollary. *If $\mathbf{p} \in \mathscr{T}$ then $\mathbf{p}^{*m}$ is not unimodal for any $m \geq 1$.*

Ushakov has also given a method of constructing distributions $\mathbf{p}$ such that $\mathbf{p}$ assigns positive mass to all integer points and $\mathbf{p}^{*m}$ is not unimodal for any m. Suppose $\mathbf{p}_1$, $\mathbf{p}_2$ are *symmetric* discrete distributions with characteristic functions φ_1, φ_2, respectively. Let $k \geq 2$ be a fixed integer and write

$$\varphi(t) = a\varphi_1(t) + (1-a)\varphi_2(kt),$$

where $0 < a < \frac{1}{2}$. Suppose φ_1 is differentible everywhere whereas φ_2 is not differentiable at 0. Let $\bar{D}$ and $\underline{D}$ denote the upper and lower derivatives, respectively. Since φ_2 is periodic with period 2π, φ_2 is not differentiable at 2π and so $\bar{D}\varphi_2(2\pi) > \underline{D}\varphi_2(2\pi)$. Now $\varphi(2\pi/k) > 0$, because

$$\varphi\left(\frac{2\pi}{k}\right) = a\varphi_1\left(\frac{2\pi}{k}\right) + (1-a) \geq 1 - a - a > 0.$$

Therefore, for $m \geq 1$,

$$\bar{D}\varphi^m\left(\frac{2\pi}{k}\right) = m\varphi^{m-1}\left(\frac{2\pi}{k}\right)\left[a\varphi_1'\left(\frac{2\pi}{k}\right) + (1-a)\bar{D}\varphi_2(2\pi)\right]$$

and

$$\underline{D}\varphi^m\left(\frac{2\pi}{\mathrm{k}}\right) = m\varphi^{m-1}\left(\frac{2\pi}{k}\right)\left[a\varphi_1'\left(\frac{2\pi}{k}\right) + (1-a)\underline{D}\varphi_2(2\pi)\right].$$

It follows that $\bar{D}\varphi^m(2\pi/k) > \underline{D}\varphi^m(2\pi/k)$ and so φ^m is not differentiable at $(2\pi/k)$. Now we have seen earlier (see the discussion following Theorem 4.4) that the characteristic function of a unimodal distribution on the integers is always differentiable on the interval $(0, 2\pi)$. We therefore see that φ^m does not correspond to a unimodal distribution for any m.

The discrete counterexamples given by Brockett and Kemperman (1982) involve distributions whose supports have progressively larger gaps. We present their first example.

Example 4.2. Suppose $\{a_n, n \geq 0\}$ is a sequence of integers such that

$$0 = a_0 < a_1 < a_2 < \ldots \quad \text{and} \quad \frac{a_{n+1}}{a_n} \to \infty \quad \text{as} \quad n \to \infty.$$

Let $S_m = \sum_1^m X_i$, where $X_1, X_2, \ldots$ are independent random variables such that $P(X_i = a_n) = p_n$, where $\{p_n\}$ is a sequence of positive numbers such that $\sum_0^\infty p_n = 1$. Since the support of the distribution of S_m includes 0 as well as arbitrarily large positive integers, S_m can have a unimodal distribution only if $P(S_m = x) > 0$ for all $x = 0, 1.2, \ldots$. We now show this to be impossible. Keep m fixed. Choose n so large that $a_{n+1} > ma_n + 1$. Suppose $ma_n < x < a_{n+1}$. To get x as a possible value of S_m, we must have $x = \sum_0^\infty k_j a_j$, where the k_j are nonnegative integers satisfying $\sum_0^\infty k_j = m$. Now *either* $k_j = 0$ for all $j \geq (n+1)$ in which case $x \leq a_n \sum_0^\infty k_j = ma_n$ *or* $k_j \geq 1$ for some $j \geq (n+1)$ in which case $x \geq k_j a_j \geq a_{n+1}$. Thus $P(S_m = x) = 0$ and S_m does not have a unimodal distribution.

A continuous version of the above example is obtained by distributing the mass p_n at a_n uniformly over the interval $(a_n - \frac{1}{2}, a_n + \frac{1}{2})$. To get a distribution whose support does not have gaps, Brockett and Kemperman take the convolution of the distribution of Example 4.2 with, say, a standard normal distribution. Thus there exist continuous densities f with support $(-\infty, \infty)$ such that f^{*m} is not unimodal for any m. The following is a simpler example of this procedure.

Example 4.3. Let φ_1 be the characteristic function of symmetric distribution on the integers such that φ_1 is not differentiable at 0. Let φ_2 be the characteristic function of the standard normal distribution. Write $\varphi(t) = \varphi_1(t) \cdot \varphi_2(t)$. Then φ is a characteristic function. Since φ_1 is periodic with period 2π, we know that $\varphi_1(2\pi) = \varphi_1(0) = 1$ and that φ_1 is not differentiable at 2π. Since φ_2 has derivatives of all orders everywhere, we see that φ^m does not have a derivative at 2π. But we have seen in Chapter 1 (see the remark preceding Theorem 1.4) that the characteristic function of a unimodal distribution is differentiable on $(0, \infty)$. Thus φ^m does not correspond to a unimodal distribution for any m. It is clear that φ corresponds to a continuous density whose support is $(-\infty, \infty)$.

Brockett and Kemperman have also used a result on trigonometric series to construct a continuous density f such that f^{*m} is not differentiable anywhere for any $m \geq 1$. Such a convolution cannot therefore be unimodal. As stated by them: "The central limit effect is much too weak for the property of *exact* unimodality of high convolutions."

Finally, we mention, without proof, a positive result on the unimodality of high convolutions. This result was proved recently by Odlyzko and Richmond (1985).

Theorem 4.9. *Suppose* $\mathbf{p} = \{p_j\}$ *is a discrete distribution with* $p_j = 0$ *for* $j < 0$ *and for* $j > d$, *while* $p_0 > 0$, $p_1 > 0$, $p_{d-1} > 0$ and $p_d > 0$. *Then the n-fold convolution* $\mathbf{p}^{*n}$ *of* $\mathbf{p}$ *with itself is strongly unimodal for all sufficiently large n.*

5 Unimodality of Infinitely Divisible Distributions

5.0. Summary

This chapter studies the unimodality of infinitely divisible laws on R and also on R^n, $n \geq 2$. The simpler case of distributions on R is treated first. In Section 1 we recall the relevant basic facts on infinitely divisible distributions. Section 2 presents results on the unimodality of symmetric infinitely divisible distributions. Yamazato's (1978) proof of the unimodality of the distributions of class L is given in Section 3. The chapter concludes with a discussion of the unimodality of higher dimensional infinitely divisible laws.

5.1. Infinitely Divisible Distributions

Let F be the distribution function of a real random variable X. Let φ be the characteristic function of X. We call F (or X or φ) *infinitely divisible* if, for every $n \geq 1$, there is a characteristic function φ_n such that $[\varphi_n(t)]^n = \varphi(t)$, for all $t \in R$. An equivalent requirement is that, for every $n \geq 1$, we can find n independent and identically distributed random variables X_{nj}, $j = 1, \ldots, n$, such that $\sum_1^n X_{nj}$ has the same distribution as X. The class of infinitely divisible distributions has been extensively studied by probabilists. Gnedenko and

Kolmogorov (1954), Loéve (1977), Lukacs (1960) and Feller (1971) are some of the works the reader may consult for information on infinitely divisible laws. In this section we list a few results that we need in our discussion of the unimodality of such distributions.

I. It is convenient to make the following definition.

Definition 5.1. A measure μ on the Borel σ-field in R is called a *Lévy measure* if $\mu(\{0\}) = 0$, $\mu(B) < \infty$ for every closed set B not containing 0 and, for every $\varepsilon > 0$,

$$\int_{-\varepsilon}^{\varepsilon} x^2 \, d\mu(x) < \infty.$$

A characteristic function φ is infinitely divisible if, and only if, $\varphi(t) = \exp[\psi(t)]$ for all $t \in R$, where ψ has the form

$$\psi(t) = i\gamma t - \frac{\sigma^2 t^2}{2} + \int_R A(t, x) \, d\Lambda(x),$$

where $\gamma \in R$, $\sigma^2 \geqq 0$, Λ is a Lévy measure and

$$A(t, x) = e^{itx} - 1 - \left[\frac{itx}{(1 + x^2)}\right].$$

The measure Λ is called the Lévy measure corresponding to φ. The correspondence between φ and $(\gamma, \sigma^2, \Lambda)$ is one-to-one. This representation of φ is called the *Lévy representation*. The measure Λ is clearly determined by the function M defined on $R - \{0\}$ as follows.

$$M(x) = \begin{cases} \Lambda(-\infty, x], & x < 0 \\ -\Lambda(x, \infty), & x > 0. \end{cases} \tag{5.1}$$

We call M the *Lévy function* corresponding to φ. Our definition of M makes M continuous from the right.

II. Suppose that φ is the characteristic function of a distribution symmetric about 0 and suppose that φ is also infinitely divisible. Then φ is real and so is $\psi = \log \varphi$. Consequently, in the Lévy representation of φ, the measure Λ is symmetric and the imaginary part of $\log \varphi$ vanishes. Thus

$$\log \varphi(t) = -\frac{\sigma^2 t^2}{2} + \int_R (\cos tx - 1) \, d\Lambda(x). \tag{5.2}$$

III. Let $\{X_n, n \geq 1\}$ be a sequence of independent random variables and write $S_n = \sum_1^n X_j$. Suppose we can find a sequence $\{a_n\}$ of centering constants and a sequence $\{b_n\}$ of positive scaling constants such that the distribution function of $(S_n - a_n)/b_n$ converges weakly to some distribution function F. Suppose also that the random variables $Y_{nj} = X_j/b_n$, $1 \leq j \leq n < \infty$ are infinitestimal. That is, for every $\varepsilon > 0$,

$$\sup_{1 \leq j \leq n} P(|Y_{nj}| \geq \varepsilon) \to 0$$

as $n \to \infty$. Then F is infinitely divisible. The class of all such limit distributions (when we vary the distributions of the X_n's and the constants a_n and b_n) is known as the class L.

Distributions in the class L can be described in terms of their Lévy measures. Let φ be an infinitely divisible characteristic function with Lévy measure Λ. Let M be the Lévy function defined by (5.1). Then φ corresponds to a distribution in L if, and only if, for $0 < a < b < \infty$, $-\infty < c < d < 0$ and $0 < \alpha < 1$, we have

$$M(b) - M(a) \geq M\left(\frac{b}{\alpha}\right) - M\left(\frac{a}{\alpha}\right),$$

and (5.3)

$$M(d) - M(c) \geq M\left(\frac{d}{\alpha}\right) - M\left(\frac{c}{\alpha}\right).$$

Conditions (5.3) are equivalent to the conditions that M has right and left derivatives everywhere on $R - \{0\}$ and $xM'(x)$ is non-increasing on $(0, \infty)$ and on $(-\infty, 0)$, where M' denotes either the right or the left derivative and one may use different derivatives at different points.

The following lemma which is a consequence of (5.3) was essentially proved by Fisz (1963). However, it does not seem to have been explicitly stated in the literature.

Lemma 5.1. *The Lévy function M of a distribution in the class L is convex on $(-\infty, 0)$ and concave on $(0, \infty)$.*

Proof. Let $0 < a < c < \infty$ and $\delta > 0$. Write $\alpha = a/c$. Then $0 < \alpha < 1$. So (5.3) shows that

$$M(a + \delta) - M(a) \geq M\left[\frac{c(a + \delta)}{a}\right] - M(c). \tag{5.4}$$

Now

$$\left[\frac{c(a + \delta)}{a}\right] = c + \left(\frac{\delta c}{a}\right) > c + \delta.$$

Therefore, $M[c(a+\delta)/a] \geq M(c+\delta)$ and (5.4) gives

$$M(a+\delta) - M(a) \geq M(c+\delta) - M(c). \tag{5.5}$$

The concavity of M on $(0, \infty)$ is an easy consequence of (5.5). The convexity of M on $(-\infty, 0)$ follows in a similar way. ■

The following corollary was given, with a slightly longer proof, by Fisz (1963).

Corollary. *The Lévy measure of a distribution in the class L is absolutely continuous.*

Proof. Lemma 5.1 shows that M is convex on $(-\infty, 0)$ and concave on $(0, \infty)$. Therefore M is absolutely continuous on $R - \{0\}$. Since $\Lambda(\{0\}) = 0$, (5.1) shows that Λ is absolutely continuous. ■

The next lemma is again due to Fisz (1963).

Lemma 5.2. *Let M be the Lévy function of a distribution in the class L.*

(a) *If* $M(x_0) > 0$ *for some* $x_0 < 0$, *then* $M(0-) = \infty$.
(b) *If* $M(x_0) < 0$ *for some* $x_0 > 0$, *then* $M(0+) = -\infty$.

Proof. Suppose $M(x_0) < 0$ for some $x_0 > 0$. Since $M(\infty) = 0$, we can find $a \in (0, \infty)$ such that $M(a) - M(a/2) = \delta > 0$. Use (5.3) with $\alpha = 1/2^n$ to obtain

$$M\left(\frac{a}{2^n}\right) - M\left(\frac{a}{2^{n+1}}\right) \geq M(a) - M\left(\frac{a}{2}\right) = \delta,$$

for $n = 0, 1, 2, \ldots$ It follows that

$$M(a) - M\left(\frac{a}{2^n}\right) \geq n\delta.$$

Letting $n \to \infty$ we get $M(0+) = -\infty$. This proves (a) and the proof of (b) is similar. ■

Lemma 5.2 and the above corollary enabled Fisz and Varadarajan (1963) to prove the following useful theorem.

Theorem 5.1. *Every nondegenerate distribution in the class L is absolutely continuous.*

Proof. Let F be an infinitely divisible distribution function and let $(\gamma, \sigma^2, \Lambda)$ be the Lévy representation of F. If $\sigma^2 > 0$, then F is the convolution of a nondegenerate normal distribution with some other distribution and so F must be absolutely continuous. So let $\sigma^2 = 0$. Let M be the Lévy function of F. According to a theorem of Fisz and Varadarajan (1963), F is absolutely continuous if, for some $\delta > 0$, M is both unbounded and absolutely continuous on either $(0, \delta)$ or $(-\delta, 0)$. Now suppose that F belongs to L and F is not degenerate. Then $M(x)$ must be nonzero for some x in $R - \{0\}$. Suppose, for instance, (in view of equation (5.1)), that $M(x_0) < 0$ for some $x_0 > 0$. Lemma 5.2 shows that M is unbounded on $(0, \delta)$ for every $\delta > 0$. But the above corollary shows that M is also absolutely continuous on $(0, \delta)$. We therefore conclude that F is absolutely continuous. ■

IV. Suppose $\{X_n,\ n \geq 1\}$ is a sequence of independent and *identically distributed* random variables. Write $S_n = \sum_1^n X_j$. Suppose we can find a sequence $\{a_n\}$ of centering constants and a sequence $\{b_n\}$ of positive scaling constants such that the distribution function of $(S_n - a_n)/b_n$ converges weakly to some distribution function F. From subsection III above, we know that F belongs to the class L. But since the X_n's are identically distributed, F belongs to a class much narrower than L, namely, the class of *stable distributions*. The following theorem gives the explicit form of stable characteristic functions.

Theorem 5.2. *A characteristic function φ corresponds to a stable distribution if, and only if, $\varphi = \exp(\psi)$, where*

$$\psi(t) = i\gamma t - c\,|t|^{\alpha}\{1 + i\beta\ \mathrm{sgn}(t)\varsigma(t, \alpha)\},$$

where $\gamma \in R$, $-1 \leq \beta \leq 1$, $0 < \alpha \leq 2$ and

$$\varsigma(t, \alpha) = \begin{cases} \tan(\pi\alpha/2), & \alpha \neq 1 \\ (2/\pi)\log|t|, & \alpha = 1. \end{cases}$$

The number α appearing in Theorem 5.2 is generally known as the *index* of the stable law corresponding to φ. If $\alpha = 2$, we get the normal distribution with mean γ and variance c. If $\alpha = 1$, and $\beta = 0$, we get the Cauchy distribution with median γ and scale parameter c.

The Lévy representation $(\gamma, \sigma^2, \Lambda)$ of a stable law can also be written down explicitly. This has one of the following two forms:

(i) $\alpha = 2$, $\sigma^2 \geq 0$, $\Lambda \equiv 0$;
(ii) $0 < \alpha < 2$, $\sigma^2 = 0$ and

$$M(x) = \begin{cases} -c_1/x^\alpha, & x > 0 \\ c_2/|x|^\alpha, & x < 0, \end{cases}$$

where $c_1 \geq 0$, $c_2 \geq 0$ and $c_1 + c_2 > 0$.

5.2. Unimodality of Symmetric Infinitely Divisible Distributions

The unimodality of symmetric stable laws was proved by Wintner (1936). Later, in 1956, he proved that all symmetric distributions in the class L are unimodal. Medgyessy (1967) showed that a symmetric infinitely divisible distribution is unimodal if its Lévy function is concave on $(0, \infty)$. In view of Lemma 5.1, we see that Medgyessy's result is more general than Wintner's (1956) result. However, one of the interesting features of Wintner's proof is that it uses Khintchine's representation [Formula (1.5)] directly to prove unimodality. For this reason, we present both the proofs. We repeatedly use Wintner's classical result (Theorem 1.6) that the convolution of two symmetric unimodal distributions is unimodal.

Let φ, Λ and M, respectively, be the characteristic function, the Lévy measure and the Lévy function of a *symmetric* infinitely divisible distribution. Then Λ is symmetric and so $M(x) = -M(-x)$ for all $x \neq 0$. Suppose further that φ corresponds to a distribution in the class L. Then the discussion in Subsection III of Section 5.1 shows that M is absolutely continuous on $(0, \infty)$ and $xM'(x)$ is nonincreasing on $(0, \infty)$. If we write $h(x) = 2xM'(x)$, then (5.2) gives

$$\log \varphi(t) = -\frac{\sigma^2 t^2}{2} - \int_0^\infty \frac{(1 - \cos tx)}{x} h(x)\, dx. \tag{5.6}$$

In formula (5.6) we know that h is nonnegative and nonincreasing on $(0, \infty)$. Further, the fact that Λ is a Lévy measure yields the following conditions for every $\varepsilon > 0$.

$$\int_0^\varepsilon xh(x)\, dx < \infty \quad \text{and} \quad \int_\varepsilon^\infty \frac{h(x)}{x}\, dx < \infty. \tag{5.7}$$

We want to show that a characteristic function φ given by (5.6) corresponds to a unimodal distribution. Suppose, for the moment, that we have proved the required result when $\sigma^2 = 0$. Then, for $\sigma^2 > 0$, (5.6) shows that φ corresonds to the convolution of two symmetric unimodal distributions (one being normal). This convolution is unimodal because both the components involved are symmetric. Therefore, without loss of generality, we may and do assume that $\sigma^2 = 0$. The right side of (5.6) then depends only on the function h and the real argument t. We therefore write $\varphi(t, h)$ for $\varphi(t)$. That is,

$$\log \varphi(t, h) = -\int_0^{\infty} \frac{(1 - \cos tx)}{x} h(x)\, dx. \tag{5.8}$$

Theorem 5.3. *Every symmetric distribution in the class L is unimodal.*

Proof. Let $F(\cdot, h)$ be the distribution function corresponding to the characteristic $\varphi(\cdot, h)$. We want to show that $F(\cdot, h)$ is unimodal for every acceptable choice of h. We recall that h is nonnegative and nonincreasing on $(0, \infty)$ and satisfies (5.7).

The problem can be reduced to the case $h = h_q$, where $0 < q < \infty$ and

$$h_q(x) = \begin{cases} q, & 0 < x < 1, \\ 0, & \text{elsewhere.} \end{cases} \tag{5.9}$$

To see this, observe that any given h can be approximated by a linear combinaton of the form

$$\sum_{j=1}^{J} h_{q_j}(b_j x). \tag{5.10}$$

Now (5.8) easily shows that $F(\cdot,\ h + h^*)$ is the convolution of $F(\cdot, h)$ and $F(\cdot, h^*)$. So $F(\cdot, h + h^*)$ is unimodal as soon as $F(\cdot, h)$ and $F(\cdot, h^*)$ are unimodal. Further, if $F(\cdot, h)$ is the distribution function of a random variable X, $c > 0$ and $\tilde{h}_c(x) = h(cx)$, then $F(\cdot, \tilde{h}_c)$ is the distribution function of X/c. We thus see from (5.10) that it is sufficient to establish the unimodality of $F(\cdot, h)$ for $h = h_q$, $0 < q < \infty$.

Once again, $h_{q+q'} = h_q + h_{q'}$. Therefore, $F(\cdot, h_{q+q'})$ is unimodal as soon as $F(\cdot, h_q)$ and $F(\cdot, h_{q'})$ are unimodal. We have thus further reduced our problem to the case where $h = h_q$ and $0 < q < 1$.

Let $0 < q < 1$ and let F_q be the distribution function with characteristic function φ_q given by $\varphi_q(t) = \varphi(t, h_q)$. Then (5.9) and a simple change of variables in (5.8) show that

$$\log \varphi_q(t) = -q \int_0^{t} \frac{(1 - \cos v)}{v}\, dv \tag{5.11}$$

By Khinchine's representation, F_q is unimodal if $d[t\varphi_q(t)]/dt$ is a characteristic function. From (5.11),

$$t\varphi_q'(t) = -q(1 - \cos t)\varphi_q(t).$$

Therefore

$$\frac{d}{dt}[t\varphi_q(t)] = t\varphi_q'(t) + \varphi_q(t)$$

$$= [(1-q) + q \cos t]\varphi_q(t) \tag{5.12}$$

Now $[(1-q) + q \cos t]$ is the characteristic function of the distribution which puts mass $q/2$ at ± 1 and mass $(1-q)$ at 0. Consequently, the right side of (5.12) is a characteristic function. This proves that F_q is unimodal for $0 < q < 1$. The proof of the theorem is now complete. ■

Let $\mathscr{G}$ denote the class of all symmetric infinitely divisible distribution functions F such that the Lévy measure of F is concave on $(0, \infty)$. Lemma 5.1 shows that the class $\mathscr{G}$ is at least as wide as the class of symmetric distributions in the class L. But $\mathscr{G}$ is actually wider. This is because, by Lemma 5.2, the Lévy function of a distribution in L must be either identically zero or unbounded. On the other hand, the Lévy function of a distribution in $\mathscr{G}$ can be both nonzero and bounded. We may take, for instance $M(x) = -e^{-x}$, $0 < x < \infty$.

Medgyessy (1967) proved that the distributions in $\mathscr{G}$ are all unimodal. The discussion in the preceding paragraph shows that Medgyessy's result is an improvement over Wintner's result (Theorem 5.3). Medgyessy's method is also different from and somewhat simpler than Wintner's method. We need a lemma, whose proof is straightforward.

Lemma 5.3. *Let $\xi(t)$ be the characteristic function of a symmetric unimodal distribution and let $\mathbf{p} = \{p_n, n \geq 0\}$ be a distribution on the set of nonnegative integers. If ς is defined by*

$$\varsigma(t) = \sum_{n=0}^{\infty} p_n[\xi(t)]^n, \qquad t \in R,$$

then ς is the characteristic function of a symmetric unimodal distribution. Moreover, if $\mathbf{p}$ is infinitely divisible, then so is ς.

Theorem 5.4. *Every distribution in the class $\mathscr{G}$ is unimodal.*

Proof. Let φ be the characteristic function of a distribution in $\mathscr{G}$. That is, the Lévy measure M of φ is concave on $(0, \infty)$. Let $g(x) = M'(x)$. Then (5.2) shows that

$$\log \varphi(t) = -\frac{\sigma^2 t^2}{2} - \int_{-\infty}^{\infty} (1 - \cos tx) g(x)\, dx.$$

$$= -\frac{\sigma^2 t^2}{2} + \psi(t, g),$$

where

$$\psi(t, g) = -\int_{-\infty}^{\infty} (1 - \cos tx) g(x)\, dx. \tag{5.13}$$

Since $\exp(-t^2\sigma^2/2)$ is the characteristic function of a symmetric unimodal distribution, the theorem would follow if we prove that $\psi(t, g)$ is the logarithm of the characteristic function of a unimodal distribution.

Now M is concave on $(0, \infty)$ and $M(+\infty) = 0$. It is possible that $M(0+) = -\infty$. But it is clearly sufficient to prove the theorem in the case where $M(0+) > -\infty$. To see this, we can replace $g(x)$ by $g_n(x)$, where $g_n(x) = \min[g(1/n), g(x)]$ and then let $n \to \infty$. So, let $2M(0+) = -A > -\infty$. Then $g(x) = Au(x)$, where u is a unimodal and symmetric density on $(0, \infty)$. Let ξ be the characteristic function of u. Then (5.13) gives $\psi(t, g) = -A + A\xi(t)$. Therefore

$$\exp[\psi(t, g)] = \exp[-A + A\xi(t)]$$

$$= \sum_{n=0}^{\infty} p_n [\xi(t)]^n, \tag{5.14}$$

where $p_n = e^{-A} A^n/(n!)$. Since ξ corresponds to a symmetric unimodal distribution, Lemma 5.3 shows that the right side of (5.14) is the characteristic function of a symmetric unimodal distribution. This completes the proof of the theorem. ■

Theorem 5.4 will be generalized later to higher dimensional distributions under a suitable definition of unimodality.

Symmetric infinitely divisible distributions which are not unimodal can be constructed fairly easily. We give two examples.

Example 5.1. Let $X = Y_1 - Y_2$ where Y_1 and Y_2 are independent random variables having the Poisson distribution with a positive mean λ. Then X has a symmetric infinitely divisible distribution which cannot be unimodal because it is discrete and nondegenerate. If we require an absolutely continuous distribution, we can consider $X_n = Y_1 - Y_2 + Z_n$, where Y_1, Y_2 are as above, Z_n is $N(0, 1/n)$ and Z_n is independent of (Y_1, Y_2). Now $X_n \to X$ in distribution as $n \to \infty$ and the distribution of X is not unimodal. Therefore, for large n, the distribution of X_n is absolutely continuous but not unimodal.

Example 5.2. Hartman and Wintner (1942) have given an example of a symmetric infinitely divisible distribution which is continuous but singular. Such a distribution cannot be unimodal because a continuous unimodal distribution must be absolutely continuous. The Hartman and Wintner example takes the Lévy function M to be purely discrete with a jump of size $1+N^{-2j}$ at the point $\pm N^{-j}$, where $j = 1, 2, \ldots$ and $N \geq 2$ is a fixed integer. The verification that the resulting distribution is continuous singular is somewhat involved and is omitted. Other examples of the same type have been given by Tucker (1964).

Wolfe (1978a) has given a method of constructing a symmetric and unimodal infinitely divisible distribution function which is not in Medgyessy's class $\mathscr{G}$. We present it in the next example.

Example 5.3. Suppose that F is a distribution function on R satisfying the following conditions:

(a) F is symmetric but not unimodal;
(b) The convolution $F * F$ is unimodal;
(c) If G_v denotes the normal distribution function with zero mean and variance v, then the convolution $F * G_v$ is unimodal for some v.

Assume, for the moment, that such an F is available. Let ξ be the characteristic function of F and, for $\lambda > 0$, let H_λ be the distribution with characteristic function ς_λ given by

$$\varsigma_\lambda(t) = \exp[\lambda(\xi(t) - 1)]. \tag{5.15}$$

Conditions (b) and (c) easily show that, for some v, $G_v * F^{*n}$ is unimodal for all n and hence $G_v * H_\lambda$ is also unimodal. From (5.15), the Lévy function of H_λ is $\lambda F(x)$, $x < 0$. But the Lévy function of G_v is identically zero. Therefore, the Lévy function of $G_v * H_\lambda$ is also $\lambda F(x)$, $x < 0$. Since F is not convex on $(-\infty, 0)$, it follows that $G_v * H_\lambda$ is not in the class $\mathscr{G}$.

It remains to show that conditions (a), (b) and (c) can be met. Let g_v denote the density of G_v. Then $\sqrt{v}\, g_v(x) = g_1(x/\sqrt{v})$ which is increasing in v. Note also that $(d/dx)g_1(cx) = -c^2 x g_1(cx)$. Consider the density f defined by

$$f(x) = K[\sqrt{c}g_c(x) - bg_1(x)], \tag{5.16}$$

where $0 < b < 1 < c$ and K is a normalizing constant. One easily gets

$$\lim_{x \to 0} \left[\frac{f'(x)}{x} \right] = K_1 \left[b - \left(\frac{1}{c} \right) \right].$$

where K_1 is a positive constant. Therefore F is not unimodal if

$$b > \left(\frac{1}{c} \right). \tag{5.17}$$

Now

$$(f * f)(x) = K^2[cg_{2c}(x) + b^2 g_2(x) - 2b\sqrt{c}\, g_{1+c}(x)].$$

Therefore

$$(f * f)'(x) = -K^2 x \left[\frac{g_1(x/\sqrt{2c})}{\sqrt{8c}} + \frac{b^2 g_1(x/\sqrt{2})}{\sqrt{8}} - \frac{2b\sqrt{c}\, g_1(x/\sqrt{1+c})}{(1+c)^{3/2}} \right].$$

Now $2c > 1 + c$. Therefore, in the last expression, the first term in the bracket dominates the third term if

$$\sqrt{8c} < \frac{(1+c)^{3/2}}{(2b\sqrt{c})},$$

which is equivalent to

$$32b^2 < \frac{(1+c)^3}{c^2}. \tag{5.18}$$

Thus $(f * f)$ is unimodal if (5.18) holds. It is clear that, for fixed b, (5.17) and (5.18) hold if c is sufficiently large. Now

$$(f * g_v)(x) = K[\sqrt{c}\, g_{c+v}(x) - b g_{1+v}(x)].$$

Therefore,

$$(f * g_v)'(x) = -Kx \left[\frac{\sqrt{c}\, g_1(x/\sqrt{c+v})}{(c+v)^{3/2}} - \frac{b g_1(x/\sqrt{1+v})}{(1+v)^{3/2}} \right]$$

Again, $c + v > 1 + v$. Therefore $(f * g_v)$ is unimodal as soon as

$$\frac{(c+v)^{3/2}}{\sqrt{c}} < \frac{(1+v)^{3/2}}{b},$$

which reduces to

$$b^2 < \frac{c(1+v)^3}{(c+v)^3}. \tag{5.19}$$

Now, for fixed c and as $v \to \infty$, the right side of (5.19) tends to c and $b^2 < 1 < c$. Therefore, once b and c are chosen to satisfy (5.17) and (5.18), (5.19) must hold for all sufficiently large v. Thus, for suitable choices of b and c, the density (5.16) satisfies the required conditions.

5.3. Unimodaity of all Distributions in L

The problem of proving the unimodality of all distributions in the class L has had a very interesting history. The original Russian edition of the book by Gnedenko and Kolmogorov (1954) contained a "theorem" which asserted that all distributions in L are unimodal. But the "proof" used a "theorem" of Lapin (1947) which claimed that the convolution of two unimodal distributions is again unimodal. While translating the Gnedenko-Kolmogorov book into English, K. L. Chung obtained counterexamples to Lapin's "theorem." The validity of the original proposition on the unimodality of L distributions thereby became open to further research. Soon Ibragimov (1957) gave "examples" of distributions in L which were *not* unimodal. However, Sun (1967) showed that Ibragimov's "examples" were, in fact, unimodal. Wolfe (1971) proved that every distribution in L whose Lévy function has its support on the positive axis is unimodal. As a consequence, every distribution in L is obtainable as the convolution of at most two unimodal distributions in L. Finally, Yamazato (1978) settled the problem by proving that all distributions in L are indeed unimodal. In this section we first present Wolfe's result and then present Yamazato's proof.

Definition 5.2. A Lévy function M on $R-\{0\}$ is said to be *one-sided* if either $M(0-)=0$ or $M(0+)=0$, that is, if the support of M is contained in $(0,\infty)$ or in $(-\infty,0)$.

The following theorem was proved by Wolfe (1971).

Theorem 5.5. *Let F be a distribution function in the class L. If F has a one-sided Lévy function, then F is unimodal.*

Proof. Let F have characteristic function φ with Lévy representation (α, σ^2, M). We assume first that $\sigma^2=0$. Write $h(x)=xM'(x)$. We know from subsection III of Section 5.1 that h is nonincreasing on $(-\infty,0)$ and on $(0,\infty)$. Suppose now that M is one-sided. We may assume that $M(0-)=0$. Then

$h(x) = 0$ for $x < 0$ and

$$\log \varphi(t) = it\alpha + \int_0^\infty \left[\frac{A(t,x)}{x}\right] h(x)\, dx,$$

where $A(t,x) = e^{itx} - 1 - [itx/(1+x^2)]$. We want to show that φ corresponds to a unimodal distribution. Since weak limits of unimodal distributions are unimodal (Theorem 1.1), we may assume that h is a step function with bounded support. So, let

$$h(x) = \sum_{j=1}^{k} \lambda_j w(p_j x), \tag{5.20}$$

where $\lambda_j > 0$ for all j, $0 < p_1 < \cdots < p_k < \infty$ and $w(x)$ is the indicator function of the interval $(0, 1)$. By adjusting α, we can write

$$\log \varphi(t) = \int_0^\infty \left[\frac{(e^{itx} - 1)}{x}\right] h(x)\, dx.$$

Substituting for $h(x)$ from (5.20) and carrying out a routine change of variables, we get

$$\log \varphi(t) = \sum_{j=1}^{k} \lambda_j \psi(p_j t), \tag{5.21}$$

where

$$\psi(t) = \int_0^t \left[\frac{(e^{iy} - 1)}{y}\right] dy.$$

Now, for $t > 1$,

$$\begin{aligned}\operatorname{Re}[\psi(t)] &= \int_0^t \left[\frac{(\cos y - 1)}{y}\right] dy \\ &= \int_0^1 \left[\frac{(\cos y - 1)}{y}\right] dy + \int_1^t \left[\frac{(\cos y)}{y}\right] dy - \log t.\end{aligned}$$

Since the middle term on the right side of the last expression is bounded in $t \in [1, \infty)$, we have, for some constant A, $\operatorname{Re}[\psi(t)] \le A - \log t$ for all $t \ge 1$. Thus (5.21) shows that, as $|t| \to \infty$, $|\varphi(t)| = O(|t|^{-\lambda})$, where $\lambda = \sum_1^k \lambda_j$. Consequently,

$$F(x) - F(0) = \frac{1}{2\pi} \int_{-\infty}^{\infty} \frac{e^{-itx} - 1}{(-it)} \varphi(t)\, dt,$$

where the right side is a convergent Lebesgue integral. Let $x > 0$. The change of variables $tx = s$ shows that

$$F(x) - F(0) = (2\pi)^{-1} \int_{-\infty}^{\infty} \left[\frac{(e^{-is} - 1)}{(-is)}\right] \varphi\left(\frac{s}{x}\right) ds. \tag{5.22}$$

By Theorem 5.1, F is absolutely continuous. Let f be the density of F. Differentiating (5.22) and using (5.21), one gets

$$f(x) = (2\pi)^{-1} \int_{-\infty}^{\infty} \left[\frac{(e^{-is} - 1)}{(-is)}\right] \varphi\left(\frac{s}{x}\right) \left[\sum_{j=1}^{k} \left(\frac{\lambda_j}{x}\right)(1 - e^{ip_j s/x})\right] ds$$

Again, the change of variables $s = tx$ gives

$$\begin{aligned} f(x) &= (2\pi)^{-1} \sum_{j=1}^{k} \left(\frac{\lambda_j}{x}\right) \int_{-\infty}^{\infty} \left[\frac{(e^{-itx} - 1)(1 - e^{ip_j t})}{(-it)}\right] \varphi(t)\, dt \\ &= \sum_{j=1}^{k} \left(\frac{\lambda_j}{x}\right) [\{F(x) - F(x - p_j)\} - \{F(0) - F(-p_j)\}] \\ &= \frac{[B(x) - B(0)]}{x}, \end{aligned} \tag{5.23}$$

where $B(x) = \sum_{j=1}^{k} \lambda_j [F(x) - F(x - p_j)]$. The same formula holds for $x < 0$. We now claim that $B(x) = 0$ for all $x \le 0$. To see this, suppose that $B(0) > 0$. Then, letting $x \to \infty$, we see from (5.23) that $f(x)$ becomes negative for large x. So $B(0)$ must vanish. But then (5.23) again shows that, for a given $x < 0$, $f(x)$ would be negative if $B(x)$ is positive. Thus $B(x) = 0$ for all $x \le 0$. Consequently, $f(x) = 0$ for $x < 0$. Formula (5.23) also shows that f is continuous everywhere except possibly at 0 and, for $x > 0$,

$$xf'(x) = \left[\sum_{j=1}^{k} \lambda_j - 1\right] f(x) - \sum_{j=1}^{k} \lambda_j f(x - p_j). \tag{5.24}$$

We see from (5.24) that $f'(x)$ is continuous everywhere except perhaps at $x - 0, p_1, \ldots, p_k$.

Recall that $\lambda = \sum_{j=1}^{k} \lambda_j$. Two cases arise.

Case 1. Suppose $\lambda \le 1$. Then (5.24) shows that $f'(x) \le 0$ for all $x > 0$. Since $f(x) = 0$ for $x < 0$, we see that f is unimodal with zero as a mode.

Case 2. Suppose $\lambda > 1$. Again (5.24) shows that $xf'(x) = (\lambda - 1) f(x)$ for $0 < x < p_1$. Therefore, for some $c > 0$,

$$f(x) = cx^{\lambda - 1}, \qquad 0 < x < p_1. \tag{5.25}$$

From (5.25) and (5.24), we conclude that f is continuous at 0 and f' is continuous everywhere except perhaps at 0. Now f is strictly increasing on $(0, p_1)$ and (5.23) shows that $f(x) \to 0$ as $x \to \infty$. Therefore f has a relative maximum in the interval (p_1, ∞). Let

$$A = \{x > 0 : f \text{ has a relative maximum at } x\}.$$

Let $x_0 = \inf A$. We now consider two subcases.

Subcase (i): Suppose x_0 is an isolated point of A. Then $x_0 = \min A$. Here we set $x_1 = x_0$ and write $B = \{x > x_0 : f \text{ has a relative minimum at } x\}$. If possible, suppose $B \neq \varnothing$. Let $x_2 = \inf B$. Then $p_1 < x_1 < x_2$, $f'(x_1) = 0$, $f'(x_2) = 0$, f is strictly increasing on $(0, p_1)$ and f is strictly decreasing on (x_1, x_2).

Write $S = \{1, \ldots, k\}$, $S_1 = \{j \in S : x_2 - p_j \geq x_1\}$ and $S_2 = S - S_1 = \{j \in S : x_2 - p_j < x_1\}$. If $S_2 = \varnothing$, then $x_2 > x_2 - p_j \geq x_1$ and $f(x_2 - p_j) > f(x_2)$ for all $j \in S$. Consequently

$$\sum_{j \in S} \lambda_j f(x_2 - p_j) > f(x_2) \sum_{j \in S} \lambda_j > f(x_2) \left[\sum_{j \in S} \lambda_j - 1 \right]$$

and so $f'(x_2) < 0$ which is a contradiction. Thus $S_2 \neq \varnothing$. Note that

$$j \in S_1 \Rightarrow f(x_2 - p_j) > f(x_2) \tag{5.26}$$

and

$$j \in S_2 \Rightarrow f(x_1 - p_j) < f(x_2 - p_j). \tag{5.27}$$

From (5.26) we get

$$\sum_{j \in S_1} \lambda_j f(x_2 - p_j) > \sum_{j \in S_1} \lambda_j f(x_2). \tag{5.28}$$

But

$$\sum_{j \in S} \lambda_j f(x_2 - p_j) = \left(\sum_{j \in S} \lambda_j - 1 \right) f(x_2). \tag{5.29}$$

Subtracting (5.28) from (5.29) we get

$$\sum_{j \in S_2} \lambda_j f(x_2 - p_j) < \left(\sum_{j \in S_2} \lambda_j - 1 \right) f(x_2). \tag{5.30}$$

Now use (5.27) on the left side of (5.30) and the inequality $f(x_2) < f(x_1)$ on the right side of (5.30) to obtain

$$\sum_{j \in S_2} \lambda_j f(x_1 - p_j) < \left(\sum_{j \in S_2} \lambda_j - 1 \right) f(x_1). \tag{5.31}$$

Again $x_1 - p_j < x_1$ for all j. Therefore

$$\sum_{j\in S_1} \lambda_j f(x_1 - p_j) < \sum_{j\in S_1} \lambda_j f(x_1). \tag{5.32}$$

If we add (5.31) and (5.32), we conclude that $f'(x_1) > 0$. This contradiction shows that $B = \varnothing$ and thus f is unimodal with x_1 as a mode.

Subcase (ii). Suppose x_0 is a limit point of A. Then we can find points x_1, x_2 such that $x_0 < x_1 < x_2$, $x_2 - p_1 < x_0$, f has a relative maximum at x_1, f has a relative minimum at x_2 and $f(x_1) \geq f(x_2)$. Now $x_1 - p_j < x_2 - p_j < x_0$ for all j. Therefore

$$\left(\sum \lambda_j - 1\right) f(x_2) = \sum \lambda_j f(x_2 - p_j) > \sum \lambda_j f(x_1 - p_j)$$
$$= \left(\sum \lambda_j - 1\right) f(x_1).$$

Thus $f(x_2) > f(x_1)$, which is again a contradiction. So this subcase is vacuous.

We have thus shown that F is unimodal when $\sigma^2 = 0$. If $\sigma^2 > 0$, then F is a convolution of a unimodal distribution function with a normal distribution function. But the latter is strongly unimodal. Therefore F is unimodal. The proof of the theorem is now complete. ■

As noted earlier, Theorem 5.5 implies that an arbitrary distribution F in the class L is the convolution of two unimodal distribution functions G and H in the class L. Indeed, if F has Lévy function M, then G has Lévy function M_1, where $M_1(x) = M(x)$ for $x > 0$ and $M_1(x) = 0$ for $x < 0$. Similarly H has Lévy function M_2, where $M_2(x) = M(x)$ for $x < 0$ and $M_2(x) = 0$ for $x > 0$. If G or H is strongly unimodal, we could conclude that F is also unimodal. But there will be situatons where neither G nor H is strongly unimodal. To see this, observe that if F is symmetric, and G is strongly unimodal, then H will be strongly unimodal and hence F would also be strongly unimodal. But L does contain symmetic distributions which are not strongly unimodal; (e.g., the Cauchy distribution). Thus one needs to use the unimodaity and some additional properties of G and H to conclude the unimodality of $G * H$. These additional properties were identified by Yamazato (1978) who proved, once and for all, that every distribution in L is unimodal. We now proceed to present Yamazato's results.

Lemma 5.4. *Let g be a bounded and continuous probability density on $(0, \infty)$. Let h be a probability density on $(-\infty, 0)$ such that h has finite limits at $-\infty$ and $0-$. Assume that h is absolutely continuous on $(-\infty, 0)$ with Radon-*

Nikodym derivative h_0. Let $f = g * h$. Then f has a continuous derivative on $R - \{0\}$ and

$$f'(x) = \int_{-\infty}^{\min(x,\,0)} h_0(y) g(x-y)\, dy - g(x) h(0-), \qquad x \neq 0,$$

where we set $g(x) = 0$ for $x < 0$.

Proof. First suppose that $x > 0$. Then

$$f(x) = \int_x^{\infty} h(x-y) g(y)\, dy. \tag{5.33}$$

Since $h(-\infty) = 0$ and h is absolutely continuous on $(-\infty, 0)$, we have

$$h(z) = \int_{-\infty}^{z} h_0(t)\, dt, \qquad z < 0. \tag{5.34}$$

Now (5.33) and (5.34) yield

$$f(x) = \int_x^{\infty} \left[\int_{-\infty}^{x} h_0(t-y)\, dt \right] g(y)\, dy. \tag{5.35}$$

Let $f_0(x)$ be the function obtained by a formal differentiation of the right side of (5.35) w.r.t. x. That is,

$$\begin{aligned} f_0(x) &= \int_x^{\infty} h_0(x-y) g(y)\, dy - g(x) \int_{-\infty}^{x} h_0(t-x)\, dt \\ &= \int_x^{\infty} h_0(x-y) g(y)\, dy - g(x) h(0-) &\quad (5.36) \\ &= \int_{-\infty}^{0} h_0(y) g(x-y)\, dy - g(x) h(0-) &\quad (5.37) \end{aligned}$$

The integrability of h_0 over $(-\infty, 0)$ and the continuity and boundedness of g on $(0, \infty)$ show that the expression (5.37) for f_0 represents a continuous function on $(0, \infty)$. The lemma would, therefore, follow if we show that f_0 is the Radon–Nikodym derivative of f. From (5.36) we have

$$\int_x^{\infty} f_0(t)\, dt = \int_x^{\infty} \left[\int_t^{\infty} h_0(t-y) g(y)\, dy \right] dt - h(0-) \int_x^{\infty} g(t)\, dt.$$

If we interchange the order of integration, the first term on the right side becomes

$$\int_x^{\infty} \left[\int_x^{y} h_0(t-y)\, dt \right] g(y)\, dy = \int_x^{\infty} [h(0-) - h(x-y)] g(y)\, dy.$$

Therefore,

$$\int_x^\infty f_0(t)\,dt = -\int_x^\infty h(x-y)g(y)\,dy = -f(x).$$

The continuity of f_0 implies that $f'(x)$ exists and equals $f_0(x)$ for $x>0$.

Suppose now that $x<0$. Then

$$f(x) = \int_0^\infty h(x-y)g(y)\,dy. \tag{5.38}$$

Again, let $f_0(x)$ be the function obtained by a formal differentiation of the right side of (5.38). That is,

$$\begin{aligned} f_0(x) &= \int_0^\infty h_0(x-y)g(y)\,dy \\ &= \int_{-\infty}^x h_0(y)g(x-y)\,dy. \end{aligned} \tag{5.39}$$

It is easy to show that f_0 is the Radon–Nikodym derivative of f on $(-\infty,0)$. Further (5.39) shows that f_0 is continuous on $(-\infty,0)$. Therefore $f'(x)$ exists and equals $f_0(x)$ for $x<0$. Now the expressions (5.37) and (5.39) for f_0 agree with the expressions for $f'(x)$ in the statement of the lemma. The proof of the lemma is thus complete. ■

Corollary. *Under the conditions of Lemma 5.4, f' is continuous at 0 if $h(0-)=0$ or $g(0+)=0$.*

Lemma 5.5. *Let G and H be absolutely continuous distribution functions with unimodal densities g and h and with modes a and b respectively. Assume that $g(x)=0$ for $x\le 0$ and $h(x)=0$ for $x\ge 0$. If $a>0$, assume that g is positive and logconcave on $(0,a]$, $g(0+)=0$ and $g(a-)=g(a)\ge g(a+)$. If $b<0$, assume that h is positive and logconcave on $[b,0)$, $h(0-)=0$ and $h(b+)=h(b)\ge h(b-)$. Then the convolution $F=G*H$ is unimodal.*

Proof. A density f of F is given by

$$f(x) = \int_0^\infty h(x-y)g(y)\,dy = \int_{-\infty}^0 h(y)g(x-y)\,dy.$$

These expressions show easily that f is nondecreasing on $(-\infty,b]$ and nonincreasing on $[a,\infty)$. Thus the unimodality of F follows in the case

$a = b = 0$ and we may assume that $a - b > 0$. Further, we only need to analyze the behavior of f over the interval $[b, a]$.

Without loss of generality, let $a + b \geq 0$. Then, the condition $(a - b) > 0$ shows that $a > 0$. Therefore, $g(0+) = 0$. Assume, for the moment, that g and h are absolutely continuous on $(0, \infty)$ and $(-\infty, 0)$ respectively with g_0 and h_0 denoting the Radon–Nikodym derivatives. Suppose also that $h(0-) < \infty$. Then g and h satisfy the hypotheses of Lemma 5.4 and its corollary. Therefore, f has a continuous derivative and, for all $x \in R$,

$$f'(x) = \int_{-\infty}^{\min(x,0)} h_0(y)g(x-y)\,dy - h(0-)g(x). \tag{5.40}$$

For $\varepsilon > 0$, let

$$A_\varepsilon(x) = \begin{cases} g(x+\varepsilon)/g(x) & \text{if } g(x) > 0 \\ 0 & \text{if } g(x) = 0. \end{cases}$$

The continuity of g shows that A_ε is continuous on $(0, \infty)$. The logconcavity of g on $(0, a]$ shows that A_ε is nonincreasing on $(0, a - \varepsilon]$. But a is a mode of g. Therefore, A_ε is nonincreasing on $[a - \varepsilon, a]$ also, $A_\varepsilon(x) \geq 1$ *for* $0 < x \leq a - \varepsilon$ and $A_\varepsilon(x) \leq 1$ for $x \geq a$. We now state and prove three assertions [numbered (i), (ii), (iii) below] from which the unimodality of f will follow immediately.

(i) If $x_0 \in [0, a + b)$ and $f'(x_0) \leq 0$, then $f'(x) \leq 0$ for all $x \in (x_0, a + b)$.

Suppose x_0 has the specified properties and suppose first that $x_0 > 0$. Let $0 < \varepsilon < a + b - x_0$. Then (5.40) gives

$$\begin{aligned} f'(x_0+\varepsilon) &= \int_{-\infty}^{0} h_0(y)A_\varepsilon(x_0-y)g(x_0-y)\,dy - h(0-)g(x_0+\varepsilon) \\ &= \left(\int_{-\infty}^{b} + \int_{b}^{0}\right) - h(0-)g(x_0)A_\varepsilon(x_0). \end{aligned}$$

Now $h_0(y)g(x_0 - y)$ is integrable and h_0 does not change sign on $(-\infty, b)$ or on $(b, 0)$. Therefore, by the continuity of A_ε on $(0, \infty)$, we have

$$\begin{aligned} f'(x_0+\varepsilon) = A_\varepsilon(\xi_1) &\int_{-\infty}^{b} h_0(y)g(x_0-y)\,dy \\ &+ A_\varepsilon(\xi_2)\int_{b}^{0} h_0(y)g(x_0-y)\,dy \\ &- h(0-)g(x_0)A_\varepsilon(x_0), \end{aligned} \tag{5.41}$$

where $x_0 < \xi_2 < x_0 - b < \xi_1$. Now $x_0 - b < a - \varepsilon$. Therefore, $\xi_2 < a - \varepsilon$ and $A_\varepsilon(x_0) \geq A_\varepsilon(\xi_2)$. Now either $\xi_1 \leq a$ in which case $A_\varepsilon(\xi_2) \geq A_\varepsilon(\xi_1)$ or $\xi_1 > a$ in which case $A_\varepsilon(\xi_2) \geq 1 \geq A_\varepsilon(\xi_1)$. Thus $A_\varepsilon(x_0) \geq A_\varepsilon(\xi_2) \geq A_\varepsilon(\xi_1)$ in all cases. Therefore (5.41) shows that

$$f'(x_0 + \varepsilon) \leq A_\varepsilon(\xi_2)\left[\int_{-\infty}^{0} h_0(y)g(x_0 - y)\,dy - h(0-)g(x_0)\right]$$
$$= A_\varepsilon(\xi_2)f'(x_0) \leq 0.$$

This proves (i) in the case $x_0 > 0$. If $x_0 = 0$, then we have

$$f'(x_0 + \varepsilon) \leq \int_{-\infty}^{0} h_0(y)A_\varepsilon(x_0 - y)g(x_0 - y)\,dy,$$

because the ignored term $h(0-)g(x_0 + \varepsilon)$ is nonnegative. Proceeding as above, we get

$$f'(x_0 + \varepsilon) \leq A_\varepsilon(\xi_2)\int_{-\infty}^{0} h_0(y)g(x_0 - y)\,dy$$
$$= A_\varepsilon(\xi_2)f'(x_0), \qquad \text{because } g(x_0) = 0.$$

Thus (i) is established.

If $b = 0$, the unimodality of f follows from assertion (i) because f is continuous, nonincreasing on (a, ∞) and nondecreasing on $(-\infty, b)$. So, in the rest of the proof, we assume that $b < 0$. Therefore, $h(0-) = 0$.

(ii) *If* $x_1 \in (b, 0]$ *and* $f'(x_1) \geq 0$, *then* $f'(x) \geq 0$ *for* $x \in (b, x_1)$.

Suppose x_1 has the specified properties and let $0 < \varepsilon < x_1 - b$. Let $K = \sup\{x : g(x) > 0\}$. Then, $K \geq a$ and so $x_1 - K \leq x_1 - a \leq x_1 + b \leq b$. Therefore, (5.40) yields

$$f'(x_1 - \varepsilon) = \int_{x_1 - K - \varepsilon}^{x_1 - \varepsilon} h_0(y)g(x_1 - \varepsilon - y)\,dy$$
$$= \int_{x_1 - K - \varepsilon}^{x_1 - K} + \int_{x_1 - K}^{b} + \int_{b}^{x_1 - \varepsilon}.$$

In the last expression, the first integral is nonnegative because $x_1 - K \leq b$. In the remaining two integrals, we can write $g(x_1 - \varepsilon - y)$ as $[A_\varepsilon(x_1 - \varepsilon - y)]^{-1}g(x_1 - y)$. Therefore, the mean value theorem again shows that

$$f'(x_1 - \varepsilon) \geq [A_\varepsilon(\xi_1)]^{-1}\int_{x_1 - K}^{b} h_0(y)g(x_1 - y)\,dy$$
$$+ [A_\varepsilon(\xi_2)]^{-1}\int_{b}^{x_1 - \varepsilon} h_0(y)g(x_1 - y)\,dy$$

where $0 < \xi_2 < x_1 - \varepsilon - b < \xi_1 < K - \varepsilon$. Since $x_1 \le 0 \le a + b$, we have $x_1 - \varepsilon - b \le a - \varepsilon$. Thus, either $\xi_1 \le a$ in which case $A_\varepsilon(\xi_2) \ge A_\varepsilon(\xi_1)$ or $\xi_1 > a$ in which case $A_\varepsilon(\xi_2) \ge 1 \ge A_\varepsilon(\xi_1)$. Therefore,

$$f'(x_1 - \varepsilon) \ge [A_\varepsilon(\xi_2)]^{-1}\left[\int_{x_1 - K}^{b} h_0(y)g(x_1 - y)\,dy + \int_{b}^{x_1 - \varepsilon} h_0(y)g(x_1 - y)\,dy\right]$$

$$\ge [A_\varepsilon(\xi_2)]^{-1}\int_{x_1 - K}^{x_1} h_0(y)g(x_1 - y)\,dy, \tag{5.42}$$

where the last step follows because the added part in the second integral is negative. Now the integral on the right side of (5.42) equals $f'(x_1)$. Thus $f'(x_1 - \varepsilon) \ge 0$ and (ii) is proved.

(iii) *If* $x_2 \in [a + b, a)$ *and* $f'(x_2) \le 0$ *then* $f'(x) \le 0$ *for all* $x \in (x_2, a)$.

The proof of (iii) follows the same lines as the proof of (ii). We omit it except to note that the roles of g and h are to be reversed and one uses a function B_ε similar to A_ε but defined in terms of h.

It remains to drop the smoothness assumptions on g and h. Suppose $a > 0$. Since g is logconcave on $(0, a]$ and $g(a-) \ge g(a)$, we can find an absolutely continuous function g_n on $(0, \infty)$ such that g_n coincides with g on $(0, a]$, g_n is unimodal with mode a and $g_n(x) \to g(x)$ a.e. as $n \to \infty$. To see this, let $x_{n,m} = m/n$, $g_n(x_{n,m}) = g(x_{n,m})$ for all integers $m \ge na$ and define g_n to be linear between $x_{n,m}$ and $x_{n,m+1}$. With this choice, $g_n(x) \to g(x)$ everywhere except at the countable number of discontinuity points of g. We can find a similar approximating sequence h_n of absolutely continuous densities on $(-\infty, 0)$ such that $h_n(0-) < \infty$ and $h_n \to h$ a.e. In addition, if $b < 0$, then we can arrange h_n to agree with h on $[b, 0)$ and to have mode b. Let G_n and H_n be the distribution functions with densities g_n and h_n, respectively. Then the above proof shows that $G_n * H_n$ is unimodal. But $G_n \to G$ and $H_n \to H$ weakly. Since limits of unimodal distributions are unimodal, it follows that $G * H$ is unimodal. The lemma is thus proved. ■

Theorem 5.6. *Every distribution in the class L is unimodal.*

Proof. Let F be a distribution function in L with Lévy representation (α, σ^2, M). Assume first that $\sigma^2 = 0$. Let $v(x) = xM'(x)$. We know that v is nonincreasing on $(0, \infty)$ and on $(-\infty, 0)$. We now recall some results from the proof of Theorem 5.5. Let w be the indicator of the interval $(0, 1)$. Let

$$h_1(x) = \sum_{j=1}^{k} \lambda_j w(p_j x),$$

where $\lambda_j > 0$ for all j and $0 < p_1 < \cdots < p_k < \infty$. Let G be the distribution function with characteristic function φ_1 given by

$$\log \varphi_1(t) = \int_0^\infty \left[\frac{(e^{itx} - 1)h_1(x)}{x}\right] dx.$$

Write $\lambda = \sum \lambda_j$. It was shown in the proof of Theorem 5.5 that:

(a) G is absolutely continuous with a density g which is continuous on $R - \{0\}$. Further $g(x) = 0$ for $x < 0$ and g satisfies the equation

$$xg'(x) = (\lambda - 1)g(x) - \sum_{j=1}^{k} \lambda_j g(x - p_j). \tag{5.43}$$

(b) If $\lambda \leq 1$, then $g(x)$ is nonincreasing on $(0, \infty)$.

(c) If $\lambda > 1$, then g is unimodal with mode $a > 0$, $g(0+) = 0$ and g' is continuous on $(0, \infty)$.

To be able to apply Lemma 5.5, we need to show that g is logconcave on $(0, a]$ when $\lambda > 1$. Two cases arise.

Case 1. Suppose $1 < \lambda \leq 2$. By (c), g' is continuous on $(0, \infty)$. Therefore, (5.43) shows that $g''(x)$ exists everywhere except at $x = 0, p_1, \ldots, p_k$ and further

$$xg''(x) = (\lambda - 2)g'(x) - \sum_{j=1}^{k} \lambda_j g'(x - p_j). \tag{5.44}$$

Now $g'(x) \geq 0$ for $0 < x < a$. So (5.44) shows that $g''(x) \leq 0$ for all $x \in (0, a)$ except perhaps at $x = 0, p_1, \ldots, p_k$. Thus g is concave and hence logconcave on $(0, a]$.

Case 2. Suppose $\lambda > 2$. For $0 < x < p_1$, we have $xg'(x) = (\lambda - 1)g(x)$. Therefore $g(x) = cx^{\lambda - 1}$ for $x \in (0, p_1)$. The conditon $\lambda > 2$ ensures that g' is continuous on R and so, by (5.44), g'' is continuous on $(0, \infty)$. Now (5.43) and (5.44) yield

$$\begin{aligned} x[(g'(x))^2 - g(x)g''(x)] &= g'(x) \cdot xg'(x) - g(x)xg''(x) \\ &= g'(x)[(\lambda - 1)g(x) - \textstyle\sum \lambda_j g(x - p_j)] \\ &\quad - g(x)[(\lambda - 2)g'(x) - \textstyle\sum \lambda_j g'(x - p_j)] \\ &= g(x)g'(x) + \sum_{j=1}^{k} \lambda_j A_j(x), \end{aligned} \tag{5.45}$$

where $A_j(x) = g(x)g'(x - p_j) - g'(x)g(x - p_j)$. Write $B(x) = [g'(x)]^2 - g(x)g''(x)$. We want to show that $B(x) \geq 0$ for $x \in (0, a]$. We know that $B(x) > 0$ for

$x \in (0, p_1]$. Suppose $B(x) < 0$ for some $x \in (p_1, a)$ and let $x_0 = \inf\{x > p_1 : B(x) < 0\}$. Then the continuity of B on $(0, \infty)$ shows that $B(x_0) = 0$, $x_0 > p_1$ and $B(x) > 0$ for $x \in (0, x_0)$. Thus $g'(x)/g(x)$ is strictly decreasing on $(0, x_0]$. If $x_0 - p_j \le 0$, then $g(x_0 - p_j) = g'(x_0 - p_j) = 0$ and so $A_j(x_0) = 0$. If $x_0 - p_j > 0$, then the decreasing character of g'/g shows that

$$\frac{g'(x_0 - p_j)}{g(x_0 - p_j)} > \frac{g'(x_0)}{g(x_0)}$$

and so $A_j(x_0) > 0$. Thus $A_j(x_0) \ge 0$ in all cases and $A_1(x_0) > 0$. Since $g(x_0)g'(x_0) \ge 0$, (5.45) shows that

$$x_0 B(x_0) \ge \lambda_1 A_1(x_0) > 0$$

This contradicts the fact that $B(x_0) = 0$. Thus $B(x) \ge 0$ for all $x \in (0, a]$ and g is logconcave on $(0, a]$.

Consider the function

$$h_2(x) = \sum_{j=1}^{m} \mu_j w(q_j x),$$

where $\mu_j < 0$ for all j and $-\infty < q_m < q_{m-1} < \cdots < q_1 < 0$. Let H be the distribution function with characteristic function φ_2 given by

$$\varphi_2(t) = \int_{-\infty}^{0} \left[\frac{(e^{itx} - 1)h_2(x)}{x}\right] dx.$$

By arguments similar to the ones made with reference to G, we can show that:

(i) H is absolutely continuous with a density h satisfying $h(x) = 0$ for $x > 0$.
(ii) If $\sum \mu_j \ge -1$, then h is nondecreasing on $(-\infty, 0)$.
(iii) If $\sum \mu_j < -1$, then h is continuous on $(-\infty, 0)$ and unimodal with mode $b < 0$, $h(0-) = 0$ and h is logconcave on $[b, 0)$.

Thus G and H satisfy all the hypotheses of Lemma 5.5. We conclude that$G * H$ is unimodal. Now apart from the centering constant α, F can be easily seen to be a limit of a sequence $G_n * H_n$, where G_n and H_n have the same properties as G and H. Thus F is unimodal. Finally, if $\sigma^2 > 0$, then the unimodality of F follows from the strong unimodality of the normal distribution. The theorem is thus proved. ■

We conclude this section by presenting an example which brings out an important difference between the symmetric case (treated in Section 2) and the general case (treated in the present section). In the symmetric case, Medgyessy's (1967) theorem (see Theorem 5.4) shows that an infinitely divisible distribution is unimodal if its Lévy function M is concave on $(0, \infty)$.

The proof of the unimodality of all distributions in L, however, used the stronger condition that $xM'(x)$ is nonincreasing on $(0, \infty)$ and on $(-\infty, 0)$. The following example due to Wolfe (1978a) shows that an asymmetric distribution may not be unimodal even if its Lévy function is convex on $(-\infty, 0)$ and concave on $(0, \infty)$.

Example 5.4 Let G be a unimodal distribution function with a nonzero mean μ, variance σ^2, mode 0 and characteristic function ς. For $a > 0$, let H_a be the distribution function with characteristic function φ_a given by

$$\varphi_a(t) = \exp[a(\varsigma(t) - 1)].$$

Then H_a is infinitely divisible with Lévy function M_a, where

$$M_a(x) = \begin{cases} aG(x), & x > 0 \\ a[G(x) - 1], & x < 0. \end{cases}$$

Since G is unimodal about 0, M_a is convex on $(-\infty, 0)$. and concave on $(0, \infty)$. However, it was shown in Chapter 1 (see the remark following Example 1.2) that H_a is not unimodal if $a > [3(\mu^2 + \sigma^2)/\mu^2]$.

5.4. Unimodality of Multivariate Infinitely Divisible Distributions

In this section we present a few results that have been proved regarding the unimodality of multivariate infinitely divisible distributions. All these results involve centrally symmetric distributions. The main contributions are due to Kanter (1977) and Wolfe (1978b).

The definitions given in Section 5.1 can be easily generalized to distributions on R^k. Thus a k-dimensional random vector $\mathbf{X}$ is said to be *infinitely divisible* if, for every $n \geq 1$, there are n independent and identically distributed random variables $\mathbf{X}_{n1}, \ldots, \mathbf{X}_{nn}$ such that $\sum_1^n \mathbf{X}_{ni}$ and $\mathbf{X}$ have the same distribution. Similarly, a *Lévy measure* on R^k is a measure Λ such that $\Lambda(\{\mathbf{0}\}) = 0$ and, for every $\varepsilon > 0$,

$$\int_{\|\mathbf{x}\| < \varepsilon} \|\mathbf{x}\|^2 \, d\Lambda(\mathbf{x}) < \infty \quad \text{and} \quad \int_{\|\mathbf{x}\| > \varepsilon} d\Lambda(\mathbf{x}) < \infty.$$

Lévy (1954) showed that a distribution function F on R^k is infinitely divisible if, and only if, its characteristic function φ has the form $\varphi = \exp[\psi]$, where

$$\psi(\mathbf{t}) = i\gamma'\mathbf{t} - \tfrac{1}{2}\mathbf{t}'\Sigma\,\mathbf{t} + \int A(\mathbf{t}, \mathbf{x})\, d\Lambda(\mathbf{x}), \tag{5.46}$$

where $\gamma \in R^k$, $\boldsymbol{\Sigma}$ is a nonnegative definite symmetric $k \times k$ matrix, Λ is a Lévy measure on R^k and

$$A(\mathbf{t}, \mathbf{x}) = e^{i\mathbf{t}'\mathbf{x}} - 1 - \frac{i\mathbf{t}'\mathbf{x}}{1 + \|\mathbf{x}\|^2}.$$

The representation (5.46) is unique. The matrix $\boldsymbol{\Sigma}$ is called the *Gaussian component* of F and Λ is called the *Lévy spectral measure* of F. We will write (5.46) for short as $F = (\gamma, \boldsymbol{\Sigma}, \Lambda)$. If F is centrally symmetric, then $\gamma = \mathbf{0}$, Λ is also centrally symmetric and we will write $F = (\boldsymbol{\Sigma}, \Lambda)$.

We recall some notation from Chapter 2. For a centrally symmetric compact convex set K in R^k, the uniform distribution (of unit mass) over K will be denoted by W_K. The set of all such distributions W_K will be denoted by Θ_k. A distribution P on R^k is called *central convex unimodal* (CCUM) if P is in the closed convex hull of Θ_k. It was shown in Chapter 2 (Theorem 2.11) that P is CCUM if, and only if, P has a representation

$$P = \int_{\Theta^k} W_K \, d\nu(W_K), \tag{5.47}$$

where ν is a probability measure on Θ_k. Since we want to make representations of the type (5.47) applicable to Lévy measures, we introduce the following definition.

Definition 5.3. A measure μ on R^k is called *generalized central convex unimodal* (GCCUM) if μ has a representation

$$\mu = \int_{\Theta_k} W_K \, d\eta(W_K),$$

where η is a measure on Θ_k.

The following important theorem was proved by Kanter (1977).

Theorem 5.7. *Let F be a centrally symmetric infinitely divisible distribution function on R^k whose Lévy spectral measure is GCCUM. Then F is CCUM.*

Proof. Let $(\boldsymbol{\Sigma}, \Lambda)$ be the Lévy representation of F. Assume first that the Gaussian component $\boldsymbol{\Sigma}$ vanishes. Since Λ is *GCCUM*, we can write

$$\Lambda = \int_{\Theta_k} W_K \, d\eta(W_K), \tag{5.48}$$

for some measure η on Θ_k. Let $c = \Lambda(R^k)$. Three cases arise.

Case 1. If $c = 0$, then (5.46) shows that F is degenerate at $\mathbf{0}$ and is thus CCUM.

Case 2. Let $0 < c < \infty$. From (5.48), we see that $\eta(\Theta_k) = c$. Therefore $\eta = c\mu$, where μ is a probability measure on Θ_k. Thus (5.48) and (5.47) show that $\Lambda = cv$ where v is a CCUM distribution. Let φ and ς, respectively, denote the characteristic functions of F and v. Then (5.46) gives

$$\begin{aligned}\varphi(\mathbf{t}) &= \exp[c(\varsigma(\mathbf{t}) - 1)] \\ &= e^{-c} \sum_{n=0}^{\infty} \left[\frac{c^n \varsigma^n(\mathbf{t})}{(n!)}\right]\end{aligned}$$

By the results of Chapter 2, ς^n corresponds to a CCUM distribution and thus φ also corresponds to a CCUM distribution.

Case 3. Let $c = \infty$. Let $A_n = \{\mathbf{x} : \| x \| > 1/n\}$ and $V_n = \{W_K : W_K(A_n) \geq \frac{1}{2}\}$. Then (5.48) and Chebyshev's inequality yield

$$\eta(V_n) \leq 2\Lambda(A_n). \tag{5.49}$$

Let η_n be the restriction of η on V_n. That is $\eta_n(B) = \eta(B \cap V_n)$. Define Λ_n by

$$\Lambda_n = \int_{\Theta_k} W_K \, d\eta_n(W_K).$$

Since Λ is a Lévy measure, $\Lambda(A_n) < \infty$ for all n and so (5.49) shows that η_n is a finite measure on Θ_k. Consequently, Λ_n is a *finite* Lévy measure which is GCCUM. From Case 2, we conclude that the distribution function F_n with Lévy representation $(\mathbf{0}, 0, \Lambda_n)$ is CCUM. But is is clear that $\Lambda_n \to \Lambda$ weakly on the complement of every neighborhood of $\mathbf{0}$. Thus $F_n \to F$ weakly. Since each F_n is CCUM, we conclude that F is CCUM.

Suppose now that the Gaussian component $\mathbf{\Sigma}$ does not vanish. Then F is the convolution of two distribution functions G and H, where G is normal with covariance matrix $\mathbf{\Sigma}$ and H is CCUM (by the earlier part of the proof). Since G is CCUM, it follows that $G * H$ is CCUM. This proves the theorem. ■

Remark. Theorem 5.7 is a generalization of Medgyessy's univariate theorem (Theorem 5.4) to the multivariate case. This is because a GCCUM Lévy measure Λ on R is precisely one whose Lévy function M is concave on $(0, \infty)$ and odd. To see this, suppose that

$$\Lambda = \int_0^{\infty} H_z \, dv(z),$$

where H_z is the uniform distribution on $(-z, z)$ and ν is a measure on $(0, \infty)$. Then, for $x > 0$,

$$M(x) = -\Lambda(x, \infty) = \tfrac{1}{2} \int_x^\infty \left(\frac{x}{z} - 1\right) d\nu(z)$$

$$= \tfrac{1}{2} \int_0^\infty \min\left(\frac{x}{z} - 1, 0\right) d\nu(z),$$

which is clearly concave on $(0, \infty)$. Conversely, suppose M is concave on $(0, \infty)$. For $x > 0$, let $g(x) = M'(x)$, where we take the right derivative for the sake of definiteness. Since g is nonincreasing on $(0, \infty)$, we can define a measure ν on $(0, \infty)$ by setting $d\nu(x) = -2x\, dg(x)$. Then $dg(x) = [-1/(2x)]\, d\nu(x)$ and so

$$g(x) = \tfrac{1}{2} \int_x^\infty \left(\frac{1}{z}\right) d\nu(z).$$

The last relation means that

$$\Lambda = \int_0^\infty H_z\, d\nu(z).$$

Thus the remark is justified. We also note that the measure ν is uniquely determined by Λ.

Theorem 5.7 can be used to prove that certain important multivariate infinitely divisible laws are CCUM. Before presenting these results, we define distributions of class L and stable distributions on R^k.

Let $\{\mathbf{Z}_n\}$ be a sequence of random k-vectors of the form

$$\mathbf{Z}_n = \frac{\sum_1^n \mathbf{X}_j - \mathbf{a}_n}{b_n}.$$

where $\{\mathbf{X}_n\}$ is a sequence of independent random k-vectors, $\{\mathbf{a}_n\}$ is a sequence of k-vectors, $\{b_n\}$ is a sequence of positive numbers and the system $\{\mathbf{X}_j/b_n, j \le n\}$ is infinitesimal. The class of all possible limit distributions of the sequence $\{\mathbf{Z}_n\}$ is called the *class* L of distributions in R^k. If we restrict the random vectors $\mathbf{X}_n$ to be not only independent but also identically distributed, then the class of all possible limit distributions of $\{\mathbf{Z}_n\}$ is called the class of *stable* distributions in R^k.

Ghosh (1974) gave a theorem which can be easily modified to show that all centrally symmetric stable laws in R^k are linear unimodal (LUM). Wolfe (1975) showed that all *spherically* symmetric stable laws in R^k are CCUM. Kanter (1977) used Theorem 5.7 to show that all *centrally* symmetric stable distributions in R^k are CCUM. Wolfe (1978b) generalized Kanter's result by proving that all centrally symmetric distributions of class L in R^k are CCUM.

Since CCUM$\Rightarrow$LUM, Ghosh's result is contained in the results of Kanter (1977) and Wolfe (1978b). We now present a slightly expanded version of Wolfe's 1978 results.

It will be convenient to write sets in their polar form. Let $S = \{\mathbf{x} \in R^k: \|\mathbf{x}\| = 1\}$ be the unit sphere in R^k and let $\mathscr{F}$ denote the σ-algebra of Borel sets in S. For $B \subset S$ and $A \subset (0, \infty)$, we use the notation $B \times A$ for the set $\{\mathbf{x} \in R^k: \|\mathbf{x}\| \in A \text{ and } \mathbf{x}/\|\mathbf{x}\| \in B\}$. Let Λ be a Lévy measure on R^k. For $B \in \mathscr{F}$ and $r > 0$, let $N_B(r) = \Lambda[B \times (r, \infty)]$.

Lemma 5.6. *Suppose $N_B(\cdot)$ is convex on $(0, \infty)$ for every $B \in \mathscr{F}$ and suppose Λ is centrally symmetric. Then Λ is* GCCUM.

Proof. Let $\xi_B(r) = -N_B'(r)$, where the prime denotes (say) the right derivative. Then ξ_B is nonnegative and nonincreasing on $(0, \infty)$. Therefore the remark following Theorem 5.7 shows that there is a unique measure ν_B on $(0, \infty)$ such that

$$\xi_B(r) = \int_r^\infty \frac{1}{2z} \, d\nu_B(z), \qquad r > 0.$$

Thus if H_z denotes the uniform distribution on $(-z, z)$ and Λ_B denotes the measure on $(0, \infty)$ with density ξ_B, then

$$\Lambda_B = \int_0^\infty H_z \, d\nu_B(z). \tag{5.50}$$

We note that $\Lambda_B(r, \infty) = N_B(r) = \Lambda[B \times (r, \infty)]$. Thus, $\Lambda_B(A) = \Lambda(B \times A)$ and consequently, for every fixed Borel set A in $(0, \infty)$, $\Lambda_B(\mathrm{A})$ is a measure in $B \in \mathscr{F}$. Now the uniqueness of ν_B shows that $\nu_B(A)$ is a measure in $B \in \mathscr{F}$. We can therefore define a measure η on R^k by

$$\eta(B \times A) = \nu_B(A) \quad \text{and} \quad \eta\{\mathbf{0}\}) = 0.$$

From this definition, one easily verifies the relation

$$\int_0^\infty f(z) \, d\nu_B(z) = \int_{R^k} f(z) I_B(\mathbf{b}) \, d\eta(\mathbf{x}), \tag{5.51}$$

for every bounded measurable function f on $(0, \infty)$, where $(z, \mathbf{b})$ is the polar representation of $\mathbf{x}$ in the integrand.

For $\mathbf{x} \in R^k - \{0\}$, let $W_\mathbf{x}$ denote the uniform distribution on the line segment $(-\mathbf{x}, \mathbf{x})$. We claim that

$$\Lambda = \int_{R^k} W_\mathbf{x} \, d\eta(\mathbf{x}). \tag{5.52}$$

In the following verification, all integrals are over R^k unless indicated otherwise.

$$\begin{aligned}\int W_{\mathbf{x}}(B \times A)\, d\eta(\mathbf{x}) &= \int W_{z\mathbf{b}}(B \times A)\, d\eta(\mathbf{x}) \\ &= \int I_B(\mathbf{b}) W_{z\mathbf{b}}(\{b\} \times A)\, d\eta(\mathbf{x}) \\ &= \int I_B(\mathbf{b}) H_z(A)\, d\eta(\mathbf{x}) \\ &= \int_0^\infty H_z(A)\, d\nu_B(z) \\ &= \Lambda_B(A) = \Lambda(B \times A),\end{aligned}$$

where we have used (5.51) in going from η to ν_B and used (5.50) on the penultimate step. We have thus verified (5.52) and proved that Λ is GCCUM. ■

Lemma 5.6 and Theorem 5.7 immediately yield the following theorem.

Theorem 5.8. *Let F be a centrally symmetric infinitely divisible distribution functon on R^k with Lévy spectral measure Λ. For $B \in \mathscr{F}$ and $r > 0$, let $N_B(r) = \Lambda[B \times (r, \infty)]$. If $N_B(\cdot)$ is convex on $(0, \infty)$ for every fixed $B \in F$,* then *F is* CCUM.

The most important application of Theorem 5.8 is in proving that all centrally symmetric distributions in the class L are CCUM. As in the univariate case, distributions in L can be characterized in terms of their Lévy spectral measures. Let F be an infinitely divisible distribution function on R^k with Lévy spectral measure Λ. Then F belongs to L if, and only if, for every Borel set A in R^k and for $\alpha \in (0, 1)$, we have

$$\Lambda(A) - \Lambda\left(\frac{A}{\alpha}\right) \geq 0. \tag{5.53}$$

As before, let $N_B(r) = \Lambda[B \times (r, \infty)]$, where B is a Borel subset of the unit sphere and $r > 0$. Then (5.53) is equivalent to

$$N_B(r) - N_B\left(\frac{r}{\alpha}\right) \geq N_B(s) - N_B\left(\frac{s}{\alpha}\right) \tag{5.54}$$

for $0 < r < s < \infty$. Condition (5.54) is the multivariate version of condition

(5.3). Note that, for fixed B, $N_B(r)$ is a function of a single variable. Therefore we can show, as in Lemma 5.1, that condition (5.54) implies that $N_B(\cdot)$ is convex on $(0, \infty)$ for every fixed B. The following theorem now follows immediately from Theorem 5.8.

Theorem 5.9. *Every centrally symmetric distribution of class L in R^k is* CCUM.

Recall that the class L of distributions in R^k is the class of all possible limit distributions of normed sums of the form $(\sum_1^n \mathbf{X}_j - \mathbf{a}_n)/b_n$, where the random vectors $\mathbf{X}_n$ are independent and the system $\{\mathbf{X}_j/b_n, j \leq n\}$ is infinitesimal. If norming is done by $k \times k$ matrices B_n instead of scalars b_n, that is, if we replace $(1/b_n)$ above by B_n, then the resulting limit distributions are called *Lévy distributions*. These have been studied by Urbanik (1972) and Wolfe (1980). Again, if the $\mathbf{X}_n$ are not only independent but also indentically distributed and norming is done by matrices, the resulting limit distributions are called *operator stable*. For $k \geq 2$, the class of Lévy (respectively, operator stable) distributions in R^k is strictly wider than the class of L (respectively, stable) distributions in R^k. The discusson of Lévy distributions and operator stable distributions is beyond the scope of this book. But we note that, for $k \geq 2$, a centrally symmetric operator stable distribution in R^k may not even be k-unimodal. Thus the nice results on the unimodality of centrally symmetric distributions of class L do not carry over to Lévy distributions.

6 Unimodality and Notions of Dependence

6.0. Summary

In this chapter, we show how some concepts of dependence can be combined with suitable concepts of unimodality to yield useful probability inequalities. The relevant concepts of dependence are reviewed in the first section. The second section discusses the role played by unimodality in verifying some of the dependence conditions in a few situations.

6.1. Preliminaries on Notions of Dependence

In this section, we review some notions of dependence for multivariate distributions. These will be found useful in the next section where we will discuss some probability inequalities via some concepts of multivariate unimodality.

Let $\mathbf{X} = (X_1, \ldots, X_n)$ be a random vector. We say that $\mathbf{X}$ has *positive upper orthant dependence* (PUOD) if

$$P[X_i \geq c_i, i = 1, \ldots, n] \geq \prod_{i=1}^{n} P(X_i \geq c_i) \tag{6.1}$$

for every $(c_1, \ldots, c_n) \in R^n$. If, on both sides of (6.1), we replace "$X_i \geq c_i$" by "$X_i \leq c_i$," we get the concept of *positive lower orthant dependence* (PLOD). Under both PUOD and PLOD, the random variables $X_1, \ldots, X_n$ have a tendency to hang together to a greater extent than they would under independence. Again, we can reverse the middle inequality in (6.1) to obtain the definitions of *negative upper orthant dependence* (NUOD) and *negative lower orthant dependence* (NLOD).

The following two observations follow immediately from the above definitions.

(a) If **X** has PUOD (respectively, PLOD, NUOD or NLOD) then every subvector of **X** has PUOD (respectively, PLOD, NUOD or NLOD).

(b) In the bivariate case, PUOD and PLOD are equivalent and NUOD and NLOD are also equivalent. [To see this, it is sufficient to observe that $P(A^c \cap B^c) = 1 - P(A) + P(B) + P(A \cap B)$.] For this reason, in the bivariate case, PUOD and PLOD can be merged into a single concept called *positive quadrant dependence* (PQD).

The word "positive" in the definition of PUOD (or PLOD) is natural because this dependence condition *does* imply nonnegativity for the correlation coefficients. To see this, let $\mathbf{X} = (X_1, \ldots, X_n)$ have PUOD or PLOD. Let F_i denote the distribution function of X_i and let F_{ij} be the joint distribution function of X_i and X_j. By observations (a) and (b) above, the pair (X_i, X_j) has both PUOD and PLOD. Consequently,

$$F_{ij}(u, v) \geq F_i(u)F_j(v). \tag{6.2}$$

By a formula of Hoeffding (1940),

$$\operatorname{Cov}(X_i, X_j) = \iint [F_{ij}(u, v) - F_i(u)F_j(v)]\, du\, dv. \tag{6.3}$$

It now follows from (6.2) and (6.3) that $\operatorname{Cov}(X_i, X_j) \geq 0$. Moreover, we also see that X_i and X_j are independent as soon as they are uncorrelated. One justifies the word "negative" in the definitions of NUOD and NLOD in a similar fashion by proving that $\operatorname{Cov}(X_i, X_j) \leq 0$ for every pair (i, j) whenever **X** has NUOD or NLOD.

Another important concept of dependence can be based on the following lemma.

Lemma 6.1. *A random vector (X_1, X_2) has positive quadrant dependence* (PQD) *if, and only if,*

$$\operatorname{Cov}[g_1(X_1), g_2(X_2)] \geq 0 \tag{6.4}$$

for every pair (g_1, g_2) of nondecreasing measurable functions.

Proof. The "if" part follows by taking $g_i = I_{[c_i, \infty)}$, $i = 1, 2$. Conversely, suppose (X_1, X_2) has PQD and let g_1, g_2 be nondecreasing functions on R. If we write $Y_i = g_i(X_i)$, then (Y_1, Y_2) also has PQD. Therefore, formulas (6.2) and (6.3) again show that $\text{Cov}(Y_1, Y_2) \geq 0$. ■

Property (6.4) and its generalizations and applications were studied by Lehmann (1966). One important generalization, called *association* was discussed by Esary, Proschan and Walkup (1967). A random vector $\mathbf{X} = (X_1, \ldots, X_n)$ is said to be *associated* if

$$\text{Cov}[g_1(\mathbf{X}), g_2(\mathbf{X})] \geq 0 \tag{6.5}$$

for every pair (g_1, g_2) of functions on R^n which are nondecreasing in each coordinate.

The following properties follow immediately from the definition of association. Property (i) is known as Chebyshev's covariance inequality.

(i) A real random variable is associated.
(ii) A vector of independent random variables is associated.
(iii) If $\mathbf{X}$ is associated, then every subvector of $\mathbf{X}$ is also associated.
(iv) For $i = 1, 2$, let $\mathbf{X}_i$ be an associated random vector. If $\mathbf{X}_1$ and $\mathbf{X}_2$ are independent, then the combined vector $(\mathbf{X}_1, \mathbf{X}_2)$ is associated.
(v) If $\mathbf{X}$ is associated, then $\mathbf{X}$ has PUOD and PLOD.
(vi) If $\mathbf{X}$ is associated and $Y_j = g_j(\mathbf{X})$, $j = 1, \ldots, m$, where each g_j is nondecreasing in each coordinate, then $(Y_1, \ldots, Y_m)$ is associated.

The verification that a given random vector is associated can sometimes be quite difficult. For distributions having densities, there is a condition known as *total positivity of order* 2 (TP_2) which implies association. It should be mentioned in passing that a condition of the type TP_2 was used by Fortuin, Kasteleyn and Ginibre (1971) to derive (6.5) in the setting of a distributive lattice. Their result is now well known as the FKG inequality.

To show that the TP_2 condition (given below) implies association, it is convenient to introduce an intermediate condition. A random variable Y is said to be *stochastically increasing* (SI) in a random variable X, if

$$P(Y > y \mid X = x)$$

can be chosen to be nondecresing in x for every fixed y. It is very easy to show that, if Y is SI in X, then $E[h(Y) \mid X = x]$ can be chosen to be nondecreasing in x for every nondecreasing function h; see Lehmann (1986, p. 116).

Suppose now that X, Y are random variables having joint density $f(x, y)$. We write $f_1(x)$ and $f_2(y)$ for the marginal densities of X and Y, respectively.

The density f is said to be *totally positive of order* 2 (TP_2) if, for every choice of points (x_1, y_1), (x_2, y_2) with $x_1 < x_2$ and $y_1 < y_2$, we have

$$f(x_1, y_1)f(x_2, y_2) \geq f(x_1, y_2)f(x_2, y_1). \tag{6.6}$$

The TP_2 condition is closely related to logconcavity. Indeed, it is not difficult to show that a univariate density p is logconcave if, and only if, $p(x-y)$ is TP_2 in (x, y). A logconcave density is also sometimes called a Polya frequency function of order 2.

Theorem 6.1. *Let X, Y be random variables. Then*

(a) *(X, Y) satisfies $\mathrm{TP}_2 \Rightarrow X$ and Y are SI in each other,*
(b) *Y SI in $X \Rightarrow (X, Y)$ is associated.*

Proof. (i) Suppose (X, Y) satisfies TP_2. Then (6.6) holds. Write

$$A(x, v) = \int_{-\infty}^{v} f(x, y)\,dy \quad \text{and} \quad B(x, v) = \int_{v}^{\infty} f(x, y)\,dy.$$

Then (6.6) easily yields

$$A(x_1, v)B(x_2, v) \geq B(x_1, v)A(x_2, v). \tag{6.7}$$

But $A(x, v) + B(x, v) = f_1(x)$. Therefore, (6.7) can be written as

$$B(x_2, v)[f_1(x_1) - B(x_1, v)] \geq B(x_1, v)[f_1(x_2) - B(x_2, v)].$$

Thus

$$\frac{B(x_2, v)}{f_1(x_2)} \geq \frac{B(x_1, v)}{f_1(x_1)} \tag{6.8}$$

Since $P(Y \geq v \mid X = x) = B(x, v)/f_1(x)$, we see from (6.8) that $P(Y \geq v \mid X = x)$ is nondecreasing in x for every fixed v. Thus Y is SI in X. Similarly X is SI in Y.

(ii) Let Y be SI in X. Let g_1 and g_2 be functions on R^2 which are nondecreasing in each coordinate. Let $U = g_1(X, Y)$ and $V = g_2(X, Y)$. As noted earlier, for every nondecreasing function h, $E[h(Y) \mid X = x]$ is nondecreasing in x. Therefore, $E(U \mid X = x)$ and $E(V \mid X = x)$ are both nondecreasing in x. Since a single random variable X is associated, we conclude that

$$\mathrm{Cov}[E(U \mid X), E(V \mid X)] \geq 0. \tag{6.9}$$

Again, for fixed x, $g_1(x, y)$ and $g_2(x, y)$ are nondecreasing in y. Therefore the conditional covariance of $g_1(x, Y)$ and $g_2(x, Y)$ given $X = x$ is nonnegative.

Consequently

$$E[\mathrm{Cov}(U, V) \mid X] \geq 0. \tag{6.10}$$

If we add (6.9) and (6.10) and use the formula

$$\mathrm{Cov}(U, V) = \mathrm{Cov}[E(U \mid Z), E(V \mid Z)] + E[\mathrm{Cov}(U, V) \mid Z], \tag{6.11}$$

with Z being any random vector, we see that $\mathrm{Cov}(U, V) \geq 0$. Thus (X, Y) is associated. ■

The multivariate analog of the TP_2 condition is obtained by requiring a density $f(x_1, \ldots, x_n)$ to be TP_2 in every pair (x_i, x_j) when the remaining $(n-2)$ variables are fixed. This condition is known as MTP_2. By repeated applications of (6.11), one can show that a random vector $\mathbf{X}$ is associated as soon as its density satisfies the MTP_2 condition. An example of an MTP_2 density is that of a multivariate normal density with a covariance matrix Σ such that Σ^{-1} has all its off-diagonal entries nonpositive. Karlin and Rinott (1981) and Bølviken (1982) have shown that, if a multivariate normal vector $\mathbf{X}$ has mean vector $\mathbf{0}$ and satisfies the MTP_2 condition and we write $|\mathbf{X}|$ for the vector $(|X_1|, \ldots, X_n|)$, then $|\mathbf{X}|$ is associated.

6.2. Monotonicity Properties of Probabilities of Rectangular Regions

Some concepts of dependence were introduced in the previous section. In this section we use some arguments involving unimodality to prove that certain distributions satisfy some of these dependence conditions. The concepts of unimodality are found especially useful in problems of dimension ≥ 3.

Let $\mathbf{X}$ be a random n-vector with mean vector $\boldsymbol{\mu}$. A probability of the type $P(|\mathbf{X} - \boldsymbol{\mu}| \leq \mathbf{c})$ plays an important role in the construction of simultaneous confidence interval estimates for the component means. When the precise knowledge of dependence between $X_1, \ldots, X_n$ is lacking or does not readily permit the evaluation of the above probability, it is often useful to obtain a lower bound for it. In particular, if the random deviates $|X_i - \mu_i|$ satisfy PLOD (see Section 6.1), then we would have

$$P(|\mathbf{X} - \boldsymbol{\mu}| \leq \mathbf{c}) \geq \prod_{i=1}^{n} P(|X_i - \mu_i) \leq c_i).$$

Such an inequality can be established by using certain monotonicity properties of probabilities of rectangular regions. We now proceed to discuss these properties.

We begin with a well known result of Slepian (1962). Let $\mathbf{X}$ have the n-variate normal distribution with mean vector $\boldsymbol{\mu}$ and covariance matrix $\boldsymbol{\Sigma}$. We denote the density of $\mathbf{X}$ by $h_n(\mathbf{x}; \boldsymbol{\mu}, \boldsymbol{\Sigma})$. For $\mathbf{c} \in R^n$, the notation $\mathbf{X} \leq \mathbf{c}$ will mean $X_j \leq c_j, j = 1, \ldots, n$. Write σ_{jk} for the (j, k)th entry in $\boldsymbol{\Sigma}$. The characteristic function of $\mathbf{X}$ is given by

$$\psi_n(\mathbf{t}) = \exp[i\mathbf{t}'\boldsymbol{\mu} - \tfrac{1}{2}\mathbf{t}'\boldsymbol{\Sigma}\mathbf{t}].$$

The multivariate inversion formula for characteristic functions then gives

$$h_x(\mathbf{x}; \boldsymbol{\mu}, \boldsymbol{\Sigma}) = (2\pi)^{-n} \int_{R^n} \exp[i\mathbf{t}'(\boldsymbol{\mu} - \mathbf{x}) - \tfrac{1}{2}\mathbf{t}'\boldsymbol{\Sigma}\mathbf{t}]\, d\mathbf{t}. \tag{6.12}$$

It follows easily from (6.12) that, for $k \neq j$,

$$\frac{\partial}{\partial \sigma_{kj}} h_n(\mathbf{x}, \boldsymbol{\mu}, \boldsymbol{\Sigma}) = \frac{\partial^2}{\partial x_j\, \partial x_k} h_n(\mathbf{x}, \boldsymbol{\mu}, \boldsymbol{\Sigma}). \tag{6.13}$$

Identity (6.13) is usually attributed to Plackett (1954) although it seems to have been known earlier as evidenced by its use by Hotelling and Pabst (1936). Slepian (1962) proved that, for $k \neq j$,

$$\frac{\partial}{\partial \sigma_{jk}} P[\mathbf{X} \leq \mathbf{c}] \geq 0. \tag{6.14}$$

To see that (6.14) follows immediately from (6.13), observe that

$$\begin{aligned}\frac{\partial}{\partial \sigma_{12}} P(\mathbf{X} \leq \mathbf{c}) &= \int_{-\infty}^{c_1} \cdots \int_{-\infty}^{c_n} \frac{\partial^2}{\partial x_1\, \partial x_2} h_n(\mathbf{x}; \boldsymbol{\mu}, \boldsymbol{\Sigma})\, d\mathbf{x} \\ &= \int_{-\infty}^{c_3} \cdots \int_{-\infty}^{c_n} h_n(c_1, c_2, x_3, \ldots, x_n; \boldsymbol{\mu}, \boldsymbol{\Sigma})\, dx_3 \ldots dx_n.\end{aligned}$$

The last expression is clearly nonnegative. The same argument shows that

$$\frac{\partial}{\partial \sigma_{jk}} P[\mathbf{X} \geq \mathbf{c}] \geq 0. \tag{6.15}$$

Indeed, if $\mathbf{X}^{(kj)}$ denotes the $(n-2)$-vector obtained from $\mathbf{X}$ by deleting X_j and X_k, then the above proof shows that

$$\frac{\partial}{\partial \sigma_{jk}} E[g_1(\mathbf{X}^{(kj)}, X_j) g_2(\mathbf{X}^{(kj)}, X_k)] \geq 0,$$

where g_i, g_2 are both nondecreasing (or both nonincreasing) in their last argument; see Das Gupta, Eaton, Olkin, Perlman, Savage and Sobel (1972).

The technique of applying (6.13) to get monotonicity of the type given by (6.14) or (6.15) can also be applied to probabilities of rectangular regions.

Theorem 6.2. *Let* (X_1, X_2) *have the bivariate normal distribution with mean vector* $\mathbf{0}$ *and covariance matrix* $\mathbf{\Sigma} = (\sigma_{jk})$. *Then* $P(|X_1| \le c_1,\ |X_2| \le c_2)$ *is nondecreasing in* $|\sigma_{12}|$.

Proof. Formula (6.13) shows that

$$\frac{\partial}{\partial \sigma_{12}} P(|X_1| \le c_1, |X_2| \le c_2) = \int_{-c_1}^{c_1} \int_{-c_2}^{c_2} \frac{\partial^2}{\partial x_1\, \partial x_2} h_2(x_1, x_2;\, \mathbf{0}, \mathbf{\Sigma})\, dx_1\, dx_2$$
$$= 2[h_2(c_1, c_2; \mathbf{0}, \mathbf{\Sigma}) - h(c_1, -c_2, \mathbf{0}, \mathbf{\Sigma})]$$
$$\gtrless 0 \text{ according as } \sigma_{12} \gtrless 0.$$

Thus $P(|X_1| \le c_1,\ |x_2| \le c_2)$ is nondecreasing in $|\sigma_{12}|$. ■

Corollary. *Under the conditions of Theorem* 6.2,

$$P(|X_1| \le c_1, |X_2| \le c_2) \ge P(|X_1| \le c_1) \cdot P(|X_2| \le c_2). \qquad (6.16)$$

Proof. From the theorem, the left side of (6.16) is nondecreasing in $|\sigma_{12}|$. Therefore a lower bound for its value is obtained by setting $\sigma_{12} = 0$. The corollary follows. ■

Both Theorem 6.2 and its corollary can be suitably extended to higher dimensions. These ideas were initiated by Dunn (1958) and the basic results were given by Sidak (1967, 1968). Jogdeo (1970) used concepts of multivariate unimodality to give direct simple proofs of Sidak's results. We now present these proofs.

Recall that a distribution P on R^n is called monotone unimodal if, for every centrally symmetric convex set $C \subset R^n$ and every $\mathbf{y} \in R^n$, $P(C + k\mathbf{y})$ is symmetric unimodal in $k \in R$.

Lemma 6.2. *Suppose that a probability density* f *on* R^n *is continuously differentiable and corresponds to a monotone unimodal distribution. Then for every* $\mathbf{a} \in R^n$ *and every centrally symmetric convex set* $C \subset R^n$, *we have*

$$\int_C \sum_{i=1}^{n} a_i \frac{\partial}{\partial x_i} f(\mathbf{x} + \mathbf{a})\, d\mathbf{x} \le 0. \qquad (6.17)$$

Proof. Let $h(k) = P(C + k\mathbf{a})$. Since h is nonincreasing on $(0, \infty)$, $h'(k) \le 0$ for $k > 0$. Now

$$h(k) = \int_{C + k\mathbf{a}} f(\mathbf{x})\, d\mathbf{x} = \int_C f(\mathbf{x} + k\mathbf{a})\, d\mathbf{x}.$$

Therefore

$$h'(k) = \int_C \sum_{i=1}^{n} a_i \frac{\partial}{\partial x_i} f(\mathbf{x} + k\mathbf{a})\, d\mathbf{x} \le 0 \qquad \text{if } k > 0.$$

The lemma follows by setting $k = 1$. ■

We can now state a generalization of Theorem 6.2.

Theorem 6.3. *Let* $\mathbf{X}$ *be a multinormal random vector with mean vector* $\mathbf{0}$ *and covariance matrix* (σ_{ij}) *with* $\sigma_{ii} = 1$ *for all i and* $\sigma_{1j} = \lambda\rho_{1j}, j > 1$. *Then for every* $\mathbf{c} = (c_1, \ldots, c_n)$ *with* $c_i > 0$ *for all i, the probability*

$$P(|X_i| \le c_i,\ i = 1, \ldots, n)$$

is a nondecreasing function of $\lambda > 0$.

Proof. It is clearly sufficient to prove the theorem for the case where $\mathbf{X}$ has a density. The assertion of the theorem is equivalent to

$$\sum_{j=2}^{n} \rho_{1j} \frac{\partial}{\partial \sigma_{1j}} P(|X_i| \le c_i,\ i = 1, \ldots, n) \ge 0. \tag{6.18}$$

Let C be the $(n-1)$-dimensional rectangle

$$[-c_2, c_2] \times [-c_3, c_3] \times \cdots \times [-c_n, c_n].$$

Further, let h_n denote the density of $\mathbf{X}$. Below we write $\mathbf{x}^* = (x_2, \ldots, x_n)$. The left side of (6.18) is seen to be equal to

$$2 \sum_{j=2}^{n} \rho_{1j} \frac{\partial}{\partial \sigma_{1j}} \int_0^{c_1} \int_C h_n(\mathbf{x})\, d\mathbf{x} = 2 \sum_{j=2}^{n} \rho_{1j} \int_0^{c_1} \int_C \frac{\partial^2}{\partial x_1\, \partial x_j} h_n(\mathbf{x})\, d\mathbf{x}$$

$$= 2 \sum_{j=2}^{n} \rho_{1j} \int_C \frac{\partial}{\partial x_j} h_n(c_1, \mathbf{x}^*)\, d\mathbf{x}^*. \tag{6.19}$$

The integral in (6.19) can be written in terms of the conditional density of $(X_2, \ldots, X_n)$ given $X_1 = c_1$. Write $\boldsymbol{\rho} = (\rho_{12}, \ldots \rho_{1n})$. Let h^* be the density of the $(n-1)$-variate normal distribution with mean vector $\mathbf{0}$ and covariance matrix (σ^*_{jk}), where

$$\sigma^*_{jk} = \sigma_{jk} - \lambda^2 \rho_{1j}\rho_{1k}, \qquad j, k = 2, \ldots, n.$$

We note that h^* corresponds to a monotone unimodal distribution. Therefore,

(6.19) can be written as

$$(\text{const})\cdot\int_C \sum_{j=2}^{n} \rho_{1j} \frac{\partial}{\partial x_j} h^*(\mathbf{x}^* - \lambda c_1 \boldsymbol{\rho})\, d\mathbf{x}^*. \tag{6.20}$$

It follows from (6.17) that expression (6.20) is nonnegative. The theorem is thus proved. ■

Corollary. *Let* $\mathbf{X}$ *be a multinormal random vector with mean vector* $\mathbf{0}$ *and covariance matrix* (σ_{ij}). *Then, for all* $\mathbf{c} = (c_1, \ldots, c_n)$ *with* $c_i > 0$ for all i, we have

$$P(|X_i| \le c_i,\ i = 1, \ldots, n) \ge \prod_{i=1}^{n} P(|X_i| \le c_i). \tag{6.21}$$

Proof. Use the notation of Theorem (6.3). That is, we assume the covariance structure (σ_{ij}) with $\sigma_{1j} = \lambda\rho_{1j}$, $j \ge 2$. From the theorem, the value of $P(|X_i| \le c_i, i = 1, \ldots, n)$ for $\lambda = 1$ must be $\ge$ its value for $\lambda = 0$. That is

$$\begin{aligned} P(|X_i| \le c_i,\ i = 1, \ldots, n) &\ge P(|X_1| \le c_1)\cdot P(|X_j| \le c_j,\ j > 1) \\ &\ge \prod_{i=1}^{n} P(|X_i| \le c_i). \end{aligned}$$

This proves the corollary. ■

Remark. Theorem 6.3 shows that the result of Theorem 6.2 is really a result on directional monotonicity. Indeed, it can be shown that if a particular $|\rho_{ij}|$ is increased while keeping other correlations fixed, then the probability of a rectangular region may *decrease*. In terms of the notation introduced in Section 6.1, the vector $(|X_1|, \ldots, |X_n|)$ has PLOD as soon as $\mathbf{X}$ is multinormal with zero mean vector. On the other hand, it can be shown that, even in the multinormal case, the vector $(|X_1|, \ldots, |X_n|)$ may not satisfy PUOD. Examples illustrating the comments of this remark have been given by Sidak (1968).

It is of some interest to look for some simple conditions (of course, less stringent than the condition of independence) which will yield PUOD or even the stronger property of association for $(|X_1|, \ldots, |X_n|)$. The next theorem considers a "contaminated independence model" for which the property of association holds.

Theorem 6.4. *Let* $\mathbf{X} = \mathbf{Z} + \mathbf{U}$, *where*

(i) *the n components of* $\mathbf{Z}$ *are independent random variables each with a symmetric unimodal distribution;*

(ii) **Z** *and* **U** *are independent; and*
(iii) $|\mathbf{U}|$ *is associated.*

Then $|\mathbf{X}|$ *is associated.*

Proof. Let I_1 and I_2 be indicators of sets in R^n which are nonincreasing in absolute values of the coordinates. Then I_1 and I_2 are axially unimodal functions (see Section 3.1). Therefore, by Theorem 3.2, $E(I_1(\mathbf{X})\,|\,\mathbf{U})$ and $E(I_2(\mathbf{X})\,|\,\mathbf{U})$ are nonincreasing in $|\mathbf{U}|$. But $|\mathbf{U}|$ is associated. Therefore

$$\text{Cov}[E(I_1(\mathbf{X})\,|\,\mathbf{U}),\ E(I_2(\mathbf{X})\,|\,\mathbf{U})] \geq 0. \tag{6.22}$$

Now given **U**, **X** has independent components and so $|\mathbf{X}|$ is associated. Therefore, given **U**, the conditional covariance of $I_1(\mathbf{X})$ and $I_2(\mathbf{X})$ is nonnegative. Therefore,

$$E[\text{Cov}\{I_1(\mathbf{X}),\ I_2(\mathbf{X})\,|\,\mathbf{U}\}] \geq 0. \tag{6.23}$$

Adding (6.22) and (6.23), we get

$$\text{Cov}[I_1(\mathbf{X}),\ I_2(\mathbf{X})] \geq 0.$$

Thus $|\mathbf{X}|$ is associated and the theorem is proved. ∎

The following is an example illustrating the use of Theorem 6.4.

Example 6.1. Suppose **X** is multinormal with mean vector **0** and covariance matrix (σ_{ij}) where $\sigma_{ij} = \lambda_i \lambda_j$, $i \neq j$ and $\sigma_{ii} \geq \lambda_i^2$ where $\lambda_1, \ldots, \lambda_n$ are some constants. Then **X** has the representation $X_i = Z_i + \lambda_i V$, where $V, Z_1, \ldots, Z_n$ are independent, V is $N(0, 1)$ and Z_i is $N(0, \sigma_{ii} - \lambda_i^2)$. In terms of Theorem 6.4, $\mathbf{U} = (\lambda_1 V, \ldots, \lambda_n V)$. Since a single real random variable is associated, $|\mathbf{U}|$ is trivially associated. Since each Z_i has a symmetric unimodal distribution, the hypotheses of Theorem 6.4 are satisfied. Thus $|\mathbf{X}|$ is associated.

Remark. It can be seen that the random vector **X** in Example 6.1 need not have mean zero. If $E(V) = c$, then we could have $E(X_i) = c\lambda_i$, $\text{Var}(X_i) \geq \lambda_i^2$ and $\text{Cov}(X_i, X_j) = \lambda_i \lambda_j$. There are some other ways of obtaining a larger class of examples giving association for $|\mathbf{X}|$. For details, see Jogdeo (1977) or Tong (1980).

We conclude this section by mentioning a property which is stronger than PLOD for $(|X_1|, \ldots, |X_n|)$. This stronger property requires that

$$P(\mathbf{X} \in C_1 \cap C_2) \geq P(\mathbf{X} \in C_1) \cdot P(\mathbf{X} \in C_2), \tag{6.24}$$

for every choice of centrally symmetric convex sets C_1, C_2 in R^n. It had been conjectured that a multivariate normal random vector $\mathbf{X}$ with mean vector $\mathbf{0}$ satisfies (6.24). Pitt (1977) verified this conjecture in the bivariate case and we understand that the multivariate conjecture has also been settled in the affirmative by W. Beckner.

Property (6.24) indicates a type of unimodality in the sense that the distribution of $\mathbf{X}$ puts more mass near the origin than away from it. Now the uniform distribution on a centrally symmetric convex set was unimodal under every definition of multivariate unimodality considered in Chapter 2. Consequently, it is of interest to note that such uniform distributions may not satisfy (6.24). This is exhibited by the next example.

Example 6.2. Let $\mathbf{X} = (X_1, X_2)$ be uniformly distributed on the square S with vertices $(\pm\frac{1}{2}, \pm\frac{1}{2})$. Let T be the square formed by the midpoints of the sides of S, namely, $(\pm\frac{1}{2}, 0)$, $(0, \pm\frac{1}{2})$. The set $S - T$ consists of four triangles which can be written as A_i, $i = 1, 2, 3, 4$, where the subscript i denotes the quadrant in which A_i is situated. Let $C_1 = T \cup A_1 \cup A_3$ and $C_2 = T \cup A_2 \cup A_4$. Then

$$P(\mathbf{X} \in C_1 \cap C_2) = \tfrac{1}{2} < \tfrac{9}{16} = P(\mathbf{X} \in C_1) \cdot P(\mathbf{X} \in C_2).$$

Thus (6.24) fails. This example also shows that (6.24) is strictly stronger than the PLOD property for $|\mathbf{X}|$.

7 Ordering of Distributions by Peakedness

7.0. Summary

Many useful partial orders have been introduced in the literature for the class of all probability distributions on R^n. For instance, Lehmann (1986) has discussed the well known stochastic ordering of distributions on R. Some orderings of interest in reliability theory will be discussed in Chapter 9. In this chapter, we consider an ordering of distributions on R^n by "peakedness." The first section defines the ordering and gives the basic properties resulting from the definition. Section 2 studies the peakedness of logconcave distributions, multivariate normal distributions and other elliptically contoured distributions. The last section presents an application where considerations of unimodality and peakedness help in deciding the recurrence of symmetric random walks.

7.1. Basic Results on Peakedness Ordering

We begin by recalling the definition of stochastic ordering. Let X_1, X_2 be random variables with distribution functions F_1 and F_2, respectively. Then X_1 is said to be *stochastically larger than* X_2 if $P(X_1 \geq t) \geq P(X_2 \geq t)$ or, equivalently, if $F_1(t) \leq F_2(t)$ for all $t \in R$. If this condition holds, we write $X_1 \geq^{\text{st}} X_2$. It will be seen in this section that some of the properties of peakedness ordering are closely connected with the stochastic ordering defined above and with suitable unimodality properties.

The partial ordering by peakedness results from the following definition.

Definition 7.1. Suppose $\mathbf{X}_1$, $\mathbf{X}_2$ are random n-vectors with centrally symmetric probability distributions P_1, P_2, respectively. Then P_1 is said to be *less peaked than* P_2 if, for every centrally symmetric convex set $C \subset R^n$, we have

$$P_1(C) \leq P_2(C). \tag{7.1}$$

If (7.1) holds, we write $P_1 \leq^p P_2$ or $\mathbf{X}_1 \leq^p \mathbf{X}_2$. For distributions on R, (7.1) is equivalent to the property that $|X_1|$ is stochastically larger than $|X_2|$.

Definition 7.1 was given for the univariate case by Birnbaum (1948) and the generalization to the multivariate case was considered by Sherman (1955). The basic result on peakedness orderings states that the ordering is preserved under convolutions with distributions which are "unimodal" in a suitable sense. Once again the univariate result is due to Birnbaum (1948) and the multivariate result is due to Sherman (1955). Kanter (1977) uses the term "dominance" in place of peakedness ordering and gives results on orderings of general measures.

Remark. In Definition 7.1, we can restrict attention to centrally symmetric *compact* convex *bodies* $C \subset R^n$. This is because P_1 and P_2 are probability measures and are therefore regular.

We recall that a nonnegative function g on R^n into R is called *quasi-concave* if the set $\{\mathbf{x} : g(\mathbf{x}) \geq s\}$ is convex for every $s \geq 0$. The simplest example of such a function is the indicator of a convex set. The following theorem is easy to establish.

Theorem 7.1. *Let* $\mathbf{X}$ *and* $\mathbf{Y}$ *be random n-vectors. Then* $\mathbf{X} \leq^p \mathbf{Y}$ *if, and only if,* $g(\mathbf{X}) \leq^{st} g(\mathbf{Y})$ *for every centrally symmetric quasi-concave function* g *on* R^n.

We need a notion of distance between a point and a set. Suppose $\|\mathbf{x}\|$ denotes the usual Euclidean norm of $\mathbf{x} \in R^n$. If $\mathbf{x} \in R^n$ and $A \subset R^n$, write

$$d(\mathbf{x}, A) = \inf\{\|\mathbf{x} - \mathbf{y}\| : \mathbf{y} \in A\}.$$

It is well known that $d(\mathbf{x}, A)$ is continuous in $\mathbf{x}$. More importantly, if A is convex, then $d(\mathbf{x}, A)$ can be easily shown to be convex in $\mathbf{x}$.

Let $\mathscr{G}_n$ denote the set of all bounded, centrally symmetric quasi-concave functions on R^n. We note that $\mathscr{G}_n$ includes all centrally symmetric logconcave functions on R^n.

Theorem 7.2. *Let P_1, P_2 be centrally symmetric probability distributions on R^n. The following five statements are equivalent.*

(i) $P_1 \leq^p P_2$.

(ii) $\int g\, dP_1 \leq \int g\, dP_2$ *for all* $g \in \mathscr{G}_n$.

(iii) $\int g\, dP_1 \leq \int g\, dP_2$ *for all continuous* $g \in \mathscr{G}_n$.

(iv) $\int g\, dP_1 \leq \int g\, dP_2$ *for all centrally symmetric logconcave functions* g *on* R^n.

(v) $\int g\, dP_1 \leq \int g\, dP_2$ *for all centrally symmetric continuous logconcave functions* g *on* R^n.

Proof. The assertions (ii)$\Rightarrow$(iii), (ii)$\Rightarrow$(iv), (iii)$\Rightarrow$(v) and (iv)$\Rightarrow$(v) are trivial. We show that (i)$\Rightarrow$(ii) and (v)$\Rightarrow$(i).

(a) Suppose (i) holds and $g \in \mathscr{G}_n$. For $i = 1, 2$, *let* $\mathbf{X}_i$ be a random vector with distribution P_i. By Theorem 7.1, $g(\mathbf{X}_1) \leq^{st} g(\mathbf{X}_2)$. Therefore $E[g(\mathbf{X}_1)] \leq E[g(\mathbf{X}_2)]$. This shows that (ii) holds. Thus (i)$\Rightarrow$(ii).

(b) Suppose (v) holds. Let C be a nonempty centrally symmetric compact convex set in R^n. Let D_k be the set $\{\mathbf{x} : d(\mathbf{x}, C) \leq 1/k\}$. Write

$$g(\mathbf{x}) = [1 - kd(\mathbf{x}, C)] \cdot I_{D_k}.$$

Since $d(\mathbf{x}, C)$ is convex in $\mathbf{x}$ and D_k is also convex, it is easily checked that g is logconcave. Since g can be written as

$$g(\mathbf{x}) = \max\{1 - kd(\mathbf{x}, C), 0\},$$

g is also seen to be continuous. In addition, $0 \leqq g(\mathbf{x}) \leqq 1$ for all $\mathbf{x}$ and

$$C \subset \{\mathbf{x} : g(\mathbf{x}) = 1\} \subset \{\mathbf{x} : g\{\mathbf{x}\} > 0\} \subset D_k.$$

Thus

$$P_1(C) \leq \int g\, dP_1 \leq \int g\, dP_2 \leq \int_{\{g>0\}} g\, dP_2 \leq P_2(D_k).$$

Letting $k \to \infty$ and noting that C is compact, we see that $P_1(C) \le P_2(C)$. This shows that (v)$\Rightarrow$(i) and completes the proof of the theorem. ■

Criterion (iii) of the above theorem enables us to prove that weak limits preserve peakedness ordering.

Theorem 7.3. *Let $\{P_{1,k}\}$ and $\{P_{2,k}\}$ be two sequences of probability distributions on R^n such that $P_{1,k} \le^p P_{2,k}$ for all k. Suppose that, for $i = 1, 2$, $P_{i,k} \to P_i$ weakly as $k \to \infty$. Then $P_1 \le^p P_2$.*

Proof. Let g be a bounded, continuous and quasi-concave function on R^n. By Theorem 7.2(iii),

$$\int g\, dP_{1,k} \le \int g\, dP_{2,k} \qquad \text{for all } k.$$

By weak convergence, we get

$$\int g\, dP_1 \le \int g\, dP_2.$$

Theorem 7.2(iii) again shows that $P_1 \le^p P_2$. ■

The proofs of the following two lemmas on mixtures and projections, respectively, are trivial.

Lemma 7.1. *Let $(T, \mathscr{T}, \nu)$ be a probability space. Let $\{P_{1,t}, t \in T\}$ and $\{P_{2,t}, t \in T\}$ be two sets of centrally symmetric probability measures on R^n such that $P_{1,t} \le^p P_{2,t}$ for all $t \in T$.* For *$i = 1, 2$, let*

$$P_i(B) = \int_T P_{i,t}(B)\, d\nu(t),$$

where B is a Borel set in R^n and the maps $t \to P_{i,t}(B)$ are assumed to be $\mathscr{T}$-measurable. Then $P_1 \le^p P_2$.

Lemma 7.2. *For $i = 1, 2$, let $\mathbf{Y}_i = \mathbf{L}\mathbf{X}_i$, where $\mathbf{X}_i$ is a random n-vector and $\mathbf{L}$ is a matrix of order $m \times n$. If $\mathbf{X}_1 \le^p \mathbf{X}_2$, then $\mathbf{Y}_1 \le^p \mathbf{Y}_2$.*

The result on the preservation of peakedness ordering under the operation of convolutions (with suitably unimodal distributions) will follow from Lemma 7.2 if we also show that the ordering is preserved under cartesian products. This is done in the next theorem.

Theorem 7.4. *Suppose P_1 and P_2 are centrally symmetric probability distributions on R^n such that $P_1 \leq^p P_2$. Let Q be a central convex unimodal distribution on R^m. Then $P_1 \times Q \leq^p P_2 \times Q$.*

Proof. Let g be a centrally symmetric bounded logconcave function on R^{m+n}. Write

$$\beta_i = \int_{R^m} \int_{R^n} g(\mathbf{x}, \mathbf{y}) \, dP_i(\mathbf{x}) \, dQ(\mathbf{y}), \qquad i = 1, 2.$$

We want to show that $\beta_2 \geq \beta_1$. Clearly,

$$\beta_i = \int_{R^n} h(\mathbf{x}) \, dP_i(\mathbf{x}), \tag{7.2}$$

where

$$h(\mathbf{x}) = \int_{R^m} g(\mathbf{x}, \mathbf{y}) \, dQ(\mathbf{y}).$$

(a) First assume that Q corresponds to the uniform distribution on a centrally symmetric convex *body* $C \subset R^m$. Then Q has the density αI_C where α is a normalizing constant. Therefore

$$h(\mathbf{x}) = \alpha \int_{R^m} g(\mathbf{x}, \mathbf{y}) I_C(\mathbf{y}) \, d\mathbf{y}.$$

Now the function $g(\mathbf{x}, \mathbf{y}) I_C(\mathbf{y})$ is logconcave on R^{m+n}. Therefore, by Theorem 2.16, the function h is logconcave on R^n. Theorem 7.2 and (7.2) now show that $\beta_1 \leq \beta_2$.

(b) Suppose next that Q corresponds to the point mass at $\mathbf{0}$. Then $h(\mathbf{x}) = g(\mathbf{x}, \mathbf{0})$, which is logconcave. Therefore we again see from (7.2) and Theorem 7.2 that $\beta_1 \leq \beta_2$.

(c) Suppose now that Q corresponds to the uniform distribution on a centrally symmetric convex *set* $C \subset R^m$ which is *not* a body. If L_1 is the subspace generated by C and L_2 is the orthocomplement of L_1 in R^m, then we can write $R^m = L_1 \times L_2$ and $Q = Q_1 \times Q_2$, where Q_2 is the point mass at $\mathbf{0} \in L_2$ and Q_1 is now the uniform distribution on the convex *body* C (considered as a subset of L_1). From (a) and (b), we conclude that

$$P_1 \times Q_1 \times Q_2 \leq^p P_2 \times Q_1 \times Q_2,$$

which means that $P_1 \times Q \leq^p P_2 \times Q$.

(d) Let $\mathscr{P}$ be the set of all probability distributions Q on R^m such that $P_1 \times Q$ is less peaked than $P_2 \times Q$. From (c) above, $\mathscr{P}$ includes all uniform distributions on centrally symmetric convex sets. But by Theorem 7.3 and Lemma 7.1, the set $\mathscr{P}$ is convex under mixtures and closed under weak convergence. Therefore, $\mathscr{P}$ contains all the central convex unimodal distributions. This completes the proof of the theorem. ■

By combining Theorem 7.4 and Lemma 7.2, we get the following important result on convolutions.

Theorem 7.5. *Let P_1, P_2 be centrally symmetric probability distributions on R^n such that $P_1 \leq^p P_2$. Let Q be a central convex unimodal distribution on R^n. Then $(P_1 * Q) \leq^p (P_2 * Q)$.*

Proof. By Theorem 7.4, $P_1 \times Q \leq^p P_2 \times Q$. If $\mathbf{L}$ maps the point $(\mathbf{x}, \mathbf{y}) \in R^{2n}$ into the point $(\mathbf{x} + \mathbf{y}) \in R^n$, then the distribution on R^n determined by $(P_i \times Q)$ and $\mathbf{L}$ is just $P_i * Q$. By Lemma 7.2 we see that $P_1 * Q \leq^p P_2 * Q$. ■

Corollary. *Suppose P_1, P_2, Q_1, Q_2 are centrally symmetric probability distributions on R^n such that $P_1 \leq^p P_2$ and $Q_1 \leq^p Q_2$. Assume that P_2 and Q_1 are central convex unimodal. Then $P_1 * Q_1 \leq^p P_2 * Q_2$.*

Proof. By Theorem 7.5, $(P_1 * Q_1) \leq^p (P_2 * Q_1)$, because Q_1 is central convex unimodal. Similarly, $(P_2 * Q_1) \leq^p (P_2 * Q_2)$, because P_2 is central convex unimodal. Thus the corollary follows by the transitivity of the peakedness ordering. ■

Corollary. *For $i = 1, 2$, let $\mathbf{X}_j^{(i)}, j = 1, \ldots, m$ be independent random n-vectors having central convex unimodal distributions. Suppose $\mathbf{X}_j^{(1)} \leq^p \mathbf{X}_j^{(2)}, j = 1, \ldots, m$. Then $\sum_{j=1}^m X_j^{(1)} \leq^p \sum_{j=1}^m X_j^{(2)}$.*

Proof. Immediate from the earlier corollary. ■

It should be noted that one cannot completely dispense with the unimodality assumptions in Theorems 7.4 and 7.5. This is exhibited by the following simple example due to Birnbaum (1948).

Example 7.1. Let P_1 be the uniform distribution on $[-1, 1]$ and let P_2 put mass $\frac{1}{2}$ at ± 1. Then $P_2 \leq^p P_1$. We note that P_2 is not unimodal. Let $C_a = (-a, a)$, where $0 < a < 2$. Then $(P_2 * P_2)(C_a) = \frac{1}{2}$. Since $P_1 * P_2$ is just the uniform distribution on $[-2, 2]$ we see that $(P_1 * P_2)(C_a) = a/2$. Therefore, $(P_1 * P_2)$ and $(P_2 * P_2)$ are not ordered under the peakedness ordering.

To conclude this section, let us indicate how the concept of ordering by peakedness works in the discrete case. Suppose P_1 and P_2 are symmetric distributions on the set of integers. Since P_1, P_2 are also distributions on R, we can compare them by using Definition 7.1. But since P_1, P_2 are discrete, we only need to check the inequality (7.1) for symmetric intervals with integer end points. So we get the following definition. We denote by A_m the set of all integers k satisfying $-m \le k \le m$, with $m = 0, 1, 2, \ldots$.

Definition 7.2. If P_1, P_2 are distributions on the set of integers, P_1 is said to be less peaked than P_2 and we write $P_1 \le^p P_2$ if

$$P_1(A_m) \le P_2(A_m) \tag{7.3}$$

for all $m = 0, 1, 2, \ldots$.

Remark. If (7.3) holds then it is also true that, for every $m \ge 0$ and every $\theta \in (0, 1)$,

$$P_1(A_m) + \theta P_1(\{m+1\}) \le P_2(A_m) + \theta P_2(\{m+1\}). \tag{7.4}$$

The truth of (7.4) is seen by writing the left side as $(1-\theta)P_1(A_m) + \theta P_1(A_{m+1})$ and then writing the right side in a similar fashion.

Since Definition 7.2 cannot be easily extended to the bivariate discrete case, we do not have discrete analogs of Theorem 7.4 and Lemma 7.2. However, convolutions can still be defined and Theorem 7.5 and its corollaries have discrete analogs. We state the new version of the first corollary only.

Theorem 7.6. *Let P_1, P_2, Q_1, Q_2 be symmetric discrete distributions such that $P_1 \le^p Q_1$ and $P_2 \le^p Q_2$. If P_2 and Q_1 are unimodal, then $P_1 * P_2 \le Q_1 * Q_2$.*

7.2. Peakedness Comparisons for the Multivariate Normal and Other Suitably Unimodal Distributions

In this section we apply the results of the previous section to compare the peakedness of a few distributions. Anderson's (1955) results for the multivariate normal case are obtained by an easy application of Theorem 7.5. For the generalization to the elliptically contoured case we present the results of Fefferman, Jodeit and Perlman (1972). The section concludes with a discussion of peakedness comparisons for certain logconcave distributions. These last results are due to Proschan (1965) and Olkin and Tong (1985).

We begin with a result of Anderson (1955).

Theorem 7.7. *For $i = 1, 2$, let $\mathbf{X}_i$ have an n-variate normal distribution with mean vector* $\mathbf{0}$ *and covariance matrix $\mathbf{\Sigma}_i$. Suppose $\mathbf{\Sigma}_1 - \mathbf{\Sigma}_2$ is nonnegative definite. Then $\mathbf{X}_1 \leq^p \mathbf{X}_2$.*

Proof. Without loss of generality we may assume that $\mathbf{X}_1 = \mathbf{X}_2 + \mathbf{Z}$, where $\mathbf{Z}$ is independent of $\mathbf{X}_2$ and has the normal distribution with mean vector $\mathbf{0}$ and covariance matrix $\mathbf{\Sigma}_1 - \mathbf{\Sigma}_2$. Let $\mathbf{Y}$ be a random n-vector which is $\mathbf{0}$ with probability one. Then $\mathbf{Z} \leq^p \mathbf{Y}$ and the distribution of $\mathbf{X}_2$ is central convex unimodal. Therefore, by Theorem 7.5, $(\mathbf{Z} + \mathbf{X}_2) \leq^p (\mathbf{Y} + \mathbf{X}_2)$ or $\mathbf{X}_1 \leq^p \mathbf{X}_2$ as was to be shown. ■

Remark. For the univariate case, Theorem 7.7 says, as expected, that, for a normal distribution with mean zero, the peakedness decreases as the variance increases.

The result of Theorem 7.7 was used by Anderson (1955) to prove the following theorem. A function $K(s, t)$ on a product set $B \times B$ will be called *nonnegative definite* if $\sum_{i=1}^n \sum_{j=1}^n a_i a_j K(t_i, t_j)$ is nonnegative for every n and for every choice of $t_1, \ldots, t_n$ in B and $a_1, \ldots, a_n$ in R.

Theorem 7.8. *For $i = 1, 2$, let $\{X_i(t),\ 0 \leq t \leq T\}$ be a separable Gaussian process with mean function* 0 *and covariance function $\sigma_i(s, t)$, assumed to be continuous on $[0, T] \times [0, T]$. Suppose that $\sigma_2(s, t) - \sigma_1(s, t)$ is nonnegative definite. Then $\int_0^T X_2^2(t)\,dt$ is stochastically larger than $\int_0^T X_1^2(t)\,dt$ and $\sup_{0 \leq t \leq T} |X_2(t)|$ is stochastically larger than $\sup_{0 \leq t \leq T} |X_1(t)|$.*

Proof. Let $0 = t_0 < t_1 < \ldots < t_n = T$ be a partition of $[0, T]$ and write $h(\mathbf{x}) = \sum_{i=1}^n x_i^2(t_i - t_{i-1})$. Then, $\{\mathbf{x} : h(\mathbf{x}) \leq s\}$ is convex for every $s \geq 0$. Further, the assumptions on the σ_i and Theorem 7.7 show that $\{X_1(t_j), j = 1, \ldots, n\}$ is more peaked than $\{X_2(t_j),\ j = 1,\ \ldots,\ n\}$. Therefore by Theorem 7.1, $\sum_1^n X_1^2(t_j)(t_j - t_{j-1})$ is stochastically smaller than $\sum_1^n X_2^2(t_j)(t_j - t_{j-1})$. The first assertion of the theorem now follows by taking finer and finer partitions which become dense in $[0, T]$ as $n \to \infty$. The proof of the second assertion is similar and uses the functon $h(\mathbf{x}) = \max\{\,|x_j|, j = 1, \ldots, n\}$. The assumption of separability is used to assert that $\sup_{0 \leq t \leq T}|X_i(t)|$ is a random variable, for $i = 1, 2$. ■

Anderson (1955) has applied Theorem 7.8 to some tests of goodness of fit. Suppose F_N is the empirical distribution function obtained from a random sample of size N drawn from a continuous distribution function F on R. Let

$$U_N = N^{1/2} \sup_x |F_N(x) - F(x)|.$$

It is well known that the limiting distribution of U_N is that of $U = \sup\{|Z(t)|, 0 \leq t \leq 1\}$, where $\{Z(t)\}$ is a separable Gaussian process with mean function 0 and covariance function

$$r(s, t) = \min(t, s) - ts.$$

Now, if F is also indexed by some parameter θ and we replace θ by its maximum likelihood estimator $\hat{\theta}_N$, we would get a modified statistic U_N^* in place of U_N. Darling (1955) has shown that, under certain conditions, the limiting distribution of U_N^* is that of $U^* = \sup\{X(t), 0 \leq t \leq 1\}$, where $\{X(t)\}$ is a Gaussian process with mean function 0 and covariance matrix $r^*(s, t) = r(s, t) - \psi(s)\psi(t)$, where ψ is a well defined function. Now

$$r(s, t) - r^*(s, t) = \psi(s)\psi(t),$$

which is a nonnegative definite functon. Therefore, Theorem 7.8 applies and we see that U^* is stochastically smaller than U. Therefore, if we use the percentage points of U in conjunction with the statistic U^*, then the assumed level of significance will be preserved. A similar conclusion holds for the statistics

$$W_N = \int |F_N(x) - F(x)|^2 \, dF(x).$$

Theorem 7.8 has a somewhat restrictive setting in that it applies to the multivariate normal case only. Fefferman, Jodeit and Perlman (1972) have proved an inequality for surface measures which enabled them to extend the result of Theorem 7.8 to certain elliptically contoured distributions. The proof of their inequality involves arguments via unimodality. We present their results in Theorems 7.9 and 7.10 below.

For $\mathbf{x} \in R^n$, let $\|\mathbf{x}\| = (\Sigma\ x_i^2)^{1/2}$ be the usual Euclidean norm. Let $S = \{\mathbf{x} \in R^n : \|\mathbf{x}\| = 1\}$ be the unit sphere in R^n and let μ denote the uniform surface measure on S. For a linear transformation $\mathbf{A}$ on R^n into R^n, the norm $\|\mathbf{A}\|$ of $\mathbf{A}$ is defined by

$$\|\mathbf{A}\| = \sup\{\|\mathbf{Ax}\| : \mathbf{x} \in S\}.$$

Theorem 7.9. *Let C be a centrally symmetric closed convex subset of R^n and let $\mathbf{A}$ be a linear transformation on R^n into R^n such that $\|\mathbf{A}\| \leq 1$. Then*

$$\mu[(\mathbf{A}C) \cap S] \leq \mu(C \cap S).$$

Proof. We can take C to be a centrally symmetric, closed and convex polyhedron. Further using orthogonal transformations, we can reduce $\mathbf{A}$ to a matrix $D = \text{diag}(\lambda_1, \ldots, \lambda_n)$, with $0 \le \lambda_i \le 1$. The theorem would follow if we prove that $\mu[(DC) \cap S]$ is a nondecreasing function of each λ_i. Now D is a product of matrices $E_1, \ldots, E_n$ where E_i is a diagonal matrix whose ith diagonal entry is λ_i and the remaining diagonal entries equal 1. Thus the proof reduces to showing that the function $u(\lambda)$ given by

$$u(\lambda) = \mu[(D_\lambda C) \cap S]$$

is a nondecreasing function of $\lambda \in (0, 1)$, where $D_\lambda = \text{diag}(\lambda, 1, \ldots, 1)$. Note that

$$u(\lambda) = \int_S I_C[D_\lambda^{-1}\mathbf{x}]\, d\mu(\mathbf{x}),$$

where I_C is the indicator function of C.

Let f_ε denote the density of the n-variate normal distribution with mean vector zero and covariance matrix εI_n. Write $\varphi_\varepsilon = I_C * f_\varepsilon$ and

$$u_\varepsilon(\lambda) = \int_S \varphi_\varepsilon(D_\lambda^{-1}\mathbf{x})\, d\mu(\mathbf{x}).$$

Let ∂C denote the boundary of C. If $\mathbf{x} \in R^n - \partial C$, then $\varphi_\varepsilon(\mathbf{x}) \to I_C(\mathbf{x})$ as $\varepsilon \to 0$. But $\mu(\partial C) = 0$ because C is a polyhedron. Therefore $u_\varepsilon(\lambda) \to u(\lambda)$ as $\varepsilon \to 0$ and so it suffices to show that $u'_\varepsilon(\lambda) \ge 0$.

Set $\psi(\mathbf{x}) = \varphi_\varepsilon[D_\lambda^{-1}\mathbf{x}]$. Then

$$u'_\varepsilon(\lambda) = -\frac{1}{\lambda} \int_S x_1 \frac{\partial \psi}{\partial x_1}\, d\mu(\mathbf{x}). \tag{7.5}$$

Let B be the unit ball in R^n and consider $I_B * \psi$. Since B and C are convex and centrally symmetric, both I_B and ψ are quasi-concave functions. Therefore, by Theorem 2.20, $I_B * \psi$ is some multiple of the density of a central convex unimodal distribution. In particular $(I_B * \psi)(t\mathbf{x})$ is symmetric unimodal in $t \in R$ for every fixed $\mathbf{x} \in R^n$. Thus, setting $\mathbf{z} = (1, 0, \ldots, 0)$, we have

$$\begin{aligned} 0 \ge \frac{d^2}{dt^2}[(I_B * \psi)(t\mathbf{z})]\bigg|_{t=0} &= \frac{\partial^2}{\partial x_1^2}(I_B * \psi)(\mathbf{x})\bigg|_{\mathbf{x}=0} \\ &= \int_B \frac{\partial^2 \psi}{\partial x_1^2}\, d\mathbf{x}. \end{aligned} \tag{7.6}$$

Now by the divergence theorem,

$$\int_B \frac{\partial^2 \psi}{\partial x_1^2}\, d\mathbf{x} = \int_S x_1 \frac{\partial \psi}{\partial x_1}\, d\mu(\mathbf{x}). \tag{7.7}$$

Therefore, (7.5), (7.6) and (7.7) show that $-\lambda u_\varepsilon'(\lambda) \le 0$. Thus $u_\varepsilon'(\lambda) \ge 0$ and the theorem is proved. ■

Theorem 7.9 yields the following generalization of Theorem 7.8 to certain elliptically contoured distributions.

Theorem 7.10. *Let ν be a spherically symmetric measure on the Borel σ-algebra in R^n. For a positive definite $n \times n$ matrix $\mathbf{\Sigma}$ define the measure $\nu_{\mathbf{\Sigma}}$ by $\nu_{\mathbf{\Sigma}}(B) = \nu(\mathbf{\Sigma}^{-1/2}B)$, where B is a Borel set. Then $\nu_{\mathbf{\Sigma}_2}$ is less peaked than $\nu_{\mathbf{\Sigma}_1}$ if $\mathbf{\Sigma}_2 - \mathbf{\Sigma}_1$ is nonnegative definite.*

Proof. For $t > 0$, *let* μ_t denote the uniform surface measure on the sphere $S_t = \{\mathbf{x} : |\mathbf{x}| = t\}$. The spherical symmetry of ν shows that there is a measure λ on $(0, \infty)$ such that, for every Borel set B,

$$\nu(B) = \nu(B \cap \{0\}) + \int_{0^+}^{\infty} \mu_t(B \cap S_t)\, d\lambda(t) \tag{7.8}$$

Now assume that $\mathbf{\Sigma}_2 - \mathbf{\Sigma}_1$ is nonnegative definite and let C be a centrally symmetric, closed, convex set. Write $\mathbf{A} = \mathbf{\Sigma}_2^{-1/2}\mathbf{\Sigma}_1^{1/2}$. Then A has norm ≤ 1. Therefore, by Theorem 7.9,

$$\begin{aligned} \mu_t[(\mathbf{\Sigma}_2^{-1/2}C) \cap S_t] &= \mu_t[\mathbf{A}\mathbf{\Sigma}_1^{-1/2}C) \cap S_t] \\ &\le \mu_t[(\mathbf{\Sigma}_1^{-1/2}C) \cap S_t]. \end{aligned} \tag{7.9}$$

Observe that both C and AC contain the origin. Therefore, (7.8) and (7.9) show that

$$\begin{aligned} \nu_{\Sigma_2}(C) &= \nu(\Sigma_2^{-1/2}C) \\ &= \nu(\{0\}) + \int_{0^+}^{\infty} \mu_t[(\Sigma_2^{-1/2}C) \cap S_t]\, d\lambda(t) \\ &\le \nu(\{0\}) + \int_{0^+}^{\infty} \mu_t[(\Sigma_1^{-1/2}C) \cap S_t]\, d\lambda(t) \\ &= \nu(\Sigma_1^{-1/2}C) = \nu_{\Sigma_1}(C). \end{aligned}$$

The theorem is thus proved. ■

Note. In terms of the notation of Theorem 7.10, Theorem 7.8 corresponds to the case where ν is the Gaussian measure with mean vector $\mathbf{0}$ and covariance matrix I.

The last application to be presented in this section shows how Theorem

7.5 can be used to obtain peakedness orderings for linear functions of random vectors having logconcave densities. These results were first obtained for the univariate case by Proschan (1965) and were generalized to the multivariate case by Olkin and Tong (1985). We will use some results on majorization mentioned earlier in Section 3.1.

Theorem 7.11. *Let* $\mathbf{Z}_1, \ldots, \mathbf{Z}_k$ *be independent, identically distributed random n-vectors having a centrally symmetric logconcave distribution. Suppose* $\mathbf{a}$ *and* $\mathbf{b}$ *are vectors in* R^k *such that* $\mathbf{a}$ *majorizes* $\mathbf{b}$. *Then*

$$\sum_{i=1}^{k} a_i \mathbf{Z}_i \leq^p \sum_{i=1}^{k} b_i \mathbf{Z}_i.$$

Before we present the proof of the theorem, let us illustrate it in a simple case. Suppose Z_1, Z_2 are independent $N(0, 1)$. For $a \in R$, write $Y_a = aZ_1 + (1-a)Z_2$. Then Y_a is normal with mean zero and variance $a^2 + (1-a)^2 = g(a)$, say. Now $g(a) > g(b)$ whenever b is closer to $\frac{1}{2}$ than a. Thus $g(a) > g(b)$ whenever the vector $(a, 1-a)$ majorizes the vector $(b, 1-b)$. But the condition $g(a) > g(b)$ implies that Y_b is more peaked than Y_a. The theorem is thus verified in this simple case.

Proof of Theorem 7.11. Without loss of generality, we may assume that $a_i = b_i$ for $i \geq 3$, and $a_1 > b_1 \geq b_2 > a_2$. Write $\mathbf{V}_1 = c(\mathbf{Z}_1 + \mathbf{Z}_2)$ and $\mathbf{V}_2 = \mathbf{Z}_1 - \mathbf{Z}_2$, where c is a suitable real number. The assumptions on the $\mathbf{Z}_i$ imply that the joint distribution of $(\mathbf{Z}_1, \mathbf{Z}_2)$ has permutation and sign symmetry. In addition, this joint distribution is also logconcave. Therefore, the conditional distribution of $\mathbf{V}_1$ given $\mathbf{V}_2 = v_2$ is symmetric and logconcave and hence monotone unimodal.

Set $c = (a_1 + a_2)/2 = (b_1 + b_2)/2$, $\lambda_1 = (a_1 - a_2)/2$ and $\lambda_2 = (b_1 - b_2)/2$. Then $\lambda_1 > \lambda_2$. Further $a_1\mathbf{Z}_1 + a_2\mathbf{Z}_2 = \mathbf{V}_1 + \lambda_1\mathbf{V}_2$ and $b_1\mathbf{Z}_1 + b_2\mathbf{Z}_2 = \mathbf{V}_1 + \lambda_2\mathbf{V}_2$. Now let C be a centrally symmetric convex set in R^n. Then

$$\begin{aligned}
P[(a_1\mathbf{Z}_1 + a_2\mathbf{Z}_2) \in C] &= P[(\mathbf{V}_1 + \lambda_1\mathbf{V}_2) \in C] \\
&= \int P[\mathbf{V}_1 \in C - \lambda_1\mathbf{v}_2 | \mathbf{V}_2 = \mathbf{v}_2]\, dF_{\mathbf{v}_2}(\mathbf{v}_2) \\
&\leq \int P[\mathbf{V}_1 \in C - \lambda_2\mathbf{v}_2 | \mathbf{V}_2 = \mathbf{v}_2]\, dF_{\mathbf{v}_2}(\mathbf{v}_2) \\
&= P[(\mathbf{V}_1 + \lambda_2\mathbf{V}_2) \in C] \\
&= P[(b_1\mathbf{Z}_1 + b_2\mathbf{Z}_2) \in C],
\end{aligned}$$

where the inequality follows from the monotone unimodality of the conditional distribution of $\mathbf{V}_1$ given $\mathbf{V}_2 = \mathbf{v}_2$. We have thus shown that $(a_1\mathbf{Z}_1 + a_2\mathbf{Z}_2) \leq^p (b_1\mathbf{Z}_1 + b_2\mathbf{Z}_2)$. Now the distribution of $\sum_{i=3}^k a_i\mathbf{Z}_i$ is centrally symmetric and logconcave and hence central convex unimodal. Using Theorem 7.5 and noting that $a_i = b_i$ for $i \geq 3$, we see that

$$(a_1\mathbf{Z}_1 + a_2\mathbf{Z}_2) + \sum_3^k a_i\mathbf{Z}_i \leq^p (b_1\mathbf{Z}_1 + b_2\mathbf{Z}_2) + \sum_3^k b_i\mathbf{Z}_i.$$

The theorem is thus proved. ■

Theorem 7.11 can be used to prove a monotonicity property for confidence coefficients. Let $\mathbf{Z}_1, \ldots, \mathbf{Z}_k$ be independent random n-vectors with a common logconcave distribution which is centrally symmetric about a parameter $\boldsymbol{\mu}$. Let $C \subset R^n$ be a centrally symmetric convex set. If $\bar{\mathbf{Z}}_k = \sum_1^k \mathbf{Z}_i/k$, then we can take $C + \bar{\mathbf{Z}}_k$ as a confidence region for $\boldsymbol{\mu}$. The confidence probability $P(\boldsymbol{\mu} \in C + \bar{\mathbf{Z}}_k)$ is then a nondecreasing function of k. This follows from Theorem 7.11 if we use

$$a_1 = \cdots = a_k = \frac{1}{k}, \qquad a_{k+1} = 0,$$

and

$$b_1 = \cdots = b_k = b_{k+1} = \frac{1}{(k+1)}.$$

One can, of course, paraphrase this result about the confidence probability into another about the probability of a type I error in a hypothesis testing problem.

We conclude this section by showing that the hypothesis of logconcavity in Theorem 7.11 cannot be replaced by the weaker hypothesis of quasi-concavity.

Example 7.2. Suppose Z_1, Z_2 are independent real random variables with a common density f given by

$$f(z) = \begin{cases} \alpha, & |z| < 1 \\ \beta, & 1 \leq |z| < (N+1) \\ 0, & \text{elsewhere.} \end{cases}$$

Here $\alpha > \beta$, $N > 2$ and $2[\alpha + N\beta] = 1$. The density f is symmetric unimodal on R. Now the vector $(1, 0)$ majorizes the vector $(\frac{1}{2}, \frac{1}{2})$. Therefore, if Theorem 7.11 holds for convex unimodal distributions, Z_1 should be less peaked than

$(Z_1 + Z_2)/2$. However, we show that $P(|Z_1| \leq 1) > P(|Z_1 + Z_2| \leq 2)$ as soon as (β/α) is sufficiently small. Now

$$P(|Z_1| \leq 1) = 2\alpha = 2\alpha[2\alpha + 2N\beta] \\ = 4\alpha[\alpha + N\beta],$$

where we have used the condition $2(\alpha + N\beta) = 1$. It is easy to verify (see Fig. 7.1) that

$$P(|Z_1 + Z_2| \leq 2) = 4\alpha^2 + 8\alpha\beta + 8\beta^2(N - 1).$$

Therefore, we have $P(|Z_1 + Z_2| \leq 2) < P(|Z_1| \leq 1)$ as soon as

$$4\alpha^2 + 8\alpha\beta + 8\beta^2(N - 1) < 4\alpha(\alpha + N\beta),$$

which holds whenever

$$\frac{\beta}{\alpha} < \frac{(N-2)}{2(N-1)}.$$

To see that the last condition can hold, we can take $N = 3$, $\alpha = \frac{1}{3}$ and $\beta = 1/(18)$. Thus Theorem 7.11 does not hold under the hypothesis of quasi-concavity. Another example to this effect has been given by Proschan (1965) who used suitable modifications of Cauchy random variables.

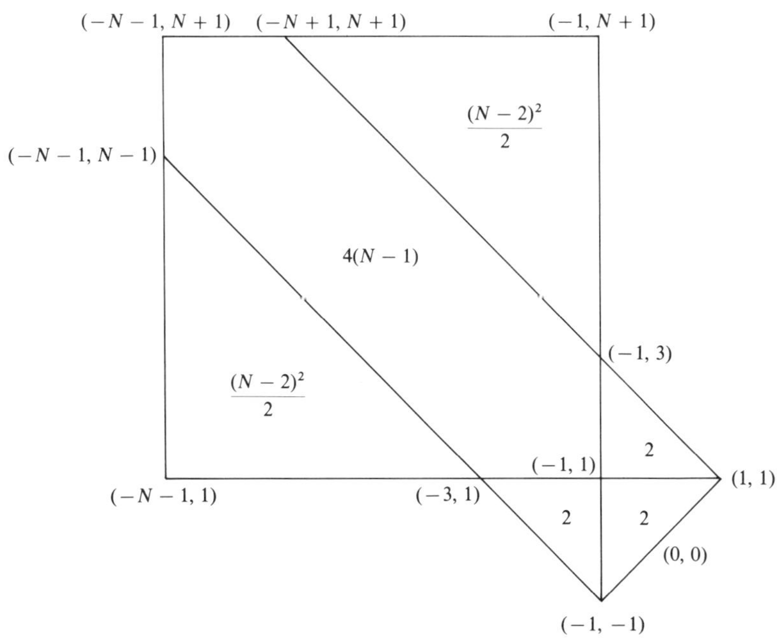

Figure 7.1. Showing the areas of regions considered in Example 7.2.

7.3. The Use of Unimodality and Peakedness Comparisons in Determining the Recurrence of Symmetric Random Walks

This section presents some results on the recurrence of symmetric random walks which involve the concepts of unimodality and peakedness comparisons. The results were given by Shepp (1962) except that, for the multivariate case, the results we present constitute an improved version of Shepp's results given by Dharmadhikari and Joag-dev (1983b).

We begin with a definition of recurrent random walks. Let $\{\mathbf{X}_n, n \geq 1\}$ be a sequence of independent random vectors taking values in R^m and having a common distribution function F. Write $\mathbf{S}_n = \sum_{j=1}^{n} \mathbf{X}_j$, where addition is by coordinates. The sequence $\{\mathbf{S}_n\}$ is called a *random walk*. Clearly, the distribution of the random walk is completely determined by F. The random walk $\{\mathbf{S}_n\}$ or, equivalently, the distribution function F is called *recurrent* if, for every open set A containing the origin, the random walk S_n visits A infinitely often with probability 1. If F is not recurrent, then it is called *transient*.

Recurrence of random walks was studied by Chung and Fuchs (1951) who proved that every genuinely three dimensional random walk is transient. For this reason, results on recurrence of random walks are vacuous in dimension ≥ 3. However, some of the lemmas presented below are applicable to all dimensions and so they are stated without any condition on the dimension m.

Suppose F is a centrally symmetric distribution function on R^m. Let φ be the characteristic function of F. Chung and Fuchs (1951) showed that either of the following two conditions is necessary and sufficient for the recurrence of F.

(a) $\sum_{n=1}^{\infty} P(\mathbf{S}_n \in C) = \infty$ for every centrally symmetric open convex set $C \subset R^m$.

(b) $\int_D [1 - \varphi(\mathbf{u})]^{-1} \, d\mathbf{u} = \infty$, where D is the unit cube in R^m.

Suppose F and G are two centrally symmetric distribution functions on R^m and suppose that F is less peaked than G. Then criterion (*a*) above seems to indicate that G would be recurrent as soon as F is so. Such a result was proved by Shepp (1962) under suitable unimodality conditions. We know that, for distributions on R, there is just one concept of symmetry and of unimodality. But in R^m, there is no unique concept of symmetry or of unimodality. Shepp used mirror symmetry about the coordinate axes. Also, his class of "unimodal" distributions was the closed convex hull of the set of all uniform distributions on symmetric rectangles with sides parallel to coordinate axes. Dharmadhikari and Joag-dev (1983b) showed that Shepp's results can be made applicable to a wider class of distributions by taking symmetry to mean central symmetry and unimodality to mean central convex unimodality. The results given below use these less restrictive notions of symmetry and unimodality.

As indicated in the preceding paragraph we will have occasion to use some of the results on peakedness comparisons proved in section 7.1. We also need the concept of a unimodal correspondent. Suppose F is the distribution function of a random vector $\mathbf{X}$. The *unimodal correspondent* (U, F) of F is defined to be the distribution function of $U\mathbf{X}$, where U is a real random variable independent of $\mathbf{X}$ and uniformly distributed on $(0, 1)$. If F is a centrally symmetric distribution function with characteristic function φ, then (U, F) has the characteristic function ψ given by

$$\psi(\mathbf{u}) = \int_{R^m} \frac{\sin(\mathbf{u}\cdot\mathbf{x})}{\mathbf{u}\cdot\mathbf{x}}\, dF(\mathbf{x}). \tag{7.10}$$

The following Lemma is important for the arguments of this section.

Lemma 7.3. *If F is centrally symmetric, then (U, F) is central convex unimodal and $F \leq^p (U, F)$.*

Proof. Let $G_{\mathbf{c}}$ be the distribution function which assigns mass $\frac{1}{2}$ to each of the points $\pm\mathbf{c}$ in R^m. Then F is the weak limit of a sequence F_n of finite mixtures of distribution functions of the form $G_{\mathbf{c}}$. Now (7.10) shows easily that the operation of taking the unimodal correspondent commutes with the operations of taking mixtures and weak limits. Therefore, the first assertion will follow if we show that $(U, G_{\mathbf{c}})$ is central convex unimodal. But $(U, G_{\mathbf{c}})$ is just the uniform distribution on the line segment $(-\mathbf{c}, \mathbf{c})$. The latter distribution is clearly central convex unimodal. This proves the first assertion.

To prove the second assertion, let F be the distribution function of a random vector $\mathbf{X}$. If C is a centrally symmetric convex set, then, for every $u \in (0, 1)$,

$$P(u\mathbf{X} \in C) \geq P(\mathbf{X} \in C).$$

Integrating over $u \in (0, 1)$, we get

$$P(U\mathbf{X} \in C) \geq P(\mathbf{X} \in C),$$

whenever U is independent of $\mathbf{X}$ and uniformly distributed on $(0, 1)$. Thus $F \leq^p (U, F)$ and the lemma is proved. ∎

We want to prove the following theorem.

Theorem 7.12. *Let F and G be centrally symmetric distribution functions on R^m and let F be central convex unimodal. Suppose $F \leq^P G$. Then the recurrence of F implies the recurrence of G.*

The proof of Theorem 7.12 is broken down into a few lemmas. First of all, the result of Theorem 7.12 follows easily from the results of section 7.1 if G is also assumed to be central convex unimodal.

Lemma 7.4. *If F and G are central convex unimodal and $F \leq^P G$, then the recurrence of F implies the recurrence of G.*

Proof. Let P and Q denote the probability measures defined on R^m by F and G, respectively. The symbol P^{*n} denotes the n-fold convolution of P with itself. Theorem 7.5 and its corollaries show that $P^{*n} \leq^P Q^{*n}$. Therefore, for every centrally symmetric open convex set C,

$$\sum_{n=1}^{\infty} P^{*n}(C) \leq \sum_{n=1}^{\infty} Q^{*n}(C). \tag{7.11}$$

If the left side of (7.11) is infinite, then so is the right side. The lemma now follows by the criterion (a) for recurrence mentioned above. ■

Lemma 7.5. *Suppose F is a centrally symmetric distribution function on R^m.*

(i) *If (U, F) is recurrent, then so is F;*
(ii) *If F is central convex unimodal, then F is recurrent if, and only if, (U, F) is recurrent.*

Proof. Let φ and ψ denote the characteristic functions of F and (U, F), respectively. It is elementary to show that, for some $\alpha > 0$,

$$1 - \cos(\mathbf{u} \cdot \mathbf{x}) \leq \alpha\left[1 - \frac{\sin(\mathbf{u} \cdot \mathbf{x})}{(\mathbf{u} \cdot \mathbf{x})}\right].$$

Integrating w.r.t. F and using (7.10), we get

$$[1 - \varphi(\mathbf{u})] \leq \alpha[1 - \psi(\mathbf{u})].$$

Therefore, if D is the unit cube in R^m, then

$$\alpha \int_D [1 - \varphi(\mathbf{u})]^{-1}\, d\mathbf{u} \geq \int_D [1 - \psi(\mathbf{u})]^{-1}\, d\mathbf{u}. \tag{7.12}$$

If the right side of (7.12) is infinite, then so is the left side. The first assertion of the lemma now follows from the criterion (b) for recurrence mentioned above.

Now suppose F is central convex unimodal and recurrent. By lemma 7.3, $F \leq^p (U, F)$. Now (U, F) is seen to be recurrent by Lemma 7.4. Thus the recurrence of F implies the recurrence of (U, F). The converse of this last statement is just the first assertion which has already been proved. The lemma is thus completely proved. ■

Proof of Theorem 7.12. The assumption $F \leq^p G$ and Lemma 7.3 show that $F \leq^p G \leq^p (U, G)$. Now F and (U, G) are both central convex unimodal. So if F is recurrent, then, by Lemma 7.4, (U, G) is also recurrent. So, by Lemma 7.5(i), G is recurrent. This completes the proof. ■

Note. Shepp (1962) has given an example to show that the unimodality assumption on F in Theorem 7.12 cannot be completely removed.

8 Applications of Unimodality in Statistical Inference

8.0 Summary

Arguments involving unimodality have been used quite often in statistical inference. This chapter brings together a few important examples of such applications. Section 1 presents results on the unimodality of the likelihood function for grouped and ungrouped observations as well as the directional unimodality of the likelihood function for certain compound multinomial samples. Some results on the estimation of the mode are presented in Section 2. Applications of Anderson-type theorems for proving the unbiasedness of some multivariate tests appear in Section 3. The chapter concludes with a discussion of the use of unimodality to construct minimum volume confidence regions.

8.1. Unimodality of the Likelihood Function

When applying the method of maximum likelihood for estimation of parameters, the unimodality of the likelihood function often facilitates the required computation. Several results on the unimodality of the likelihood function are available in the literature and some of these are presented in the current section.

A special situation in which the likelihood function is unimodal is as follows. Suppose that a density $f(x, \theta)$ is logconcave in θ for every fixed x. If $L_n(\theta \mid \mathbf{x})$ is the likelihood function based on a sample of size n from $f(x, \theta)$, then L_n is logconcave in θ and so will also be unimodal in θ. An example of this situation is where $f(x, \theta) = g(x - \theta)$ and g is logconcave.

Pratt (1981) has considered a model where the underlying density is logconcave but the observations are transformed by grouping. Specifically, suppose F is a known and differentiable distribution function on R. Let Z be a random variable taking values $1, \ldots, m$ with

$$P(Z = k \mid \mathbf{x}, \boldsymbol{\tau}, \boldsymbol{\beta}) = F(\tau_k - \mathbf{x}'\boldsymbol{\beta}) - F(\tau_{k-1} - \mathbf{x}'\boldsymbol{\beta}), \tag{8.1}$$

where $\boldsymbol{\tau}$ and $\boldsymbol{\beta}$ are vector parameters, and $-\infty = \tau_0 < \tau_1 < \cdots < \tau_m = \infty$. The original ungrouped random variable Y obeys the model $Y = \mathbf{x}'\boldsymbol{\beta} + e$, where e has distribution function F and the observed $Z = k$ if $\tau_{k-1} < Y \leq \tau_k$.

Theorem 8.1. *If F has a logconcave density f, then* (8.1) *is logconcave in $\boldsymbol{\tau}$ and $\boldsymbol{\beta}$.*

Proof. Write $w = \tau_{k-1} - \mathbf{x}'\boldsymbol{\beta}$ and $v = \tau_k - \mathbf{x}'\boldsymbol{\beta}$. Let $I(z, v, w) = 1$ if $w < z \leq v$ and 0 otherwise. Then

$$P(Z = k \mid \mathbf{x}, \boldsymbol{\tau}, \boldsymbol{\beta}) = \int I(z, v, w) f(z)\, dz. \tag{8.2}$$

Now $I(z, v, w)$ is concave and hence logconcave in (z, v, w). If f is also logconcave, then the integrand in (8.2) is logconcave in (z, v, w). Therefore, by Theorem 2.16, the left side of (8.2) is logconcave in (v, w). But v and w are linear in $\boldsymbol{\tau}$ and $\boldsymbol{\beta}$. Thus the logconcavity of f implies the logconcavity of the likelihood (8.1). ■

Pratt (1981) has also proved the converse of the above theorem. More precisely, if (8.2) is logconcave in (τ_{k-1}, τ_k) for some $\mathbf{x}$ and $\boldsymbol{\beta}$, then f is logconcave.

Burridge (1982) has a general approach where the grouping is done by partitioning the sample space into sets which may not necessarily be intervals. However, the results were obtained for the specialized model described by an exponential-type density $p(x, \theta)$ given by

$$p(x, \theta) = \exp[a(\theta) + b(x) + \theta x],$$

where x and θ are real. Burridge (1982) has also considered a regression model $Y = \mu + \mathbf{x}'\mathbf{b} + \sigma e$, where e is assumed to have a logconcave density.

Using the same techniques as above, Burridge shows that the likelihood for grouped data is logconcave in $1/\sigma$, μ/σ and $\mathbf{b}/\sigma$. These results are closely related to those of Pratt (1981).

A standard example of a unimodal density which is not logconcave is the Cauchy density. The likelihood function for the family of Cauchy distributions has been studied by several authors. In general, the likelihood equation for such a family has multiple roots. As a result, the maximum likelihood estimator may not be unique; [see Dharmadhikari and Joag-dev (1985)]. Barnett (1966) has made an extensive numerical study of locating maximum likelihood estimators. In a recent article, Reeds (1985) has shown that, if T_n is the number of local minima in θ for the likelihood for a sample of size n from the density $f(x, \theta) = \{\pi[1 + (x - \theta)^2]\}^{-1}$, then T_n is asymptotically distributed like a Poisson random variable with mean $1/\pi$.

The difficulties mentioned in the preceding paragraph do not arise if one considers the Cauchy family with both location and scale parameters. These results are due to Copas (1975). Suppose $X_1, \ldots, X_n$ is a random sample from the Cauchy density

$$p(x; \theta, \sigma) = \frac{\sigma}{\pi[\sigma^2 + (x - \theta)^2]}.$$

If L is the likelihood function, then

$$\log L = -n \log \pi + n \log \sigma - \sum \log[\sigma^2 + (x_i - \theta)^2].$$

The likelihood equations obtained by equating $\partial L/\partial\theta$ and $\partial L/\partial\sigma$ to zero are:

$$\sum_i \left[\frac{y_i}{\sigma^2 + y_i^2}\right] = 0, \tag{8.3}$$

and

$$\sum_i \left[\frac{\sigma^2}{\sigma^2 + y_i^2}\right] = \frac{n}{2}, \tag{8.4}$$

where $y_i = (x_i - \theta)$. Assume that there is no value of x at which half or more of the observations are coincident. Then (8.4) has a unique positive solution and whenever (8.4) holds, one easily gets

$$\frac{\partial^2 \log L}{\partial\sigma^2} = -4 \sum_i \left[\frac{y_i^2}{(\sigma^2 + y_i^2)^2}\right].$$

Since the last expression is negative, we conclude that for each fixed θ, L is a unimodal function of σ and attains its maximum at $\sigma = \hat{\sigma}(\theta) > 0$.

Again, it is easy to check that

$$\frac{\partial^2 \log L}{\partial \theta^2} = 2 \sum_i \frac{(y_i^2 - \sigma^2)}{(y_i^2 + \sigma^2)^2}. \tag{8.5}$$

Since, for positive numbers, a, b we always have

$$\frac{2(a-b)}{(a+b)^2} \leq \frac{(a-b)}{b(a+b)},$$

we see from (8.5) that

$$\frac{\partial^2 \log L}{\partial \theta^2} \leq \frac{1}{\sigma^2} \sum_i \frac{(y_i^2 - \sigma^2)}{(y_i^2 + \sigma^2)}. \tag{8.6}$$

By (8.4), the right side of (8.6) is zero. Thus when $(\partial \log L/\partial\sigma) = 0$, both of the second partial derivatives of $\log L$ are nonpositive. Finally, one can also show that

$$\left[\frac{\partial^2 \log L}{\partial\theta\,\partial\sigma}\right]^2 - \frac{\partial^2 \log L}{\partial\sigma^2} \cdot \frac{\partial^2 \log L}{\partial\theta^2} \leq 0.$$

It follows that there are no saddle points and there is exactly one maximum of L attained in the region $\sigma > 0$. We recall the requirement that there is no value of x at which half or more of the observations are coincident. These calculations also show that, if $\lambda(\theta \mid \mathbf{x}) = \max\{L(\theta, \sigma \mid \mathbf{x}) : \sigma \geq 0\}$ is the maximized likelihood function for θ, then λ is unimodal in θ. Thus if the likelihood function is multimodal in the location parameter θ for a fixed σ, the explanation might lie in the fact that the assumed value of σ is other than that suggested by the observations.

In an interesting paper, Levin and Reeds (1977) have considered the unimodality of the compound multinomial likelihood function. Their results verified a conjecture of Good (1965) by following the technique of Laplace transforms. Suppose $\mathbf{n} = (n_1, \ldots, n_t)$ be a multinomial random vector with parameters $N = \sum n_j$ and $\mathbf{p} = (p_1, \ldots, p_t)$. Suppose further that $\mathbf{p}$ has the Dirichlet distribution with parameter $\boldsymbol{\alpha} = (\alpha_1, \ldots, \alpha_t)$. The marginal distribution of $\mathbf{n}$ is then a compound multinomial distribution $CM(\boldsymbol{\alpha})$ given by

$$P(\mathbf{n} \mid \boldsymbol{\alpha}) = \binom{N}{n} \frac{\Gamma(\sum \alpha_j)}{\prod \Gamma(\alpha_j)} \cdot \frac{\prod \Gamma(n_j + \alpha_j)}{\Gamma(N + \sum \alpha_j)}. \tag{8.7}$$

If we omit the initial combinatorial factor and consider (8.7) as a function of $\boldsymbol{\alpha}$, we will get the likelihood function $L(\boldsymbol{\alpha} \mid \mathbf{n})$ of $\boldsymbol{\alpha}$ given the observation $\mathbf{n}$.

One easily gets

$$L(\boldsymbol{\alpha}|\mathbf{n}) = \frac{\prod_{j=1}^{t} \prod_{h=0}^{n_j-1} (\alpha_j + h)}{\prod_{h=0}^{N-1} (\sum_1^t \alpha_j + h)}. \tag{8.8}$$

Suppose now that we take a random sample $\mathbf{n}_1, \ldots, \mathbf{n}_m$ from (8.8). The likelihood of $\boldsymbol{\alpha}$ based on $\mathbf{n}_1, \ldots, \mathbf{n}_m$ will then be a product of expressions of the form (8.8). Following Good (1965), Levin and Reeds consider the "directional" unimodality of the likelihood function. Specifically, suppose $\boldsymbol{\lambda} = (\lambda_1, \ldots, \lambda_t)$ is fixed so that each $\lambda_i > 0$ and $\sum \lambda_i = 1$. The likelihood of $\boldsymbol{\alpha}$ evaluated at the point $k\boldsymbol{\lambda}$ is then

$$L(k) = \frac{\prod_{i=1}^{m} \prod_{j=1}^{t} \prod_{h=0}^{n_{ij}-1} (h + k\lambda_j)}{\prod_{h=0}^{N-1} (k+h)^m}, \tag{8.9}$$

where we write $\mathbf{n}_i$ as $(n_{i1}, \ldots, n_{it})$. Levin and Reeds (1977) have proved the following theorem.

Theorem 8.2. *The likelihood function $L(k)$ given by (8.9) has at most one local maximum. It occurs at a finite k if*

$$\chi^2 > \sum_{j=1}^{t} \left[\frac{N_j}{N\lambda_j}\right] - m$$

and for $k = \infty$ otherwise, where $N_j = \sum_{i=1}^m n_{ij}$ and

$$\chi^2 = \sum_{i=1}^{m} \sum_{j=1}^{t} \left[\frac{(n_{ij} - N\lambda_j)^2}{N\lambda_j}\right].$$

The proof of Theorem 8.2 will be given through a few lemmas. Before we present these, consider the special case where $m = 1$ and $\lambda_j = 1/t$ for all j. In this case, the theorem says that the likelihood $L(k)$ has at most one local maximum and it occurs at a finite k or at $k = \infty$ according as $\chi^2 > t-1$ or $\chi^2 \le t-1$. This was precisely the conjecture given by Good (1965).

Let us fix some notation. If M is a function of bounded variation on $[0, \infty)$ with $M(0) = 0$, we write

$$\mathscr{L}_{dM}(y) = \int_{0^-}^{\infty} e^{-uy}\, dM(u),$$

and

$$\mathscr{L}_M(y) = \int_0^\infty e^{-uy} M(u)\, du.$$

One easily gets $\mathscr{L}_{dM}(y) = y \cdot \mathscr{L}_M(y)$ for all $y > 0$ and we will use this relationship without comment. The main tool used by Levin and Reeds to prove Theorem 8.2 is the following lemma.

Lemma 8.1. *If the function M changes sign at most once on $(0, \infty)$, then the number of zeros of $\mathscr{L}_M(y)$ in $(0, \infty)$ does not exceed the number of sign changes of M in $(0, \infty)$.*

This lemma is well known; see, for instance, Lehmann (1986), p. 85. It is a special case of a more general theorem of Karlin (1968).

Differentiating (8.9) and writing $\tau_j = \lambda_j^{-1}$, we get

$$\frac{d}{dk} \log L(k) = \sum_{i=1}^{m} \sum_{j=1}^{t} \sum_{h=0}^{n_{ij}-1} (k + h\tau_j)^{-1} - m \sum_{h=0}^{N-1} (k+h)^{-1}. \tag{8.10}$$

To be able to apply Lemma 8.1, we write (8.10) as a Stieltjes transform. Let G_{ij} be the "distribution function" of the measure which puts a unit mass at each of the points $u = h\tau_j$, where $0 \le h < n_{ij}$ and h is an integer. Write $G = \sum_{i=1}^m \sum_{j=1}^t G_{ij}$. Next, let F be the "distribution function" of the measure which puts mass m at each of the points $0, 1, \ldots, (N-1)$. Then (8.10) shows that

$$\frac{d}{dk} \log L(k) = \int_{0^-}^\infty (k+u)^{-1}\, d(G-F)(u). \tag{8.11}$$

Now notice that

$$(k+u)^{-1} = \int_0^\infty e^{-(k+u)y}\, dy.$$

Therefore, (8.11) yields

$$\begin{aligned} \frac{d}{dk} \log L(k) &= \int_0^\infty e^{-ky} \int_{0^-}^\infty e^{-uy}\, d(G-F)(u)\, dy \\ &= \int_0^\infty e^{-ky} \varphi(y)\, dy = \mathscr{L}_\varphi(k), \end{aligned} \tag{8.12}$$

where

$$\varphi(y) = \int_{0^-}^\infty e^{-uy}\, d(G-F)(u) = \mathscr{L}_{d(G-F)}(y). \tag{8.13}$$

In view of Lemma 8.1, (8.12) shows that Theorem 8.2 will follow if we prove that φ has at most one change of sign in $(0, \infty)$. Again (8.13) shows that φ will have at most one change of sign in $(0, \infty)$ if $G - F$ does the same. But $(G - F)$ has many changes of sign on $(0, \infty)$ and so we have to smooth it by taking its convolution with the uniform distribution function W given by

$$W(u) = \begin{cases} 0 & \text{if} \quad u \le 0 \\ u & \text{if} \quad 0 \le u \le 1 \\ 1 & \text{if} \quad u \ge 1. \end{cases}$$

Write $F^* = W * F$ and $G^* = W * G$. Then F^* corresponds to the measure which distributes the mass mN uniformly over the interval $(0, N)$. That is

$$F^*(u) = \begin{cases} 0 & \text{if} \quad u \le 0 \\ mu & \text{if} \quad 0 \le u \le N \\ mN & \text{if} \quad u \ge N. \end{cases}$$

Similarly, G^* is a sum of functions G_β^*, where G_β^*, corresponds to a measure which distributes a total mass of β uniformly over the union of the intervals $(h\tau,\ h\tau + 1)$, where $h = 0,\ 1,\ \ldots,\ \beta$ and $\tau > 1$. A useful property of G^* is described by the next Lemma. The proof given here is slightly simpler than that given by Levin and Reeds.

Lemma 8.2. *For all* $u \ge 0$,

$$\int_{0^-}^{u} x\, dG^*(x) \le \left(\frac{u}{2}\right) \int_{0^-}^{u} dG^*(x). \tag{8.14}$$

Proof. As mentioned above G^* is a sum of functions G_β^* which corresponds to Lebesgue measure on the union of the intervals $(0, 1), (\tau, \tau + 1), (2\tau, 2\tau + 1)$, etc., where $\tau > 1$. Thus, because of linearity, we may assume that $G^* = G_\beta^*$. We note that the above intervals are disjoint because $\tau > 1$. Therefore, for $u \in (j\tau + 1,\ (j + 1)\tau)$, the left side of (8.14) stays constant whereas the right side increases. Thus it suffices to prove (8.14) for $u \in (j\tau,\ j\tau + 1)$. For such a u, the left side of (8.14) becomes

$$\begin{aligned} &\tfrac{1}{2} \sum_{h=0}^{j-1} [(h\tau + 1)^2 - (h\tau)^2] + \tfrac{1}{2}(u^2 - j^2\tau^2) \\ &= \tfrac{1}{2}[\tau j(j - 1) + j + u^2 - j^2\tau^2]. \end{aligned}$$

Similarly, the right side of (8.14) becomes

$$\frac{u}{2}[j+u-j\tau)].$$

Thus (8.14) is equivalent to

$$u[j+u-j\tau] \geq [\tau j(j-1)+j+u^2-j^2\tau^2],$$

which in turn is equivalent to

$$u(\tau-1) \leq (j\tau+1)(\tau-1)$$

which is true because $u \leq j\tau+1$ and $\tau > 1$. The lemma is thus proved. ∎

We note that $\mathscr{L}_{dW}(y) = (1-e^{-y})/y$. Therefore

$$\mathscr{L}_{d(G^*-F^*)}(y) = \frac{(1-e^{-y})}{y} \mathscr{L}_{d(G-F)}(y). \tag{8.15}$$

So, we need to show that $\mathscr{L}_{d(G^*-F^*)}$ has at most one zero on $(0, \infty)$. To this end, we take one more convolution with the "distribution function" I corresponding to the Lebesgue measure on $(0, \infty)$. That is, $I(u) = 0$ for $u \leq 0$ and $I(u) = u$ for $u \geq 0$. Let

$$H(u) = [I*(G^*-F^*)](u) = \int_0^u [G^*(x) - F^*(x)]\, dx.$$

Then $\mathscr{L}_{dH}(y) = \mathscr{L}_{dI}(y) \cdot \mathscr{L}_{d(G^*-F^*)}(y)$ and so,

$$\begin{aligned}\mathscr{L}_H(y) &= \mathscr{L}_I(y) \cdot \mathscr{L}_{d(G^*-F^*)} \\ &= \frac{1}{y^2} \mathscr{L}_{d(G^*-F^*)} \\ &= \frac{(1-e^{-y})\varphi(y)}{y^3}, \qquad \text{from (8.15).}\end{aligned}$$

Thus we need to show that the function H changes sign at most once. This will follow immediately from the following lemma.

Lemma 8.3. *If $u > 0$ and $H(u) < 0$, then $H'(u) \leq 0$.*

Proof. We know that

$$H(u) = \int_0^u [G^*(x) - F^*(x)]\, dx.$$

Therefore,

$$H'(u) = G^*(u) - F^*(u).$$

Suppose that $u > 0$ and $H(u) < 0$. If $u \geq N$, then $G^*(u) \leq mN = F^*(u)$, because the total mass for G^* is mN. Thus $H'(u) \leq 0$ for $u \geq N$. Let $u < N$. Then $F^*(x) = mx$ for $0 \leq x \leq u$ and so

$$\int_0^u F^*(x)\,dx = \frac{mu^2}{2} = \frac{u}{2}F^*(u). \tag{8.16}$$

Now, an alternative expression for H is

$$H(u) = \int_{0^-}^u (u - x)d(G^* - F^*)(x). \tag{8.17}$$

Thus $H(u) < 0$ means that

$$\int_{0^-}^u (u - x)\,dF^*(x) > \int_{0^-}^u (u - x)\,dG^*(x)$$

or,

$$\int_0^u F^*(x)\,dx > uG^*(u) - \int_{0^-}^u x\,dG^*(x). \tag{8.18}$$

Now, by (8.16), the left side of (8.18) equals $(u/2)F^*(u)$ whereas, by Lemma 8.2, the right side is larger than or equal to $(u/2)G^*(u)$. Thus

$$\frac{u}{2}F^*(u) \geq \frac{u}{2}G^*(u),$$

which means that $H'(u) \leq 0$. This proves the lemma. ■

Proof (of Theorem 8.2). We have proved so far that $(d/dk)\log L(k)$ has at most one zero on $(0, \infty)$. To study the behavior of this derivative near $k = 0$, let $c_{ij} = 0$ or 1 according as n_{ij} is zero or positive. Then (8.10) shows that, near $k = 0$, $(d/dk)\log L(k)$ behaves like $a + (b/k)$, where

$$b = \sum_{i=1}^m \sum_{j=1}^t c_{ij} - m.$$

It is clear that, for each i, there is at least one j for which $n_{ij} > 0$. Therefore $b \geq 0$. The only case where $b = 0$ is where, for each i, there is a j for which $n_{ij} = N$. This case is trivial and will not be considered. We then have $b \geq 1$. Consequently, the derivative $(d/dk)\log L(k)$ is positive near $k = 0$. If this derivative is positive on $(0, \infty)$ then $L(k)$ is increasing and becomes maximum at $k = \infty$. On the other hand, if the derivative vanishes at k_0 then $L(k)$ becomes maximum at k_0 and, moreover, $L(k_0)$ can be the only local maximum of $L(k)$.

To complete the proof of Theorem 8.2, it remains to connect the χ^2 condition with the zeros of the derivative of $L(k)$. First, around $k = \infty$, the leading term in $(d/dk) \log L(k)$ is easily seen [from (8.10)] to be

$$\frac{1}{2k^2}\,[mN(N-1) - \sum_i \sum_j \tau_j n_{ij}(n_{ij}-1)]$$

$$= \frac{N}{2k^2}\left[mN - m - \sum_i \sum_j \left(\frac{\tau_j n_{ij}^2}{N}\right) + \sum_j \left(\frac{\tau_j N_j}{N}\right)\right]$$

$$= \frac{N}{2k^2}\left[\sum_j \left(\frac{\tau_j N_j}{N}\right) - m - \chi^2\right],$$

where, on the last step, we have used the fact that

$$\chi^2 = \sum_i \sum_j \frac{\tau_j (n_{ij} - N\lambda_j)^2}{N} = \sum_i \sum_j \left(\frac{\tau_j n_{ij}^2}{N}\right) - Nm.$$

Thus the condition

$$\chi^2 > \sum \left(\frac{\tau_j N_j}{N}\right) - m \tag{8.19}$$

is equivalent to $(d/dk)\log L(k)$ being negative near $k = \infty$. Since this derivative is positive near $k = 0$, we see by continuity that (8.19) implies the existence of a k_0 at which the derivative vanishes. Now we see from (8.17) that, for all sufficiently large u,

$$H(u) = \int_{0^-}^{u} x\, dF^*(x) - \int_{0^-}^{u} x\, dG^*(u).$$

Using the calculations in the proof of Lemma 8.2, we see that

$$H(u) = \frac{mN^2}{2} - \tfrac{1}{2} \sum_i \sum_j [\tau_j n_{ij}(n_{ij} - 1) + n_{ij}]$$

$$= \frac{N}{2}\left[\sum \left(\frac{\tau_j N_j}{N}\right) - m - \chi^2\right].$$

Thus, if (8.19) fails, then $H(\infty) \geq 0$ and so, by Lemma 8.3, $H(u) \geq 0$ for all $u \geq 0$. But this means that H has no change of sign and so $(d/dk)\log L(k)$ has no zero on $(0, \infty)$. The proof of the theorem is thus complete. ■

Levin and Reeds (1977) have applied their techniques to prove the unimodality of the likelihood function for the negative binomial distribution and for mixtures of Poisson distributions.

8.2. Estimaton of a Mode

While the mode of a distribution is one of the three principal measures of location (the other two being the mean and the median), the problem of estimating the mode received very little attention until the middle of the 1960's. An interesting historical account of the literature on the subject has been given by Sager (1983). Here we will be content with presenting the main results and techniques that have been developed to attack the estimation problem for the mode.

The estimation procedures are roughly of two types: direct and indirect. The direct methods, based on the clustering of observations were proposed by Chernoff (1964) and Dalenius (1965). The indirect methods obtain estimators for the density and derive estimators for the mode as a by-product. Such indirect estimators were proposed by Parzen (1962) in a pioneering paper. A few of the other important papers on the estimation of a univariate mode are by Venter (1967) and Grenander (1965). Generalizations to the multivariate case have been given by Sager (1978, 1979).

Consider first the direct method of estimating a mode. Let $(X_1, \ldots, X_n)$ be a random sample from a density f having a unique mode θ. There are two direct ways of estimating θ.

(i) For $x \in R$ and $\delta > 0$, let $N(x, \delta)$ be the number of sample observations in the interval $[x - \delta, x + \delta]$. An estimator of θ can then be taken to be x_0, where

$$N(x_0, \delta) = \max\{N(x, \delta) : x \in R\}.$$

It is clear that x_0 will depend on δ and we would generally choose δ to depend on n.

(ii) Fix α so that $0 < \alpha < 1$. For $x \in R$, let $\delta(x)$ be the smallest $\delta > 0$ such that the interval $[x - \delta, x + \delta]$ contains at least $n\alpha$ observations. An estimator of θ can then be taken to be x_0 where

$$\delta(x_0) = \inf\{\delta(x) : x \in R\}.$$

Again, the estimator x_0 depends on α and we would generally choose α to depend on n.

Approach (i) has been used by Chernoff (1964) and approach (ii) has been used by Venter (1967). An estimator given by Grenander (1965) is a "weighted" version of Venter's estimator. As already observed, the indirect method of estimating the mode concentrates on estimating the density. In the following, we will present the results of Parzen (1962) and Venter (1967) in some details and make comments about generalizations. The selection of topics is done mainly on the basis of ease of presentation. However, some of the important ideas and techniques will be covered by our presentation.

Let $Y_1 < \cdots < Y_n$ be the order statistics for a random sample drawn from a population with distribution function F. Suppose that F has a density f such that

(a) the support of f is an interval (a, b);
(b) f is continuous; and
(c) f has a unique maximum at $\theta \in (a, b)$.

Let r_n be an integer satisfying $1 \leq r_n \leq (n-1)/2$. Define

$$V_j = Y_{j+r_n} - Y_{j-r_n},$$

where $j = r_n + 1, \ldots, n - r_n$. Let K_n be such that

$$V_{K_n} = \min_j V_j.$$

It is then clear that two reasonable estimators of θ are

$$\theta_{1n} = \frac{(Y_{K_n+r_n} + Y_{K_n-r_n})}{2},$$

and

$$\theta_{2n} = Y_{K_n}.$$

For use in the next theorem we need a condition on the steepness of f around θ. Write

$$\alpha_1(\delta) = \min\{f(x) : |x - \theta| \leq \delta\},$$

$$\alpha_2(\delta) = \max\{f(x) : |x - \theta| \geq 2\delta\},$$

and

$$\alpha(\delta) = \frac{\alpha_1(\delta)}{\alpha_2(\delta)}.$$

The following theorem was proved by Venter (1967).

Theorem 8.3. *Suppose f satisfies the conditions* (a), (b), (c) *above and suppose*

$\alpha(\delta) > 1$ for all small $\delta > 0$. Let $\{r_n\}$ be a sequence of positive integers such that

$$n^{-1}r_n \to 0 \tag{8.20}$$

and, for every $\lambda \in (0,1)$,

$$\sum_{n=1}^{\infty} n\lambda^{r_n} < \infty. \tag{8.21}$$

Then, with probability 1, $\theta_{in} \to \theta$ as $n \to \infty$, where $i = 1, 2$.

Proof. Let Z_i, $i = 1, 2, \ldots$ be independent exponentially distributed random variables with $E(Z_i) = 1$. Write $U_i = F(Y_i)$, $G = F^{-1}$ and $S_i = \sum_1^i Z_j$. Then it is clear that Y_i has the same distribution as $G(U_i)$ and that U_i and S_i/S_{n+1} have identical (beta) distributions. Suppose $p \in (0,1)$ is fixed and let r_n be such that

$$\left[\frac{r_n + 1}{n}\right] \leq p \leq 1 - \frac{r_n}{n}. \tag{8.22}$$

Since $(r_n/n) \to 0$, we see that (8.22) will hold for all large n. Now

$$\begin{aligned} V_{[np]} &= Y_{[np]+r_n} - Y_{[np]-r_n} \\ &\overset{d}{=} G\left[\frac{S_{[np]+r_n}}{S_{n+1}}\right] - G\left[\frac{S_{[np]-r_n}}{S_{n+1}}\right], \end{aligned}$$

where $\overset{d}{=}$ means equality in distribution. By the mean value theorem,

$$V_{[np]} \overset{d}{=} \left\{\frac{S_{[np]+r_n} - S_{[np]-r_n}}{S_{n+1}}\right\} G'[\varphi_n(p)],$$

where

$$\frac{S_{[np]-r_n}}{S_{n+1}} \leq \varphi_n(p) \leq \frac{S_{[np]+r_n}}{S_{n+1}}. \tag{8.23}$$

First we show that $\varphi_n(p) \to p$ almost surely, uniformly in p, by showing that the upper and lower bounds in (8.23) converge uniformly. Observe that

$$\frac{S_{[np]+r_n}}{S_{n+1}} - p = \frac{1}{S_{n+1}}\{T_{n,p} + ([np] - np) + p(n - S_{n+1}) + r_n\},$$

where

$$T_{n,p} = S_{[np]+r_n} - [np] - r_n. \tag{8.24}$$

Since $(S_{n+1}/n) \to 1$ with probability 1 and $(r_n/n) \to 0$, we see that it is sufficient to show that $T_{n,p}/S_{n+1} \to 0$ almost surely, uniformly in p. To prove this, let $\varepsilon > 0$ be fixed and consider

$$B_n = P\left[\sup_p |T_{n,p}| \geq n\varepsilon\right].$$

We see from (8.24) that

$$\begin{aligned} B_n &\leq P[|S_j - j| \geq n\varepsilon, \qquad \text{for some } j = 1, \ldots, n] \\ &\leq \sum_{j=1}^{n} P[S_j - j \geq n\varepsilon] + \sum_{j=1}^{n} P[S_j - j \leq -n\varepsilon]. \end{aligned} \tag{8.25}$$

Now, by Markov's inequality, every random variable W satisfies

$$P[W \geq a] \leq e^{-at} E[e^{tW}], \qquad t > 0,\ a \in R.$$

Therefore, in view of the fact that S_j has the gamma distribution, we have

$$\begin{aligned} P[S_j \geq j + n\varepsilon] &\leq e^{-t(j+n\varepsilon)}(1-t)^{-j}, \qquad 0 < t < 1. \\ &= [w(t)]^{n-j}[\beta(t,\varepsilon)]^{-n}, \end{aligned}$$

where $w(t) = (1-t)e^t$ and $\beta(t, \varepsilon) = (1-t)e^{t(1+\varepsilon)}$. It is easy to check that, near $t = 0$,

$$w(t) \sim 1 - \left(\frac{t^2}{2}\right) \quad \text{and} \quad \beta(t,\varepsilon) \sim 1 + \varepsilon t.$$

Therefore, by choosing t near zero, we can have $w(t) < 1$ and $\beta(t, \varepsilon) > 1$. Further,

$$\begin{aligned} \sum_{j=1}^{n} P(S_j \geq j + n\varepsilon) &\leq \left\{\sum_{j=1}^{n} [w(t)]^{n-j}\right\} \cdot [\beta(t,\varepsilon)]^{-n} \\ &\leq [1 - w(t)]^{-1}[\beta(t,\varepsilon)]^{-n}. \end{aligned}$$

A similar argument applies to the second term in (8.25). We thus see that B_n is bounded above by a term of the form $h_1\lambda_1^n$ where $0 < \lambda_1 < 1$. So $\sum B_n < \infty$. Consequently $[\sup_p T_{n,p}/n] \to 0$ with probability 1 as required. The lower bound for $\varphi_n(p)$ in (8.23) can be treated in the same fashion. We have thus shown that $\varphi_n(p) \to p$ almost surely, uniformly in p.

Now let $q = F(\theta)$, where θ is the mode. Let $\delta > 0$ be suitably small and choose p so that either

$$\frac{r_n + 1}{n} \leq p \leq F(\theta - 3\delta) \tag{8.26}$$

or

$$F(\theta + 3\delta) \leq p \leq 1 - \frac{r_n}{n}. \tag{8.27}$$

For the sake of definiteness we assume that (8.26) holds. Then

$$\frac{V_{[np]}}{V_{[nq]}} = \frac{S_{[np]+r_n} - S_{[np]-r_n}}{S_{[nq]+r_n} - S_{[nq]-r_n}} \cdot \frac{G'[\varphi_n(p)]}{G'[\varphi_n(q)]}. \tag{8.28}$$

It is shown below that the right side of (8.28) is larger than 1. This would yield information about K_n and in turn about the estinators θ_{in}.

Because of the uniform convergence of $\varphi_n(p)$ and $\varphi_n(q)$, there exists an n_0 such that for $n \geq n_0$,

$$\varphi_n(p) \leq F(\theta - 2\delta),$$

and

$$F(\theta - \delta) \leq \varphi_n(q) \leq F(\theta + \delta).$$

Therefore

$$\frac{G'[\varphi_n(p)]}{G'[\varphi_n(q)]} = \frac{f[G(\varphi_n(q))]}{f[G(\varphi_n(p))]} \geq \alpha(\delta) > 1.$$

Thus the second factor on the right side of (8.28) is greater than 1. To show that the first factor is close to 1, let

$$D_{n,p} = S_{[np]+r_n} - S_{[np]-r_n}.$$

We show that $D_{n,p}/(2r_n)$ converges to 1 almost surely, uniformly in p. Observe that

$$\begin{aligned} P\left[\sup_p D_{n,p} > 2r_n(1+\varepsilon)\right] &\leq \sum_{j=r_n+1}^{n-r_n} P[S_{j+r_n} - S_{j-r_n} > 2r_n(1+\varepsilon)] \\ &\leq nP[S_{2r_n} > 2r_n(1+\varepsilon)] \\ &\leq nh_2(t)[\lambda_2(t,\varepsilon)]^{r_n}, \end{aligned}$$

where the last inequality is obtained by using an exponential bound as done earlier in the proof. Again we can arrange λ_2 to be in (0, 1) by choosing t and ε suitably. A similar bound can be derived for $P[\inf_p D_{n,p} < 2r_n(1-\varepsilon)]$. The Borel–Cantelli lemma again shows that $D_{n,p}/(2r_n) \to 1$ almost surely, uniformly in p. We thus see that

$$\frac{V_{[np]}}{V_{[nq]}} > 1 \tag{8.29}$$

with probability 1 for all sufficiently large n. We had assumed above that (8.26) holds. Condition (8.27) can be treated in a similar fasion. Now when we minimize V_j over j, the minimizing subscript was denoted by K_n. If we write $K_n = [np^*]$, then (8.29) assures us that

$$F(\theta - 3\delta) < p^* < F(\theta + 3\delta).$$

In other words

$$F(\theta - 3\delta) < \frac{K_n}{n} < F(\theta + 3\delta).$$

It follows that $(K_n/n) \to q$ with probability 1. Strong convergence of the sample quantiles now implies that Y_{K_n}, $Y_{K_n+r_n}$ and $Y_{K_n-r_n}$ all converge to θ with probability one. The estimators θ_{1n} and θ_{2n} are thus strongly consistent for the mode θ. The proof of Theorem 8.3 is now complete. ■

Venter (1967) also gives some results on the rates of convergence of the estimators θ_{1n} and θ_{2n} considered in Theorem 8.3. As one can expect, the peakedness behavior of f near θ plays an important role in determining the rate of convergence. For example, suppose the ratio $\alpha(\delta)$ (defined earlier) satisfies

$$\alpha(\delta) \geq 1 + \rho\delta^k,$$

where ρ and k are both positive. Then one can take

$$r_n = \begin{cases} An^{2k/(2k+1)}, & k \geq \frac{1}{2} \\ An^k, & k < \frac{1}{2} \end{cases}$$

with $A > 0$. Venter (1967) shows that, for $i = 1, 2$, with probability one,

$$\theta_{in} = \theta + O_p(\beta_n),$$

where

$$\beta_n = \begin{cases} n^{-1/(2k+1)}(\log n)^{1/k}, & k \geq \frac{1}{2} \\ n^{-1/2}(\log n)^{1/k}, & k < \frac{1}{2}. \end{cases}$$

In particular, if $k = 1$, then with $r_n = O(n^{2/3})$, we have $\theta_{in} - \theta = O_p(n^{-1/3} \log n)$.

The estimators θ_{1n}, θ_{2n} considered above used the smallest spacing containing $2r_n$ ordered observations. Instead of giving all the weight to such shortest intervals, Grenander (1965) considered a weighted average of the mid-points of the intervals, with shorter intervals receiving larger weights.

He defined

$$\theta_{3n} = \tfrac{1}{2} \sum_{j=1}^{n-k} w_j(Y_j + Y_{j+k}),$$

where

$$w_j = \frac{(Y_{j+k} - Y_j)^{-p}}{\{\sum_{i=1}^{n-k} (Y_{i+k} - Y_i)^{-p}\}}.$$

Here $1 < p < k < n$. Grenander showed that θ_{3n} is consistent for θ^*, where

$$\theta^* = \frac{\int x f^{p+1}(x)\, dx}{\int f^{p+1}(x)\, dx}.$$

The quantity θ^* is close to the mode θ only if p is large. Thus θ_{3n} is, in general, biased for θ and also inconsistent for θ. Adriano, Gentle and Sposito (1977) have shown that, $\theta^* > \theta$ for a few standard families of distributions. Dalenius (1965) studies estimators of the type θ_{1n}, θ_{2n}, θ_{3n} by employing Monte Carlo methods. It turned out that for large p, the bias of the Grenander estimator is large whereas the standard deviation is small. Sager (1975) has extended Venter's results (including those on rates of convergence) by relaxing the steepness requirement on density. More specifically, Sager allows the density to have a modal interval.

We now proceed to see how density estimators yield estimators of the mode. The important results in this area are due to Parzen (1962). Suppose that F_n is the empirical distribution function for a sample $(X_1, \ldots, X_n)$ drawn from a distribution with a density f. A reasonable estimator of $f(x)$ is

$$f_n^*(x) = \frac{[F_n(x+h) - F_n(x-h)]}{2h},$$

where $h > 0$. If $w(y) = \frac{1}{2}$ for $-1 < y \leq 1$ and $w(y) = 0$ elsewhere, then $f_n^*(x)$ can be written as

$$f_n^*(x) = \frac{1}{h} \int_{-\infty}^{\infty} w\left[\frac{(x-y)}{h}\right] dF_n(y)$$
$$= \frac{1}{nh} \sum_{j=1}^{n} w\left[\frac{(x-X_j)}{h}\right].$$

Now $f_n^*(x)$ is seen to be just one member of a whole class of estimators of the form

$$f_n(x) = \frac{1}{h} \int_{-\infty}^{\infty} K\left[\frac{(x-y)}{h}\right] dF_n(y), \tag{8.30}$$

where K is a suitable kernel. The notations $f_n^*(x)$ and $f_n(x)$ do not bring out the fact that these estimators also depend on h. But, in practice, we will choose h to depend on n and then study the behavior of these estimators as $n \to \infty$. We will therefore write $h(n)$ in place of h when necessary. It is clear that

$$E[f_n(x)] = \frac{1}{h} \int_{-\infty}^{\infty} K\left[\frac{(x-y)}{h}\right] f(y)\, dy.$$

The next theorem proves the asymptotic unbiasedness of $f_n(x)$ when the kernel K satisfies some conditions and $h(n) \to 0$ as $n \to \infty$. The theorem is essentially given by Bochner (1955).

Theorem 8.4. *Let g be a measurable function on R which is Lebesgue integrable over R. Let K be a measurable kernel on R which is Lebesgue integrable over R. Assume that $yK(y) \to 0$ as $|y| \to \infty$. Finally, let $h = h(n) \to 0$ as $n \to \infty$.* Define

$$g_n(x) = \frac{1}{h(n)} \int_{-\infty}^{\infty} K\left[\frac{y}{h(n)}\right] g(x-y)\, dy.$$

Then, at every continuity point x of g,

$$g_n(x) \to g(x) \int_{-\infty}^{\infty} K(y)\, dy,$$

as $n \to \infty$. Morevoer, if g is uniformly continuous on R then the above convergence is also uniform in x.

Proof. Since $K \cdot g$ is the same as $(cK) \cdot (g/c)$, we can rescale K so that

$$\int_{-\infty}^{\infty} K(y)\, dy = 1. \tag{8.31}$$

Let $\varepsilon > 0$ be fixed and let x be a continuity point of g. We can find $\delta > 0$ such that

$$|y| < \delta \Rightarrow |g(x-y) - g(x)| \le \varepsilon.$$

Since $yK(y) \to 0$ as $|y| \to \infty$ and K is integrable over R, we can find M such that

$$|y| \ge M \Rightarrow |yK(y)| \le \varepsilon\delta$$

and

$$\int_{|y| \ge M} |K(y)|\, dy \le \varepsilon.$$

Next, having chosen δ and M, we can find N so that

$$n \geq N \Rightarrow h(n) \leq \frac{\delta}{M}.$$

Finally, let A and B, respectively, be the integral of $|g|$ and $|K|$ over R. Now, in view of (8.31), we have

$$g(x) = g(x) \int_{-\infty}^{\infty} K(y)\, dy = \frac{g(x)}{h(n)} \int_{-\infty}^{\infty} K\left[\frac{y}{h(n)}\right] dy.$$

Therefore,

$$\begin{aligned} g_n(x) - g(x) = [h(n)]^{-1} &\int_{|y|<\delta} K\left[\frac{y}{h(n)}\right][g(x-y) - g(x)]\, dy \\ &+ [h(n)]^{-1} \int_{|y|\geq\delta} K\left[\frac{y}{h(n)}\right] g(x-y)\, dy \\ &- g(x)[h(n)]^{-1} \int_{|y|\geq\delta} K\left[\frac{y}{h(n)}\right] dy \\ &= c_1 + c_2 + c_3, \text{ say.} \end{aligned}$$

Assume that $n \geq N$. Note that

$$\begin{aligned} |y| \geq \delta \Rightarrow \left[\frac{|y|}{h(n)}\right] \geq \left[\frac{\delta}{h(n)}\right] \geq M \\ \Rightarrow [h(n)]^{-1} \left| K\left[\frac{y}{h(n)}\right]\right| \leq \left[\frac{\varepsilon\delta}{|y|}\right] \leq \varepsilon. \end{aligned}$$

Therefore,

$$|c_2| \leq \varepsilon \int_{-\infty}^{\infty} |g(x-y)|\, dy = \varepsilon A,$$

and

$$\begin{aligned} |c_3| &\leq |g(x)| \int_{|y|\geq[\delta/h(n)]} |K(y)|\, dy \\ &\leq |g(x)| \int_{|y|\geq M} |K(y)|\, dy \leq |g(x)|\varepsilon. \end{aligned}$$

Finally,

$$|c_1| \leq \varepsilon \int_{-\infty}^{\infty} |K(y)|\, dy = \varepsilon B.$$

Thus, if $n \geq N$, then

$$|g_n(x) - g(x)| \leq \varepsilon[B + A + |g(x)|].$$

It follows that $g_n(x) \to g(x)$ as $n \to \infty$. Suppose now that g is uniformly continuous on R. Then δ and hence N can be chosen to be independent of x. Further, the integrability of g implies its boundedness. That is, there is a constant D such that $|g(x)| \leq D$ for all x. Thus, for $n \geq N$,

$$\sup_x |g_n(x) - g(x)| \leq \varepsilon[B + A + D].$$

It follows that $g_n(x) \to g(x)$ uniformly in x as $n \to \infty$. The theorem is thus proved. ■

Corollary. *The density estimator $f_n(x)$ given by* (8.30) *is asymptotically unbiased for $f(x)$ whenever x is a continuity point of f if $h = h(n)$ is chosen so that*

$$h(n) \to 0 \qquad \text{as } n \to \infty \tag{8.32}$$

and the kernel K is chosen so that

$$\int_{-\infty}^{\infty} |K(y)|\, dy < \infty, \tag{8.33}$$

$$\int_{-\infty}^{\infty} K(y)\, dy = 1, \tag{8.34}$$

and

$$yK(y) \to 0 \qquad \text{as } |y| \to \infty. \tag{8.35}$$

To simplify writing, we use the term *weighting function* to denote a *bounded, even* function K satisfying the conditions (8.33) and (8.35). The boundedness assumption implies, for instance, that K^2 is also a weighting function when K is a weighting function. The condition that K is even is convenient in practice although it is not quite necessary for the calculations to be valid.

Parzen (1962) has shown that Theorem 8.4 can also be used to obtain the limit of a suitably normed variance of $f_n(x)$. To see this, write $q(y) = h^{-1}K[(x-y)/h]$. Then (8.30) shows that

$$f_n(x) = n^{-1} \sum_{j=1}^{n} q(X_j).$$

Therefore

$$\text{Var}[f_n(x)] = n^{-1}\, \text{Var}[q(X)], \tag{8.36}$$

where X is a random variable with density f.

Theorem 8.5. *Suppose that K is a weighting function such that* (8.34) *holds. Assume that $h=h(n)\to 0$ as $n\to\infty$. Then*

$$nh \operatorname{Var}[f_n(x)] \to f(x) \int_{-\infty}^{\infty} K^2(y)\, dy$$

at all continuity points x of f.

Proof. From (8.36), we get

$$\begin{aligned} nh \operatorname{Var}[f_n(x)] &= h \operatorname{Var}[q(X)] \\ &= hE[\{q(X)\}^2] - h\{E[q(X)]\}^2. \end{aligned}$$

Using the definition of $q(y)$ and Theorem 8.4, we see that

$$\begin{aligned} hE[\{q(X)\}^2] &= h^{-1} \int_{-\infty}^{\infty} K^2\left[\frac{(x-y)}{h}\right] f(y)\, dy \\ &\to f(x) \int_{-\infty}^{\infty} K^2(y)\, dy. \end{aligned}$$

Since

$$\begin{aligned} E[q(X)] &= h^{-1} \int K\left[\frac{(x-y)}{h}\right] f(y)\, dy \\ &\to f(x), \end{aligned}$$

we also see that $h\{E[q(X)]\}^2 \to 0$ as $n\to\infty$. The theorem is thus proved. ■

Corollary. *Suppose K is a weighting function satisfying condition* (8.34). *Suppose $h=h(n)$ is such that $h(n)\to 0$ and $nh(n)\to\infty$. Then* $\operatorname{Var}[f_n(x)]\to 0$ *and, consequently,* $f_n(x)$ is consistent in quadratic mean for $f(x)$.

Note. Since $f_n(x)$ is the normed sum of a set of independent and identically distributed random variables, the standard central limit theorem for such random variables can be used to prove the asymptotic normality of $f_n(x)$ under the conditions of the preceding corollary. See Parzen (1962) for details.

Under suitable hypotheses, Parzen (1962) has proved the uniform consistency of the estimator $f_n(x)$ and has thereby obtained a consistent estimator of the populaton mode. As before, let K be a weighting function and write

$$k(u) = \int_{-\infty}^{\infty} e^{-iuy} K(y)\, dy. \tag{8.37}$$

Let φ_n be the empirical characteristic function given by

$$\varphi_n(u) = \int_{-\infty}^{\infty} e^{iux}\, dF_n(x) = n^{-1} \sum_{j=1}^{n} e^{iuX_k}.$$

Assume that $|k(u)|$ is integrable over R. Then, using the inversion formula, one can easily obtain

$$f_n(x) = (2\pi)^{-1} \int_{-\infty}^{\infty} e^{-iux}\, \varphi_n(u) k(uh)\, du. \tag{8.38}$$

We see immediately from (8.38) that $f_n(x)$ is continuous on R and tends to zero as $|x| \to \infty$. Therefore, $f_n(x)$ attains its maximum value at some point θ_n. We call θ_n the sample mode.

Theorem 8.6. *Suppose f is a uniformly continuous probability density on R. Suppose K is a weighting function satisfying* (8.34). *Assume that the Fourier transform $k(u)$ given by* (8.37) *is absolutely integrable over R. Let $h = h(n)$ be such that $h(n) \to 0$ and $nh^2(n) \to \infty$. Then*

$$\sup_x |f_n(x) - f(x)| \to 0 \text{ in probability.}$$

Moreover, if $\{\theta_n\}$ is the sequence of sample modes and f has a unique mode θ, then $\theta_n \to \theta$ in probability.

Proof. To prove the first assertion, it is sufficient to prove that

$$E\left[\sup_x |f_n(x) - f(x)|\right] \to 0. \tag{8.39}$$

But, by Theorem 8.4, $E[f_n(x)] \to f(x)$ uniformly in x. Therefore (8.39) would follow if we prove that

$$E\left[\sup_x |f_n(x) - E[f_n(x)]|\right] \to 0. \tag{8.40}$$

By (8.38),

$$\sup_x |f_n(x) - E[f_n(x)]| \leq (2\pi)^{-1} \int_{-\infty}^{\infty} |k(hu)|\, |\varphi_n(u) - E[\varphi_n(u)]|\, du.$$

But, by Minkowski's inequality,

$$E|\varphi_n(u) - E[\varphi_n(u)]| \leq \{\mathrm{Var}[\varphi_n(u)]\}^{1/2} \leq n^{-1/2}.$$

Therefore, the left side of (8.40) does not exceed

$$(2\pi)^{-1}n^{-1/2}\int_{-\infty}^{\infty}|k(hu)|\,du=(2\pi)^{-1}(n^{1/2}h)^{-1}\int_{-\infty}^{\infty}|k(u)|\,du.$$

The last quantity tends to zero as $n\to\infty$. Thus the first assertion is proved.

To prove the second assertion, observe that the uniform continuity and integrability of f show that $f(x)\to 0$ as $|x|\to\infty$. Now the uniqueness of the mode θ can be used to show easily that

$$f(x_n)\to f(\theta)\Rightarrow x_n\to\theta.$$

Therefore, the assertion that $\theta_n\to\theta$ in probability will follow as soon as we show that

$$f(\theta_n)\to f(\theta)\text{ in probability.}\tag{8.41}$$

Clearly,

$$|f(\theta_n)-f_n(\theta_n)|\le\sup_x\,|f(x)-f_n(x)|.$$

Further

$$\begin{aligned}|f_n(\theta_n)-f(\theta)|&=\left|\sup_x f_n(x)-\sup_x f(x)\right|\\&\le\sup_x|f_n(x)-f(x)|.\end{aligned}$$

Adding the last two inequalities, we get

$$|f(\theta_n)-f(\theta)|\le 2\,\sup_x|f_n(x)-f(x)|.\tag{8.42}$$

The right side of (8.42) tends to zero in probability by the first assertion. Therefore, (8.41) is established and the proof of the theorem is complete. ■

Additional results proved by Parzen (1962) include conditions for the asymptotic normality of the consistent sequence $\{\theta_n\}$ given by Theorem 8.6.

Methods of estimating a multivariate mode were studied only recently. Here also one may *either* choose a bounded convex set and find a location where it traps the maximum number of sample observations *or* find a convex set of a given shape but having the smallest volume and fetching a given proportion of the observations. Sager (1978 and 1979) has followed the second approach and has obtained results on consistency and rates of convergence.

This approach is clearly a natural one in case the underlying density f is convex unimodal (that is, $\{x: f(x) \geq c\}$ is convex for every c). An important concept concerning a multivariate density is that of an isopleth, which is the contour where the density is a constant. For a convex unimodal density, an isopleth is the boundary of a convex set. Sager (1979) gives an iterative procedure for estimating a mode and also develops procedures for estimating isopleths. He cites interesting applications to air pollution data.

We mention in passing the work on estimating the density and the mode which uses the concept of a σ-lattice. Suppose Ω is a nonempty set. A family $\mathscr{F}$ of subsets of Ω is called a *σ-lattice* if $\mathscr{F}$ is closed under countable unions and countable intersections and $\mathscr{F} \supset \{\Omega, \varnothing\}$. We note that a σ-lattice might not be closed under complementation. A nested family of intervals (or convex sets) is an example of a σ-lattice. This example is clearly suitable for studying unimodal densities. Brunk (1965) introduced the concept of a conditional expectation given a σ-lattice. Robertson (1967) and Wegman (1969, 1970a, 1970b) applied Brunk's technique to estimate unimodal densities. These authors prove results on consistency by utilizing the fact that the conditional expectation given an apropriate σ-lattice gives rise to maximum-likelihood-type estimators. The details are somewhat technical and are not directly related to the theme of this monograph.

8.3. Monotonicity of the Power Functions of Certain Multivariate Tests

Anderson's theorem and its generalizations have been used to prove the monotonicity of the power functions of several multivariate tests. For instance, consider the multivariate analysis of variance (MANOVA) problem. Here one assumes that the observations have a multivariate normal distribution. This distribution is central convex unimodal. Suppose that the acceptance region of the test under consideration is centrally symmetric and convex. Suppose further that under every fixed alternative hypothesis, the distribution corresponds to a translation of the distribution under the null hypothesis. Then Anderson's theorem immediately implies that the test is unbiased. Indeed, one can show that the power function is nonincreasing in suitable directions.

The approach described in the preceding paragraph has been considered by Das Gupta, Anderson and Mudholkar (1964) for testing the multivariate linear hypothesis. In another paper, Anderson and Das Gupta (1964) have considered tests of independence. These results were refined by Eaton and Perlman (1974) who showed that the power functions of certain tests are

Schur convex. Using the same technique, Cohen and Strawderman (1971) proved the monotonicity of the power functons of certain tests for testing the homogeneity of variances. In the following, we present some of these results emphasizing the role played by multivariate unimodality.

Consider first the multivariate linear hypothesis testing problem. In the canonical form this problem concerns a random matrix $\mathbf{X} = (\mathbf{X}_1, \mathbf{X}_2, \mathbf{X}_3)$, where

(i) $\mathbf{X}_1$, $\mathbf{X}_2$ and $\mathbf{X}_3$ have orders $p \times s$, $p \times (n - r)$ and $p \times (r - s)$, respectively;
(ii) the columns of $\mathbf{X}$ are independent p-variate normal random vectors with a common covariance matrix $\boldsymbol{\Sigma}$; and
(iii) $E(\mathbf{X}_1) = \boldsymbol{\Lambda}$, $E(\mathbf{X}_2) = \mathbf{0}$ and $E(\mathbf{X}_3) = \boldsymbol{\Gamma}$.

The null hypothesis to be tested is H_0: $\boldsymbol{\Lambda} = \mathbf{0}$ against all alternatives.

It is easy to see that the above problem remains invariant under the transformations which take $(\mathbf{X}_1, \mathbf{X}_2, \mathbf{X}_3)$ to $(\mathbf{W}\mathbf{X}_1\mathbf{F}_1, \mathbf{W}\mathbf{X}_2\mathbf{F}_2, \mathbf{W}\mathbf{X}_3\mathbf{F}_3 + G)$, where $\mathbf{W}$ is nonsingular, G is an arbitrary matrix of constants and $\mathbf{F}_1$, $\mathbf{F}_2$, $\mathbf{F}_3$ are orthogonal matrices. It can be shown that the maximal invariant under these transformations is the set of characteristic roots $c_1 \geq c_2 \geq \cdots \geq c_p$ of the matrix $\mathbf{S}_n\mathbf{S}_e^{-1}$, where

$$\mathbf{S}_n = \mathbf{X}_1\mathbf{X}_1' \quad \text{and} \quad \mathbf{S}_e = \mathbf{X}_2\mathbf{X}_2'.$$

Here $\mathbf{S}_n$ represents the "variation due to the hypothesis H_0" and $\mathbf{S}_e$ represents the "variation due to error." An invariant test is based on the maximal invariant $(c_1, c_2, \ldots, c_p)$ and the power function of such a test depends on the parameter $(\delta_1, \ldots, \delta_t)$, where $\delta_1 \geq \delta_2 \geq \cdots \geq \delta_t$ are the characteristic roots of $\boldsymbol{\Lambda}\boldsymbol{\Lambda}'\boldsymbol{\Sigma}^{-1}$ and $t = \min(p, s)$.

The following theorem is an easy consequence of Anderson's theorem.

Theorem 8.7. *Suppose* $\mathbf{X}_1, \ldots, \mathbf{X}_s$ *are independent p-variate random vectors such that* $\mathbf{X}_i$ *is* $N(k_i\boldsymbol{\mu}, \boldsymbol{\Sigma}_i)$. *Let* $\mathbf{Y}$ *be a random matrix independent of* $(\mathbf{X}_1, \ldots, \mathbf{X}_s)$. *Let E be a region in the space for* $(\mathbf{X}_1, \ldots, \mathbf{X}_s, \mathbf{Y})$ *such that for every* $i = 1, \ldots, s$ *and for every choice of values for* $\mathbf{x}_j$, $j \neq i$ *and of* $\mathbf{y}$, *E is convex and centrally symmetric in* $\mathbf{x}_i$. *Then* $P[(\mathbf{X}_1, \ldots, \mathbf{X}_s, \mathbf{Y}) \in E]$ *is nonincreasing in each* $k_i \geq 0$.

It will be shown below that the acceptance regions of the following three standard tests satisfy the conditions (on E) in Theorem 8.7.

(i) The likelihood ratio test whose acceptance region is of the form $\prod_{i=1}^{p} (1 + c_i) \leq b_1$.

(ii) The Lawley–Hotelling trace test whose acceptance region is of the form $\sum_{i=1}^{p} c_i \leq b_2$.
(iii) Roy's maximum root test whose acceptance region is of the form $c_1 \leq b_3$.

Here b_1, b_2, b_3 are constants which depend on the levels of significance.

To apply Theorem 8.7 to the acceptance regions of the above tests, we note first that the characteristic roots of $\mathbf{S}_n\mathbf{S}_e^{-1}$ are the same as the characteristic roots of $(\mathbf{UU}')(\mathbf{VV}')^{-1}$, with $\mathbf{U} = \mathbf{BX}_1\mathbf{F}_1$ and $\mathbf{V} = \mathbf{BX}_2\mathbf{F}_2$, where $\mathbf{B}$ is nonsingular and $\mathbf{F}_1, \mathbf{F}_2$ are orthogonal. Now $\mathbf{B}$, $\mathbf{F}_1$ and $\mathbf{F}_2$ can be chosen so that the joint density of $\mathbf{U}$ and $\mathbf{V}$ is given by

$$C \exp\left[-\tfrac{1}{2}\left\{\mathrm{tr}(\mathbf{vv}') + \sum_{i=1}^{t} (u_{ii} - \theta_i)^2 + \sum_{i=t+1}^{p} u_{ii}^2 + \sum_{i=1}^{p} \sum_{\substack{j=1 \\ i \neq j}}^{s} u_{ij}^2\right\}\right]. \tag{8.43}$$

In view of (8.43), the hypothesis H_0 becomes $\theta_1 = \cdots = \theta_t = 0$. Suppose $\mathbf{U}_j$ denotes the jth column vector of $\mathbf{U}$. Then (8.43) shows that

(a) $\mathbf{U}_1, \ldots, \mathbf{U}_s, \mathbf{V}$ are mutually independent,
(b) $\mathbf{U}_j$ *is* $N(\mathbf{0}, I)$ for $t < j \leq s$,
(c) if $\mathbf{e}_j$ denotes the jth coordinate vector in R^p, then $\mathbf{U}_j$ is $N(\theta_j\mathbf{e}_j, I)$ for $1 \leq j \leq t$.

Thus Theorem 8.7 can be applied to the set $\{U_1, \ldots, \mathbf{U}_t, \mathbf{V}\}$ and any test whose acceptance region is convex and centrally symmetric in each $\mathbf{U}_i$ (whenever the other $\mathbf{U}_j$ and $\mathbf{V}$ are fixed) will be such that its power function will be increasing in each $|\theta_i|$. To complete the proof that the three tests (i), (ii), (iii) mentioned above have acceptance regions satisfying the required conditions, we need two lemmas.

Lemma 8.4. *Let* $\mathbf{A}$ *be an* $m \times n$ *matrix with columns* $\mathbf{a}_1, \ldots, \mathbf{a}_n$. *Let* $W_k(\mathbf{A})$ *denote the sum of all the k-fold products of the characteristic roots of* $(\mathbf{AA}' + \mathbf{I}_m)$. *Then for* $\alpha_k \geq 0$, $k = 1, \ldots, m$, *the region defined by* $\sum_{k=1}^{m} \alpha_k W_k(\mathbf{A}) \leq \beta$ *is convex in each* $\mathbf{a}_i$ *when the other* $\mathbf{a}_j$ *are fixed.*

Proof. It is well known that $W_k(\mathbf{A})$ is the sum of all the $k \times k$ principal minors of $\mathbf{AA}' + \mathbf{I}_m$. Now a typical $k \times k$ principal minor of $\mathbf{AA}' + \mathbf{I}_m$ has the form

$$\left|(\mathbf{b}, \mathbf{C})\begin{pmatrix}\mathbf{b}' \\ \mathbf{C}'\end{pmatrix} + \mathbf{I}_k\right| = \left|\mathbf{bb}' + \mathbf{CC}' + \mathbf{I}_k\right|,$$

where $\mathbf{b}$ is a column vector. If we write $\mathbf{B} = \mathbf{CC}' + \mathbf{I}_k$, then the above minor becomes

$$|\mathbf{bb}' + \mathbf{B}| = (\mathbf{b}'\mathbf{B}^{-1}\mathbf{b} + 1)|\mathbf{B}|,$$

which is a positive definite quadratic form in $\mathbf{b}$ plus a constant. It follows that $W_k(\mathbf{A})$ is also a positive definite quadratic form in $\mathbf{a}_i$ plus a constant (when the other $\mathbf{a}_j$ are fixed). The same conclusion holds for $\sum \alpha_k W_k(\mathbf{A})$ because each α_k is nonnegative. Therefore, the region defined by $\sum \alpha_k W_k(\mathbf{A}) \leq \beta$ ellipsoidal and hence convex in each $\mathbf{a}_i$. ■

In the next lemma the maximum characteristic root of a square matrix $\mathbf{A}$ is denoted by $\mathrm{ch}_1(\mathbf{A})$.

Lemma 8.5. *Let $\mathbf{B}$ be a fixed symmetric nonnegative definite matrix of order $n \times n$. Let $\mathscr{A}$ denote the set of all matrices of order $n \times m$. Let $E = \{\mathbf{A} \in \mathscr{A}: \mathrm{ch}_1(\mathbf{AA}'\mathbf{B}) \leq \beta\}$, where β is a specified number. Then E is a convex subset of $\mathscr{A}$.*

Proof. Write $\mathbf{B} = \mathbf{TT}'$, where $\mathbf{T}$ has order $n \times n$. Then $\mathrm{ch}_1(\mathbf{AA}'\mathbf{B}) = \mathrm{ch}_1[(\mathbf{TA})(\mathbf{TA})']$. Let $\mathbf{A}_1 \in E$, $\mathbf{A}_2 \in E$, $\lambda \in (0, 1)$ and $\mathbf{A} = \lambda\mathbf{A}_1 + (1 - \lambda)\mathbf{A}_2$. Then

$$\begin{aligned}\mathbf{x}'(\mathbf{TA})(\mathbf{TA})'\mathbf{x} = {} & \lambda^2\mathbf{x}'\mathbf{TA}_1\mathbf{A}_1'\mathbf{T}'\mathbf{x} + (1 - \lambda)^2\mathbf{x}'\mathbf{TA}_2\mathbf{A}_2'\mathbf{T}'\mathbf{x} \\ & + \lambda(1 - \lambda)[\mathbf{x}'\mathbf{TA}_1\mathbf{A}_2'\mathbf{T}'\mathbf{x} + \mathbf{x}'\mathbf{TA}_2\mathbf{A}_1'\mathbf{T}'\mathbf{x}].\end{aligned}$$

Now, by the Cauchy–Schwarz inequality,

$$\mathbf{x}'\mathbf{TA}_1\mathbf{A}_2'\mathbf{T}'\mathbf{x} \leq (\mathbf{x}'\mathbf{TA}_1\mathbf{A}_1'\mathbf{T}'\mathbf{x})^{1/2}(\mathbf{x}'\mathbf{TA}_2\mathbf{A}_2'\mathbf{T}'\mathbf{x})^{1/2}.$$

Therefore

$$\begin{aligned}\mathbf{x}'\mathbf{TAA}'\mathbf{T}'\mathbf{x} & \leq [\lambda(\mathbf{x}'\mathbf{TA}_1\mathbf{A}_1'\mathbf{T}'\mathbf{x})^{1/2} + (1 - \lambda)(\mathbf{x}'\mathbf{TA}_2\mathbf{A}_2'\mathbf{T}'\mathbf{x})^{1/2}]^2. \\ & \leq [\lambda(\beta\mathbf{x}'\mathbf{x})^{1/2} + (1 - \lambda)(\beta\mathbf{x}'\mathbf{x})^{1/2}]^2 \\ & = \beta\mathbf{x}'\mathbf{x}.\end{aligned}$$

It follows that $\mathbf{A} \in E$. The lemma is thus proved. ■

We are now ready to prove the monotonicity of the power functions of the three tests mentioned earlier. Recall that $c_1 \geq c_2 \geq \cdots \geq c_p$ are the characteristic roots of $(\mathbf{UU}')(\mathbf{VV}')^{-1}$, where $(\mathbf{U}, \mathbf{V})$ has the density given by (8.43).

(i) *The likelihood ratio test*: This test has the acceptance region $\prod (1 + c_i) \leq b$. Now $(1 + c_i)$ are the characteristic roots of $(\mathbf{UU}')(\mathbf{VV}')^{-1} + \mathbf{I}_p$. After a simple linear transformation, we see that the acceptance region is of the form $W_p \leq b$. This region is thus convex in each $\mathbf{U}_i$ when the other $\mathbf{U}_j$ and $\mathbf{V}$ are fixed. Therefore, the power function is increasing in each $|\theta_i|$ by Theorem 8.7.

(ii) *The Lawley-Hotelling trace test*: This test has the acceptance region $\sum c_i \leq b$, which is equivalent to $\sum(1 + c_i) \leq b + p$. Again, in terms of Lemma 8.4, the acceptance region has the form $W_1 \leq \beta$. Therefore, the power function is again increasing in each $|\theta_i|$.

(iii) *Roy's maximum root test*: This test has the acceptance region $c_1 \leq b$. Lemma 8.5 shows that the acceptance region is convex in each $\mathbf{U}_j$. Theorem 8.7 then implies that the power function is increasing in each $|\theta_i|$.

Some of the above results can be improved if the acceptance regions satisfy convexity conditions stronger than those required by Theorem 8.7. Such a result was obtained by Eaton and Perlman (1974) and Das Gupta (1976b). To describe this result, consider again the canonical form (8.43). Suppose the acceptance region E is convex in the $(\mathbf{U}, \mathbf{V})$ space. This convexity condition is much stronger than the coordinatewise convexity condition assumed in Theorem 8.7. Let $\rho(E, \boldsymbol{\theta})$ be the probability of Type II error associated with the acceptance region E. Now the joint density of $(\mathbf{U}, \mathbf{V})$ is logconcave and also centrally symmetric when $\boldsymbol{\theta} = \mathbf{0}$. Also $\boldsymbol{\theta}$ is a location parameter. Therefore $\rho(E, \boldsymbol{\theta})$ is of the form $\rho(E - \boldsymbol{\theta}, \mathbf{0})$. Now if E is also invariant under a suitable group G of transformations, then Mudholkar's theorem (Theorem 3.3) applies and if $C(\boldsymbol{\theta}_0)$ denotes the convex hull of the G-orbit of $\boldsymbol{\theta}_0$, then we can conclude that

$$\boldsymbol{\theta} \in C(\theta_0) \Rightarrow \rho(E, \boldsymbol{\theta}) \geq \rho(E, \theta_0).$$

Schwartz (1967) showed that the acceptance region of the Lawley–Hotelling and Roy tests are convex. Using this result and the above-mentioned argument, Eaton and Perlman (1974) have shown that the power functions of these two tests are Schur convex in $\boldsymbol{\lambda} = (\lambda_1, \ldots, \lambda_t)$, where $\lambda_i = |\theta_i|$.

Theorem 8.7 can be applied to prove the monotonicity of the power functions of some additional tests. Anderson and Das Gupta (1964) have established such a monotonicity property for certain tests of independence between two sets of normal random vectors. Let $\mathbf{Z}$ be a random matrix of order $(p + q) \times (n + 1)$ such that the columns $\mathbf{z}_\alpha$ of $\mathbf{Z}$ are independent observations from a $(p + q)$-variate normal distribution $N(\boldsymbol{\xi}, \boldsymbol{\Sigma})$. The matrix $\boldsymbol{\Sigma}$ can be partitioned as

$$\boldsymbol{\Sigma} = \begin{bmatrix} \boldsymbol{\Sigma}_{11}, & \boldsymbol{\Sigma}_{12} \\ \boldsymbol{\Sigma}_{21}, & \boldsymbol{\Sigma}_{22} \end{bmatrix}.$$

The hypothesis of independence $H_0: \boldsymbol{\Sigma}_{12} = \mathbf{0}$ is to be tested against all alternatives. The sample covariance matrix $\mathbf{S}$ can be partitioned in the same way as $\boldsymbol{\Sigma}$. The problem is invariant under the transformations

$$\mathbf{z}_\alpha \to \mathbf{z}_\alpha + \mathbf{b} \quad \text{and} \quad \mathbf{Z} \to \begin{pmatrix} \mathbf{B}_1 & \mathbf{0} \\ \mathbf{0} & \mathbf{B}_2 \end{pmatrix} \mathbf{Z}\mathbf{F},$$

where $\mathbf{b}$ is an arbitrary $(p+q)$-vector, $\mathbf{B}_1$ and $\mathbf{B}_2$ are nonsingular matrices of order $p \times p$ and $q \times q$ respectively and $\mathbf{F}$ is an orthogonal matrix. Assume that $p \le q$. Then the maximal invariant statistic is $(c_1^2, \ldots, c_p^2)$, where $c_1^2 \ge c_2^2 \ge \cdots \ge c_p^2$ are the characteristic roots of $S_{11}^{-1} S_{12} S_{22}^{-1} S_{21}$. The c's are the sample canonical correlation coefficients. The maximal invariant in the parameter space is $(\rho_1^2, \ldots, \rho_p^2)$, where $\rho_1^2 \ge \cdots \ge \rho_p^2$ are the characteristic roots of $\Sigma_{11}^{-1}\Sigma_{12}\Sigma_{22}^{-1}\Sigma_{21}$. Again, the ρ's are the population canonical correlations. As in the case of the MANOVA problem, the present problem can be put in canonical form. Consider two random matrices $\mathbf{X}$ and $\mathbf{Y}$ or order $p \times n$ and $q \times n$ respectively written as follows.

$$\mathbf{X} = (X_{i\alpha}), \qquad \mathbf{Y} = (Y_{j\alpha}),$$

and satisfying the following conditions.

(a) The entire set of $(p+q)n$ random variables have a multivariate normal distribution.
(b) $E(X_{i\alpha}) = E(Y_{j\alpha}) = 0$ for all α, i, j.
(c) $\operatorname{Var}(X_{i\alpha}) = \operatorname{Var}(Y_{i\alpha}) = (1 - \rho_i^2)$ for all α and for $i = 1, \ldots, p$.
(d) $\operatorname{Var}(Y_{j\alpha}) = 1$ for all α and for $j > p$.
(e) $\operatorname{Cov}(X_{i\alpha}, Y_{i\alpha}) = \rho_i$ for all α and for $i = 1, \ldots, p$.
(f) All the remaining correlations amongst variables not mentioned in (e) vanish.

The hypothesis H_0 reduces to $\rho_1 = \cdots = \rho_p = 0$. Further, the distribution of the characteristic roots of $S_{11}^{-1} S_{12} S_{22}^{-1} S_{21}$ is the same as the distribution of the characteristic roots of $(XX')^{-1}(XY')(YY')^{-1}(YX')$, Now an analysis similar to the one involving Theorem 8.7 and Lemmas 8.4 and 8.5 show that invariant tests whose acceptance regions are suitably convex will have power functions which are monotonic in each ρ_i. We omit the details but mention that two standard tests having such a monotonicity property are:

(i) The likelihood ratio test whose acceptance region is $\prod_{i=1}^{p} (1 - c_i^2) \ge b_1$; and
(ii) Roy's maximum root test whose acceptance region is $c_1 \le b_2$.

Note. Perlman and Olkin (1980) have given a unified approach to derive the monotonicity properties of the tests in MANOVA problems. Their technique uses the FKG and related inequalities, which are outside the scope of this monograph.

As a final example of the application of unimodality to proving unbiasedness of tests we consider the results of Cohen and Strawderman (1971) regarding tests for homogeneity of variances. Let $s_i^2, i = 1, \ldots, k$ be the sample variances from k independent random samples of size $(n+1)$ drawn from normal populations with unknown means and unknown variances σ_i^2. Assume $k \geq 3$. The hypothesis to be tested is:

$$H_0 : \sigma_1 = \sigma_2 = \cdots = \sigma_k$$

against all alternatives. A two-parameter family of test statistics for this problem was defined by Laue (1965). To describe the family, let

$$M(t) = [k^{-1} \textstyle\sum_{j=1}^k s_j^2]^{1/t},$$

and

$$R(\lambda, \eta) = \frac{M(\lambda)}{M(\eta)}. \tag{8.44}$$

Then Laue's family has as members the statistics $T(\lambda, \eta)$, where

$$T(\lambda, \eta) = \left[\frac{kn}{(\lambda - \eta)}\right] \log R(\lambda, \eta). \tag{8.45}$$

Many known tests of H_0 are included in this family. For instance the likelihood ratio test statistic is equivalent to

$$\frac{(k^{-1} \sum_{i=1}^k s_i^2)}{(\prod_{i=1}^k s_i^2)^{1/k}},$$

which equals $R(1, 0)$. Thus, the likelihood ratio test statistic is equivalent to $T(1, 0)$. Using Mudholkar's results (1966), Cohen and Strawderman (1971) proved that the tests based on $T(\lambda, \eta)$ are unbiased if $\lambda \geq 0$ and $\eta \leq 0$. Since $T(\lambda, \eta) = T(\eta, \lambda)$, unbiasedness is also obtained for $\lambda \leq 0$ and $\eta \geq 0$.

The proof of unbiasedness of the tests under consideration is simplified by going to a location parameter situaton. Let $\mathbf{X}$ be a $k \times 1$ random vector with density $f(\mathbf{x} - \boldsymbol{\theta})$, where $\boldsymbol{\theta}$ is a location parameter and f satisfies the following conditions:

(i) f is permutation symmetric in its k arguments, and
(ii) the set $\{\mathbf{x} : f(\mathbf{x}) \geq c\}$ is convex for each c. (This set may not be centrally symmetric, however.)

Suppose $E \subset R^k$ satisfies the following conditions.

(a) E is convex,
(b) E is permutation symmetric, and
(c) E is translation invariant along the equiangular line; that is, $E + r\mathbf{1} = E$, where $r \in R$ and $\mathbf{1}$ is the vector with all its components equal to 1.

Now fix $\boldsymbol{\theta} \in R^k$. If $r = \Sigma\, \theta_i/k$, then $\boldsymbol{\theta} = r\mathbf{1} + \mathbf{z}$, where $\Sigma\, z_i = 0$. Therefore, $r\mathbf{1}$ is easily seen to be in the convex hull of the orbit of $\boldsymbol{\theta}$ under the permutation group. Therefore, by Theorem 3.3 (with G as the permutation group), we get

$$\int_E f(\mathbf{x} - \boldsymbol{\theta})\, d\mathbf{x} \leq \int_E f(x - r\mathbf{1})\, d\mathbf{x}. \tag{8.46}$$

Now consider a test of $H_0^*: \theta_1 = \cdots = \theta_k$ for which the acceptance region is E. Then the left side of (8.46) is the probability of accepting H_0^* when the alternative $\boldsymbol{\theta}$ holds while the translation invariance of E along the equiangular line shows that the right side of (8.46) is the probability of accepting H_0^* when H_0^* holds. Thus the test with acceptance region E is unbiased.

In order to apply the discussion of the preceding paragraph to the tests based on $T(\lambda, \eta)$, we make the transformations $x_i = \log(s_i^2)$ and $\theta_i = \log(\sigma_i^2)$. Then the joint density of f of $\mathbf{X}$ has the location parameter $\boldsymbol{\theta}$ and $f(\mathbf{x})$ easily satisfies the permutation invariance condition (i) mentioned above. Further f is easily shown to be logconcave and hence f also satisfies the unimodality condition (ii). Suppose

$$E(\lambda, \eta) = \{\mathbf{x} : T(\lambda, \eta) \leq c\}. \tag{8.47}$$

Then $E(\lambda, \eta)$ is trivially symmetric with respect to permutations and translation invariant along the equiangular line. Therefore, the test of H_0 with acceptance region $E(\lambda, \eta)$ will be unbiased as soon as we check that $E(\lambda, \eta)$ is convex.

Theorem 8.8. *For $\lambda \geq 0$ and $\eta \leq 0$, the test with acceptance region $E(\lambda, \eta)$ is unbiased.*

Proof. The discussion preceding the statement of the theorem shows that we only need to establish the convexity of $E(\lambda, \eta)$. Write $\delta = -\eta$ and

$$g(\mathbf{x}) = \left(\sum e^{\lambda x_i}\right)^{1/\lambda} \left(\sum e^{-\delta x_i}\right)^{1/\delta}. \tag{8.48}$$

Then (8.48), (8.47), (8.45) and (8.44) show that

$$E(\lambda, \eta) = \{\mathbf{x} : g(\mathbf{x}) \leq d\},$$

where d is easily determined in terms of c, λ and δ.

Suppose first that $\lambda > 0$ and $\eta < 0$. The continuity of $g(\mathbf{x})$ shows that, in order to prove the convexity of $E(\lambda, \eta)$, we may and do assume that λ and η are both rational. Then we can find a positive integer N such that $a = N/\lambda$ and $b = N/\delta$ are both positive integers. Now

$$E(\lambda, n) = \{\mathbf{x} : [g(\mathbf{x})]^N \leq d^N\},$$

with

$$[g(\mathbf{x})]^N = (\sum e^{\lambda x_i})^a (\sum e^{-\delta x_i})^b.$$

By the multinomial theorem, $[g(\mathbf{x})]^N$ is seen to be a finite sum of terms of the form $A \exp[L(\mathbf{x})]$, where A is a constant and $L(\mathbf{x})$ is linear in $\mathbf{x}$. Since each such term is convex, so is $[g(\mathbf{x})]^N$. Consequently, $E(\lambda, \eta)$ is convex, whenever $\lambda > 0$ and $\eta < 0$. The convexity of $E(\lambda, \eta)$ for the cases where λ or η or both are 0 or ∞ or $-\infty$ follows by a limiting argument whose details are omitted. The theorem is proved. ■

Theorem 8.8 yields the unbiasedness of the likelihood ratio test whose test statistic is equivalent to $T(1, 0)$. The statistic $T(\infty, -\infty)$ is equivalent to (max s_i^2)/(min s_i^2), which was proposed by Hartley (1950). Thus Hartley's test is also seen to be unbiased. Although the unbiasedness of the likelihood ratio test was established earlier by Pitman (1939) and Brown (1939) and that of the Hartley test by Ramachandran (1956), the results of Cohen and Strawderman (1971) established the unbiasedness of a wide class of tests. Also their proof actually shows the monotonicity of the power function along rays perpendicular to the equiangular line. They also show that the set $E(\lambda, \eta)$ may not be convex if λ and η are both positive. In particular $E(\infty, 1)$ is not convex. We note that $T(\infty, 1)$ is equivalent to the statistic max $s_i^2/\sum s_i^2$, which was proposed by Cochran (1941). The unbiasedness of Cochran's test is an unresolved problem.

8.4. Shortest Confidence Intervals and Minimum Volume Confidence Regions

Suppose that θ is a real parameter. When $100(1-\alpha)\%$ confidence intervals are to be constructed for θ, one has to consider the problem of splitting the mass α into the two tails. In symmetric cases (e.g., when θ is the mean of a normal distribution), it is natural to use equal tails of size $\alpha/2$ each and the resulting confidence intervals are generally unbiased and also the shortest within suitable classes. However, in asymmetric cases (e.g., when θ is the variance of a normal distribution), equal tailed confidence intervals are generally biased and may not have the shortest possible length. Unbiased confidence intervals are well discussed in textbooks like Lehmann (1986). On the other hand, shortest confidence intervals are generally discussed in an ad hoc manner. Further, the minimization problem is generally solved via calculus. In this section we show how, in many standard situations, shortest confidence intervals can be constructed in a simpler manner by using the unimodality of relevant distributions. For the case where θ is a vector parameter, we present the method given by Jeyaratnam (1985) for constructing minimum volume confidence sets for θ.

The univariate examples given here are based on the following lemma. We write $L(I)$ for the length of a interval I.

Lemma 8.6. *Let X be a random variable having a unimodal density f. Let $\gamma \in (0, 1)$ and let I be an inverval such that $P(X \in I) = \gamma$ and, for a suitable $c > 0$, $f(x) \geq c$ if $x \in I$ and $f(x) \leq c$ if $x \notin I$.*

(i) *If J is any interval for which $P(X \in J) \geq \gamma$ then $L(J) \geq L(I)$.*
(ii) *If f is continuous and I has end points a and b, then $f(a) = f(b)$.*

We omit the proof of the lemma because assertion (i) is just a version of the Neyman-Pearson lemma and assertion (ii) is trivial.

Example 8.1. Let $X_1, \ldots, X_n$ be a random sample from $N(\mu, \sigma^2)$. Let $T = \Sigma(X_i - \bar{X})^2$. The shortest confidence interval for σ^2 of the form (cT, dT) has been given by Tate and Klett (1959). We derive the same result more easily by using Lemma 8.6. In what follows, we write $h_m(x)$ for the density of the χ^2 distribution with m degrees of freedom.

Let $V = \sigma^2/T$. Since $1/V$ has the χ^2 distribution with $(n-1)$ degrees of freedom, the density of V is easily obtained as

$$f_1(v) = K \cdot h_{n+3}(v^{-1}),$$

where K is a suitable constant. Since $h_m(v)$ is unimodal in v, so is $h_m(v^{-1})$. Moreover, $f_1(v)$ is continuous. By Lemma 8.6, the shortest interval (c, d) such that $P(c < V < d) = (1 - \alpha)$ must satisfy $f_1(c) = f_1(d)$. Therefore, the shortest confidence interval for σ^2 of the form (cT, dT) must satisfy

$$P\left(c < \frac{\sigma^2}{T} < d\right) = 1 - \alpha,$$

and

$$h_{n+3}(c^{-1}) = h_{n+3}(d^{-1}).$$

It is convenient to write $a = d^{-1}$ and $b = c^{-1}$. The confidence interval then becomes $(T/b, T/a)$, where a snd b satisfy

$$P(a < \chi^2_{n-1} < b) = 1 - \alpha,$$

and

$$h_{n+3}(a) = h_{n+3}(b).$$

This result agrees with the one given by Tate and Klett (1959).

Suppose now that a confidence interval is needed for σ^k. Then we can look at the distribution of $(\sigma^2/T)^{k/2}$ and follow the same steps as above to show that the shortest confidence interval for σ^k of the form $((T/b)^{k/2}, (T/a)^{k/2})$ satisfies

$$P(a < \chi^2_{n-1} < b) = 1 - \alpha,$$

and (8.49)

$$h_{n+1+k}(a) = h_{n+1+k}(b).$$

For $k = 1$, (8.49) agrees with the calculation given by Hogg and Tanis (1983, pp. 290–291).

Example 8.2. Let $X_1, \ldots, X_n$ be a random sample from the uniform distribution on $(0, \theta)$. Let $T = \max(X_1, \ldots, X_n)$ and $V = \theta/T$. Then the density of V is easily obtained as:

$$f_2(v) = \begin{cases} nv^{-(n+1)}, & v > 1 \\ 0, & v \leq 1. \end{cases}$$

The density f_2 is unimodal about 1 but it is discontinuous at $v = 1$. By the Lemma, the shortest confidence interval (c, d) such that $P(c < V < d) = 1 - \alpha$ must satisfy $c = 1$. We can then easily show that $d = \alpha^{-1/n}$. Therefore, the shortest confidence interval for θ of the form (cT, dT) is $(T, T\alpha^{-1/n})$. This result has been given in many places; see, for example, Bickel and Doksum (1977, p. 189).

Remark. The chi-square distribution arises in contexts other than those considered in Example 8.1. The formulas given in Example 8.1 can be easily modified to obtain shortest confidence intervals for the parameters of exponential, Weibull and other distributions. Similarly, simple transformations enable us to use the result of Example 8.2 to obtain confidence intervals for the location parameter of the shifted exponential distribution. See Guenther (1969) for details.

Example 8.3. Let T_1, T_2 be independent random variables such that T_i/σ_i^2 has the chi-square distribution with n_i degrees of freedom. We want a shortest confidence interval for $\theta = \sigma_1^2/\sigma_2^2$ having a certain simple form. Write $T = n_2 T_1/(n_1 T_2)$ and $V = \theta/T$. Then V is distributed as F_{n_2,n_1}. Let $w(v; m_1, m_2)$ denote the density of F_{m_1,m_2}. We note that $w(v; m_1, m_2)$ is unimodal and it is also continuous if $m_1 > 2$.

Suppose first that $n_2 > 2$. Then our Lemma shows that the shortest interval (c, d) such that $P(c < V < d) = 1 - \alpha$ must satisfy $w(c;\ n_2,\ n_1) = w(d;\ n_2,\ n_1)$. Therefore the shortest confidence interval for σ_1^2/σ_2^2 of the form $(cT,\ dT)$ satisfies

$$P(c < F_{n_2,n_1} < d) = 1 - \alpha$$

and

$$w(c;\ n_2,\ n_1) = w(d;\ n_2,\ n_1).$$

If $n_2 \leq 2$, then the density of V is unimodal about 0 and also discontinuous at 0. It follows that the shortest confidence interval for σ_1^2/σ_2^2 of the form $(cT,\ dT)$ must have $c = 0$. The constant d then satisfies $P(F_{n_2,n_1} \geq d) = \alpha$. This example does not seem to have been discussed in the literature.

Apart from the use of unimodality, the unifying thread for the last three examples is the use of a pivotal quantity like θ/T, where θ is the parameter of interest and T is a relevant statistic. An important property that makes these examples work is that the derivative of θ/T w.r.t. θ is free of θ. This is made transparent by Jeyaratnam (1985) who has used the multivariate version of the above property to construct minimum volume confidence regions for a vector parameter. His theorem follows.

Theorem 8.9. *Let* $\mathbf{X}$ *be a random vector with density* $g(\mathbf{x}, \boldsymbol{\theta})$, *where* $\boldsymbol{\theta}$ *is a vector parameter with k components. Denote the parameter space by* Ω. *Suppose* $V(\mathbf{x}, \theta)$ *is a pivotal quantity with density* $f(v)$ *such that, for each fixed* $\mathbf{x}$, $V(\mathbf{x}, \cdot)$ *is a one-to-one function on* Ω *into* R^k *whose Jacobian* $J_{\mathbf{x}}$ *is free of* $\boldsymbol{\theta}$. *Let* $S(\mathbf{x}) \subset \Omega$ *be defined by* $S(\mathbf{x}) = \{\boldsymbol{\theta} : f[V(\mathbf{x}, \boldsymbol{\theta})] \geq c\}$ *and let* $\gamma = P_{\boldsymbol{\theta}}[\boldsymbol{\theta} \in S(\mathbf{X})]$. *Then the confidence region* $S(\mathbf{X})$ *has minimum volume among all confidence regions for* $\boldsymbol{\theta}$ *which are based on* V *and have confidence level* $\geq \gamma$.

Proof. Let $S^*(\mathbf{X})$ be another confidence region for $\boldsymbol{\theta}$ based on V. Let Γ be the range of V. The statement that S^* is based on V means that there is a set $W^* \subset \Gamma$ such that

$$\boldsymbol{\theta} \in S^*(\mathbf{x}) \Leftrightarrow V(\mathbf{x}, \boldsymbol{\theta}) \in W^*.$$

If γ^* is the confidence level for S^*, then

$$\gamma^* = \int_{W^*} f(v)\, dv.$$

We note that the confidence region $S(\mathbf{X})$ *is* based on V because

$$\theta \in S(\mathbf{x}) \Leftrightarrow V(\mathbf{x}, \boldsymbol{\theta}) \in W,$$

where $W = \{v \in \Gamma : f(v) \geq c\}$. Moreover

$$\gamma = \int_W f(v)\, dv.$$

By the argument used in the Neyman–Pearson lemma, we can see that

$$\int_W dv \leq \int_{W^*} dv. \tag{8.50}$$

Now

$$\text{Volume}[S(\mathbf{x})] = \int_{S(\mathbf{x})} d\theta = \frac{1}{|J_{\mathbf{x}}|} \int_W dv. \tag{8.51}$$

Similarly,

$$\text{Volume}[S^*(\mathbf{x})] = \frac{1}{|J_{\mathbf{x}}|} \int_{W^*} dv. \tag{8.52}$$

The theorem now follows from (8.50), (8.51) and (8.52). ■

Remark. It is clear that under suitable unimodality assumptions, the confidence region S given by Theorem 8.9 will be convex or star-shaped or at least connected.

Example 8.4. Let X and Q be independently distributed as $N(\mu, \sigma^2)$ and $\sigma^2 \chi_m^2$, respectively. (For instance, X and Q may be suitable multiples of the mean and the variance for a random sample from a normal distribution). We want to construct a confidence region for $\boldsymbol{\theta} = (\mu, \lambda)$, where $\lambda = \sigma^k$. Let

$$g_1(x, q, \mu, \lambda) = \frac{(x - \mu)}{\sigma} = \frac{(x - \mu)}{\lambda^{1/k}},$$

and

$$g_2(x, q, \mu, \lambda) = \frac{\sigma^{k+1}}{q^{(k+1)/2}} = \frac{\lambda^{(k+1)/k}}{q^{(k+1)/2}}.$$

It is easy to check that, for fixed (x, q), the Jacobian $\partial(g_1, g_2)/\partial(\mu, \lambda)$ is free of (μ, λ). It is trivial to see that for fixed (x, q), the transformation which takes (μ, λ) into $(g_1(x, q, \mu, \lambda), g_2(x, q, \mu, \lambda))$ is one-to-one. Finally, if $V_i = g_i(X, Q, \mu, \lambda)$, then (V_1, V_2) is a pivotal quantity whose density is

$$f(v_1, v_2) = \left[\frac{2}{(k+1)}\right] \varphi(v_1) v_2^{-(k+3)/(k+1)} h_m(v_2^{-2/(k+1)}),$$

where φ is the standard normal density and h_m is the density of the χ_m^2 distribution. By Theorem 8.9, the confidence region based on (V_1, V_2) which has minimum volume is $S(X, Q) = \{(\mu, \lambda) : f(V_1, V_2) \geq c\}$. This calculaton is valid for $k > 0$ and agrees with the results given by Jeyaratnam for the cases $k = 1, 2$.

9 Convexity in Reliability Theory

9.0. Summary

This chapter discusses some aspects of the theory of reliability in which considerations of convexity play an important role. Many of the results involve certain types of orderings, which are discussed in Section 1. Next, we consider the important classes of IFR and IFRA distributions, with the univariate results appearing in Section 2 and the multivariate results in Section 3. Finally, Section 4 looks at order statistics and their convexity properties, some of which involve IFR and IFRA distributions.

The standard reference on reliability theory is the book by Barlow and Proschan (1981).

9.1. Some Orderings for Distributions of Nonnegative Random Variables

Ordering of distributions by peakedness was studied in Chapter 7. It was seen that stochastic ordering is useful in discussions involving peakedness. In this section we define three orderings which are relevant for some results in reliability theory. While some of the definitions and properties considered here can be given for distributions on R, we limit ourselves to distributions on $[0, \infty)$.

Two of the partial orderings we consider involve the inverse of a distribution function G. For this purpose, we need to restrict G suitably and define G^{-1} in a special way. This and other technical reasons motivate the definitions of some special classes of distribution functions in the next paragraph.

For a nondecreasing function g on R, define the support of g by

$$\text{supp}(g) = \{x \in R : g(x+\delta) > g(x-\delta) \text{ for all } \delta > 0\}.$$

Suppose $\mathscr{I}$ is the class of all distribution functions G on R such that

(a) $G(0-) = 0$
(b) supp(G) is an interval $[l_G, u_G]$, where we take the interval to be open at u_G if $u_G = \infty$.

Next, let $\mathscr{I}_0$ be the class of all distribution functions G in $\mathscr{I}$ such that $l_G = 0$. That is, a distribution function G in $\mathscr{I}_0$ must have $G(x) > 0$ for all $x > 0$, whereas a distribution function H in $\mathscr{I}$ may have $H(y) = 0$ for some $y > 0$. Finally, let $\mathscr{I}_c$ be the class of all G in $\mathscr{I}$ such that G is continuous on R.

We now define the inverse of a distribution function in $\mathscr{I}$. Suppose $G \in \mathscr{I}$. Then supp(G) is an interval $[l_G, u_G]$. Define G^{-1}on $[0, 1]$ as follows.

$$G^{-1}(t) = \inf\{x : G(x) > t\}, \qquad 0 \le t \le 1.$$

With this definition $G^{-1}(0) = l_G$ and $G^{-1}(1) = \infty$. Further G is strictly increasing on $[l_G, u_G]$. Therefore, one easily verifies that

$$G^{-1}G(x) = \begin{cases} l_G, & x < l_G \\ x, & l_G \le x < u_G \\ \infty, & x \ge u_G. \end{cases}$$

It is thus seen that $G^{-1}G$ is convex on R. It should be noted that even though G is strictly increasing on $[l_G, u_G]$, it may have discontinuities in that interval. As a result, G^{-1} may have intervals of constancy and the relation $GG^{-1}(t) = t$ may not hold for all $t \in [0, 1]$. However, if $G \in \mathscr{I}_c$ then we do get $GG^{-1}(t) = t$ for all $t \in [0, 1]$. This is the reason for the introduction of the class $\mathscr{I}_c$ above.

For use in Definition 9.3 below, we introduce the following definition.

Definition 9.1. A function g on R into $(-\infty, \infty]$ is called *star-shaped* if $g(x)/x$ is nondecreasing on $(0, \infty)$.

Remark. Suppose g is a convex function on R into $(-\infty, \infty]$. Then g is star-shaped if, and only if, either $g(0+) \le 0$ or $g(x) = \infty$ for all $x > 0$. To see the use of this observation in our context, suppose G is a distribution function in the class $\mathscr{I}$ introduced above. Then supp $G = [l_G, u_G]$ with $0 \le l_G < \infty$ and we also know that $G^{-1}G$ is convex on R. Now if $l_G > 0$, then $(G^{-1}G)(0+) = l_G > 0$ and so $G^{-1}G$ cannot be star-shaped. This is the reason for the introduction of the class $\mathscr{I}_0$ above.

Finally, we need a simple equivalence relation. Call two distribution functions F and G on R *scale equivalent* if, for some $\alpha > 0$, $G(x) = F(\alpha x)$ for all x.

Definition 9.2. Let F and G be distribution functions in $\mathscr{I}$. We say that F *convex-precedes* G and write $F \le^c G$ if $G^{-1}F$ is convex on R.

Definition 9.3. Let F and G be distribution functions in $\mathscr{I}$. We say that F *star-precedes* G and write $F \le^* G$ if $G^{-1}F$ is star-shaped.

The definition of convex ordering was given by van Zwet (1964) and that of star ordering was given by Barlow and Proschan (1966). Some properties of convex and star orderings are summarized in the next theorem.

Theorem 9.1.

(a) *If $F \in \mathscr{I}$ and $G \in \mathscr{I}_0$, then $F \le^c G \Rightarrow F \le^* G$.*
(b) *The relations $F_1 \le^c F_2$ and $F_1 \le^* F_2$ are unaffected if, for $i = 1, 2$, we replace F_i by G_i which is scale equivalent to F_i.*
(c) *The ordering $\le^c$ is reflexive on $\mathscr{I}$ and transitive on $\mathscr{I}_c$.*
(d) *The ordering $\le^*$ is reflexive on $\mathscr{I}_0$ and transitive on $\mathscr{I}_0 \cap \mathscr{I}_c$.*
(e) *The orderings $\le^c$ and $\le^*$ are antisymmetric on classes of scale equivalent functions in $\mathscr{I}_0 \cap \mathscr{I}_c$.*

Proof. All distribution functions considered in this proof are assumed to be in $\mathscr{I}$.

(i) Suppose $F \le^c G$ and $G \in \mathscr{I}_0$. Two cases arise.

Case 1. Let $u_F > 0$. Then $F(x) < 1$ *for* $x < u_F$ and so $G^{-1}F$ is finite on $(-\infty, u_F)$. Now the convexity of $G^{-1}F$ shows that $G^{-1}F$ is continuous at 0. That is $G^{-1}F(0+) = G^{-1}F(0-) = 0$. The remark following Defnition 9.1 then shows that $G^{-1}F$ is star-shaped.

Case 2. Let $u_F = 0$. Since $l_F = 0$, we see that F is degenerate at 0 and so $G^{-1}F(x) = 0$ or ∞ according as $x < 0$ or $x \ge 0$. Again, $G^{-1}F$ is star-shaped.

Assertion (a) is now verified.

(ii) The proof of assertion (b) is trivial. So consider assertion (c). We know that $F^{-1}F$ is convex so $\le^c$ is reflexive on $\mathscr{I}$. Suppose $F \le^c G$ and $G \le^c H$.

Then $G^{-1}F$ and $H^{-1}G$ are both convex. Since $H^{-1}G$ is also nondecreasing, we see that $H^{-1}GG^{-1}F$ is also convex. Now the relation $H^{-1}GG^{-1}F = H^{-1}F$ would be true if GG^{-1} is the identity map,which holds if $G \in \mathscr{I}_c$. Thus $\leq^c$ is transitive on $\mathscr{I}_c$. This proves (c) and assertion (d) can be proved in the same way.

(iii) Assertion (a) shows that we need prove assertion (e) for the ordering $\leq^*$ only. So let F, G be distribution functions in $\mathscr{I}_0 \cap \mathscr{I}_c$ such that $F \leq^* G$ and $G \leq^* F$. Write $s(x) = G^{-1}F(x)$ and $w(x) = F^{-1}G(x)$. Then s maps $[0, u_F)$ onto $[0, u_G)$ and w maps $[0, u_G)$ onto $[0, u_F)$. Now for $x \in (0, u_G)$,

$$s[w(x)] = G^{-1}FF^{-1}G(x) = G^{-1}G(x) = x.$$

Therefore,

$$\frac{x}{w(x)} = \frac{s[w(x)]}{w(x)},$$

which is nondecreasing on $(0, u_G)$ because s is star-shaped. But $w(x)/x$ is also nondecreasing on $(0, u_G)$ because w is star-shaped. Therefore $w(x)/x$ is constant on $(0, u_G)$. Thus, for some $\alpha > 0$, $w(x) = \alpha x$ for all $x \in (0, u_G)$. That is $F^{-1}G(x) = \alpha x$ or $G(x) = F(\alpha x)$ for $0 < x < u_G$. If we take limit from the left at u_G, we see that $G(x) \to 1$ and so we conclude that $F(\alpha x) = 1$ for all $x \geq u_G$. Thus $G(x) = F(\alpha x)$ for all x and F, G are scale equivalent. The theorem is thus proved. ∎

Corollary. *The orderings $\leq^c$ and $\leq^*$ are partial orderings for classes of scale equivalent classes of distribution functions in $\mathscr{I}_0 \cap \mathscr{I}_c$.*

The survival function $\bar{F}$ corresponding to a distribution function F is defined by $\bar{F}(x) = 1 - F(x)$, $x \in R$. This is the function used most often in reliability theory. Star ordering can be characterized in terms of crossings of survival functions as follows.

Theorems 9.2. *Suppose $F \in \mathscr{I}$ and $G \in \mathscr{I}_0 \cap \mathscr{I}_c$.*

(a) *$F \leq^* G$ if, and only if, for every $\alpha > 0$, $\bar{F}(x)$ crosses $\bar{G}(\alpha x)$ at most once and from above as x goes from 0 to ∞.*
(b) *If $F \leq^* G$ and F, G differ somewhere and have the same mean, then $\bar{F}(x)$ will cross $\bar{G}(\alpha x)$ exactly once and from above as x goes from 0 to ∞.*

Proof. Suppose $F \leq^* G$, $\alpha > 0$ and $s(x) = G^{-1}F(x)$. If possible, suppose there are x, y with $x < y$ such that

$$F(x) > G(\alpha x) \quad \text{and} \quad F(y) < G(\alpha y).$$

Then $\alpha x < u_G$ and so

$$s(x) = G^{-1}F(x) > G^{-1}G(\alpha x) = \alpha x.$$

To show that $s(y) < \alpha y$, we consider two cases.

Case 1. Suppose $\alpha y < u_G$. Then

$$s(y) = G^{-1}F(y) < G^{-1}G(\alpha y) = \alpha y.$$

Case 2. Suppose $\alpha y \geq u_G$. Then $G(\alpha y) = 1$. So $F(y) < 1 = G(u_G - 0)$ and

$$s(y) = G^{-1}F(y) < G^{-1}G(u_G - 0) = u_G \leq \alpha y.$$

Thus $s(y) < \alpha y$ in both cases. We now see that

$$\frac{s(x)}{x} > \alpha > \frac{s(y)}{y}.$$

That is, s is not star-shaped. This contradiction shows that $\bar{F}(x) - \bar{G}(\alpha x)$ has the sign change property mentioned in (a). The "only if" part of (a) is thus proved.

To prove the "if" part, suppose that s is not star-shaped. Then we can find u, v such that $0 < u < v$ and $[s(u)/u] > [s(v)/v]$. Now $s(v) < \infty$ and so $0 < s(u) \leq s(v) < u_G$. Choose α in the open interval $(s(v)/v,\ s(u)/u)$ and so close to the left end that $\alpha v < u_G$. We then have

$$0 < \alpha u < s(u) < u_G \quad \text{and} \quad 0 < s(v) < \alpha v < u_G.$$

Therefore,

$$G(\alpha u) < G[s(u)] = F(u) \quad \text{and} \quad G(\alpha v) > G[s(v)] = F(v).$$

Thus the function $F(x) - G(\alpha x)$ does not have the required sign change property. This proves (a).

Assertion (b) follows easily if one observes that the total area under $\bar{F}$ is equal to the mean of F. The theorem is thus proved. ∎

An interesting example of convex ordering has been given by van Zwet (1964, section 4.3.4). We present this example here supplying some of the details of the calculations. For $m > 0$, let $f(x, m)$ denote the gamma density with parameter m. That is,

$$f(x, m) = \frac{e^{-x}x^{m-1}}{\Gamma(m)}, \qquad x > 0.$$

The distribution function defined by $f(x, m)$ is denoted by F_m. We want to show that $F_n \leq^c F_m$ if $1 < m < n$. So assume that $m > 1$ and $n = m + \delta$ where $\delta > 0$. If we write

$$\varphi(x) = F_n^{-1} F_m(x),$$

then we want to show that φ is concave on $(0, \infty)$. An equivalent assertion is that, for every pair of real numbers a, b, the function ψ defined by

$$\psi(x) = F_m(x) - F_n[b(x + a)], \qquad x > 0$$

has at most two zeros and that if ψ does vanish at x_1 and x_2, say, then ψ is positive on (x_1, x_2). We now proceed to prove this last assertion. Observe first that, if $b \leq 0$, then ψ is strictly increasing and so it has at most one change of sign. So we will assume that $b > 0$, We have

$$\psi'(x) = f(x, m) - bf[b(x + a), m + \delta].$$

Three cases arise.

Case 1. Let $a > 0$. Then $\psi'(x)$ has the same sign as

$$\begin{aligned} \varsigma(x) &= \log f(x, m) - \log f[b(x + a), m + \delta] - \log b \\ &= (b - 1)x + (m - 1)\log x - (m + \delta - 1)\log(x + a) + K, \end{aligned}$$

where K is a constant. Again

$$\varsigma'(x) = (b - 1) + \beta(x),$$

where

$$\beta(x) = \frac{[a(m - 1) - \delta x]}{[x(x + a)]}.$$

Observe that $\beta(x)$ decreases to zero as x increases from 0 to $a(m - 1)/\delta$, Beyond $a(m - 1)/\delta$, $\beta(x)$ continues to decrease until it reaches a minimum at $x = x_0$, where

$$[\delta x_0 - a(m - 1)]^2 = a^2(m - 1)(m - 1 + \delta).$$

Beyond x_0, $\beta(x)$ remains negative but increases to zero.

Suppose first that $b \leq 1$. Then $\varsigma'(x)$ is positive up to a certain point and remains negative thereafter. Since $\varsigma(0+) < 0$, we see that ς and hence ψ' has at most two sign changes and the sequence of signs must be $-, +, -$. Again $\psi(0) < 0$ and it follows that ψ must have the same sign sequence.

Suppose next that $b > 1$. Then $\varsigma'(x)$ has at most two sign changes and the sequence of signs must be $+, -, +$. Consequently, ς and hence ψ' can have the sign sequence $-, +, -, +$. The resulting sign sequence for ψ can also be $-, +, -, +$. We now argue that the last sign change is not possible. Indeed, if we can find x^* such that $\psi(x) > 0$ and $\psi'(x) > 0$ for all $x > x^*$, then ψ would remain bounded away from 0 as $x \to \infty$. But $\psi(x) \to 0$ and $x \to \infty$ and we have reached a contradiction. Thus ψ has at most two sign changes and the possible sign sequence is $-, +, -$.

Case 2. Suppose $a = 0$. Then

$$\varsigma(x) = (b-1)x - \delta \log x + K$$

and

$$\varsigma'(x) = (b-1) - \frac{\delta}{x}.$$

Near $x = 0$, ς and hence ψ' are positive. But $\psi(0) = 0$. Therefore, ψ is also positive near $x = 0$.

Again, suppose that $b \le 1$. Then $\varsigma'(x) < 0$ for all $x > 0$. Therefore ς and hence ψ' has possible sign sequence $+, -$. The same sign sequence must also hold for ψ.

Suppose next that $b > 1$. Then the sign sequence for ς' is $-, +$. Therefore, the possible sign sequence for ς or ψ' is $+, -, +$. Hence the possible sign sequence for ψ is also $+, -, +$. But, as seen in Case 1, ψ' and ψ cannot be simultaneously positive on an infinite interval (x^*, ∞). Thus ψ must have the sign sequence $+, -$.

Case 3. Suppose $a < 0$ and write $A = -a$. Then ψ and ψ' are both positive on $(0, A]$. It is also easy to show that as x goes from A to ∞, $\beta(x)$ increases from $-\infty$ to 0. Therefore, either $\varsigma'(x) < 0$ for all $x > A$ or $\varsigma'(x)$ has sign sequence $-, +$ on (A, ∞). Thus this case is similar to Case 2.

We have thus completed the verification that the gamma distribution with parameter n convex-precedes the gamma distribution with parameter m if $1 < m < n$.

A third ordering of distributions can be defined in terms of survival functions. As usual, we set $x^+ = \max(x, 0)$. Suppose X is a nonnegative random variable with distribution function F. Let $a \ge 0$. If X is considered to be the length of life of a component, then $E(X-a)^+$ is the "mean residual life after a." Now

$$E(X-a)^+ = \int_0^\infty P[X - a \ge x]\, dx = \int_a^\infty \bar{F}(t)\, dt. \tag{9.1}$$

Formula (9.1) justifies the term "mean residual life" in the next definition, which was given by Bessler and Veinott (1966). We write $\mu(F)$ for the mean of F.

Definition 9.4. Suppose $\mu(F)$ and $\mu(G)$ are finite. We say that F *precedes* G *in mean residual life* and write $F \leq^{mr} G$ if, for every $a \geq 0$,

$$\int_a^\infty \bar{F}(t)\, dt \leq \int_a^\infty \bar{G}(t)\, dt. \tag{9.2}$$

Remark. From (9.2), we see easily that $\leq^{mr}$ is transistive and reflexive. It is also antisymmetric on the class of all distribution functions on $[0, \infty)$ with finite means. Thus the MR ordering is a partial ordering on the class of distributions of nonnegative random variables with finite means. We saw in Theorem 9.1 that convex and star orderings are scale invariant. In contrast the MR ordering is not scale invariant.

It was shown earlier that convex ordering is stronger than star ordering. The next theorem shows that, under certain conditions, star ordering is stronger than MR ordering.

Theorem 9.3. *Suppose* $F \leq^* G$, G *is continuous and* $\mu(F) \leq \mu(G)$. *Then* $F \leq^{mr} G$.

Proof. If $\bar{F} \leq \bar{G}$ everywhere, there is nothing to prove. So suppose $\bar{F}(x) > \bar{G}(x)$ for some x. The condition $\mu(F) \leq \mu(G)$ then shows that, for some y, $\bar{F}(y) < \bar{G}(y)$. From Theorem 9.2 we see that there is a ξ such that

$$x > \xi \quad \Rightarrow \quad \bar{F}(x) \leq \bar{G}(x)$$

and

$$x < \xi \quad \Rightarrow \quad \bar{F}(x) \geq \bar{G}(x).$$

Now (9.2) clearly holds for $a \geq \xi$. If it fails at some $a < \xi$, it must also fail at $a = 0$. But it does hold at $a = 0$ because $\mu(F) \leq \mu(G)$. Therefore (9.2) holds for all $a > 0$ and the theorem follows. ■

Bessler and Veinott (1966) gave the following characterization of MR ordering. The proof given here is taken from Ross (1982). For a detailed discussion of the MR and other related partial orderings, we refer the reader to Stoyan (1983).

Theorem 9.4. *Let F, G be distribution functions of nonnegative random variables X, Y. Then*

(a) $F \leq^{mr} G$ *if, and only if,* $E[h(X)] \leq E[h(Y)]$ *for all nondecreasing, convex functions h on* $[0, \infty)$.
(b) $E(X) = E(Y)$ *and* $F \leq^{mr} G$ *if, and only if,* $E[h(X)] \leq E[h(Y)]$, *for all convex functions h on* $[0, \infty)$.

Proof. (i) Suppose $F \leq^{mr} G$ and assume first that h is convex, nondecreasing and twice continuously differentiable. We want to show that $E[h(X)] \leq E[h(Y)]$. If either $E[h(Y)] = \infty$ or h is a constant function, there is nothing to prove. So suppose that h is nonconstant and $E[h(Y)] < \infty$. Then h is bounded below by an increasing linear function and so $E(Y) < \infty$. The following calculation is now justified.

$$\begin{aligned}\int_0^\infty h(y)dG(y) &= h(0) + \int_0^\infty h'(y)\bar{G}(y)\,dy \\ &= h(0) + \int_0^\infty \left\{h'(0) + \int_0^y h''(a)\,da\right\}\bar{G}(y)\,dy \\ &= h(0) + h'(0)E(Y) + \int_0^\infty \left\{\int_a^\infty \bar{G}(y)\,dy\right\}h''(a)\,da.\end{aligned}$$

A similar calculation is valid with G replaced by F because of the condition (9.2). The same condition and the nonnegativity of h'' now give

$$E[h(Y) - h(X)] \geq h'(0)E(Y - X). \tag{9.3}$$

But if we set $a = 0$ in (9.2), we get $E(Y) \geq E(X)$. Since $h'(0) \geq 0$, (9.3) yields $E[h(Y)] \geq E[h(X)]$. The smoothness condition on h is removed by usual limiting arguments. This proves the "only if" part of (a). The "if" part follows by taking $h(x) = (x - a)^+$ and using (9.1).

(ii) Suppose that $E(X) = E(Y)$ and $F \leq^{mr} G$. Then (9.3) shows that $E[h(Y) - h(X)] \geq 0$ without requiring that $h'(0) \geq 0$. Therefore, the inequality $E[h(Y)] \geq E[h(X)]$ is valid for all convex functions h. This justifies the "only if" part of (b). To prove the "if" part observe that the ordering $F \leq^{mr} G$ follows from (a) and the equality $E(X) = E(Y)$ follows by taking, in turn, $h(x) = x$ and $h(x) = -x$. The proof of the theorem is now complete. ■

To end this section we present a theorem to indicate how results of the type given in Theorem 9.4 can be extended to vectors of independent random variables. The theorem appears on pp. 29–30 of Stoyan (1983). As one might expect, the proof will use conditioning and induction.

Theorem 9.5. *Suppose* $\mathbf{X}$ *and* $\mathbf{Y}$ *are independent n-vectors of independent random variables such that* $X_i \leq^{mr} Y_i$, $i = 1, \ldots, n$. *Then* $E[h(\mathbf{X})] \leq E[h(\mathbf{Y})]$ *for all functions h on* $[0, \infty)^n$ *which are coordinatewise nondecreasing and coordinatewise convex.*

Proof. The theorem is true when $n = 1$. Suppose it holds in dimension $(n - 1)$ where $n \geq 2$. Let h be as stated in the theorem. Write $\mathbf{x} = (\mathbf{x}', x_n)$. When x_n is fixed, the function $h(\mathbf{x}', x_n)$ is a function of the required type in $\mathbf{x}'$. Therefore, by the induction hypothesis,

$$E[h(\mathbf{X}', x_n)] \leq E[h(\mathbf{Y}', x_n)]. \tag{9.4}$$

Write $g(x_n)$ for the right side of (9.4). The independence hypothesis of the theorem then shows that (9.4) is the same as

$$E[h(\mathbf{X}) \mid X_n] \leq g(X_n).$$

Therefore,

$$E[h(\mathbf{X})] \leq E[g(X_n)]. \tag{9.5}$$

Again, $g(x_n)$ is nondecreasing and convex in x_n. Therefore,

$$E[g(X_n)] \leq E[g(Y_n)]. \tag{9.6}$$

But the independence hypothesis again shows that

$$E[g(Y_n)] = E\{E[h(\mathbf{Y}) \mid Y_n]\} = E[h(\mathbf{Y})]. \tag{9.7}$$

Combining (9.5), (9.6) and (9.7), one gets the assertion of the theorem. ■

9.2. Univariate IFR and IFRA Classes

In reliability theory, convexity properties of the survival function play an important role in identifying important classes of distributions. This section is devoted to a discussion of some of these properties. Of the several important classes of distributions, we consider only two, namely, IFR and IFRA. Univariate results are presented in the current section and multivariate results in the next section. After defining the IFR and IFRA classes, we consider their Laplace transforms and convolutions.

We begin with the definition of an IFR distribution.

Definition 9.5. A nonnegative random variable X (or its distribution function F or its survival function $\bar{F}$) is said to have *increasing failure rate* (IFR) if $\log \bar{F}$ is concave on R.

The motivation behind this definition is as follows. If F has a density f, then the "falure rate" for an item which has survived up to time x is

$$r(x) = \frac{f(x)}{\bar{F}(x)}. \tag{9.8}$$

The increasing failure rate property should mean that $r(x)$ is nondecreasing in x and this, in turn, means that $\log \bar{F}$ is concave. Definition 9.5 makes the IFR property applicable to distributions with or without density.

The following theorem identifies a large class of IFR distributions.

Theorem 9.6. *If F has a logconcave density f, then F has IFR.*

Proof. The logconcavity of f shows that, for $0 < t_1 < t_2$ and $x > 0$, *we have*

$$\frac{f(t_1 + x)}{f(t_1)} \geq \frac{f(t_2 + x)}{f(t_2)}.$$

That is

$$f(t_2)f(t_1 + x) \geq f(t_1)f(t_2 + x).$$

Integrating w.r.t. x over $(0, \infty)$, we have

$$f(t_2)\bar{F}(t_1) \geq f(t_1)\bar{F}(t_2).$$

Therefore,

$$\frac{f(t_2)}{\bar{F}(t_2)} \geq \frac{f(t_1)}{\bar{F}(t_1)},$$

which proves the theorem. ∎

The IFR class can be enlarged to the IFRA class using Definition 9.1 of star-shaped functions.

Definition 9.6. A random variable X (or its distribution function F or its survival function $\bar{F}$) is said to have *increasing failure rate average* (IFRA) if $(-\log \bar{F})$ is star-shaped.

It is easily shown that F has IFRA if, and only if, for every $\alpha \in (0, 1)$ and every x,

$$\bar{F}(\alpha x) \geq [\bar{F}(x)]^{\alpha}. \tag{9.9}$$

Definition 9.6 can be motivated as follows. Suppose F has a density f. The failure rate $r(x)$ is then given by (9.8). Therefore, the average failure rate up to time t is

$$a(t) = \frac{1}{t}\int_0^t r(x)\,dx = -\frac{\log \bar{F}(t)}{t}.$$

The IFRA property should mean that $a(t)$ is nondecreasing in t which, in turn, means that $[-\log \bar{F}(t)]$ should be star shaped. Definition 9.6 makes the IFRA property applicable to all distributions.

As can be expected, the IFR and IFRA concepts are closely related to the convex and star orderings given in Section 9.1. To see this, let $\mathscr{E}_\lambda$ be the exponential distribution function with mean $(1/\lambda)$. Then $\mathscr{E}_\lambda^{-1}(y) = -[\log(1-y)]/\lambda$ and so $\mathscr{E}_\lambda^{-1}F(x) = -[\log \bar{F}(x)]/\lambda$. Thus F is IFR if, and only if, $F \leq^c \mathscr{E}_\lambda$ and F is IFRA if, and only if, $F \leq^* \mathscr{E}_\lambda$.

The IFR and IFRA classes of distributions can be characterized in terms of the Laplace transforms of survival functions. For the IFR class, such a characterization was given by Vinogradov (1973) and the corresponding results for the IFRA and many other classes were given by Block and Savits (1980a). We now proceed to present some of these results.

Let F be a distribution function on $(0, \infty)$ and write $\bar{F}(x) = 1 - F(x)$ as usual. For $s > 0$, define the functions $a_n(s)$ and $b_n(s)$ as follows.

$$a_n(s) = \int_0^\infty \left[\frac{e^{-st}t^n}{n!}\right]\bar{F}(t)\,dt, \qquad n = 0, 1, 2, \ldots,$$

$$a_{-1}(s) = 1, \tag{9.10}$$

and

$$b_n(s) = \int_0^\infty \left[\frac{e^{-st}t^n}{n!}\right] dF(t), \qquad n = 0, 1, 2, \ldots.$$

The sequence $\{a_n(s)\}$ will be called the *Laplace sequence* for F. It is easy to check that

$$b_n(s) = a_{n-1}(s) - s a_n(s), \qquad n = 0, 1, 2, \ldots, \tag{9.11}$$

and

$$a_n'(s) = -(n+1)a_{n+1}(s), \quad n = -1, 0, 1, \ldots. \tag{9.12}$$

If we define

$$c_n(s) = \frac{b_n(s)}{a_n(s)}, \qquad n = 0, 1, 2, \ldots$$

then (9.11) and (9.12) easily show that

$$c_n'(s) = \frac{(n+1)[a_{n+1}(s)a_{n-1}(s) - a_n^2(s)]}{a_n^2(s)}. \tag{9.13}$$

Theorem 9.7. *A distribution function F on $(0, \infty)$ belongs to the IFR class if, and only if, its Laplace sequence $\{a_n(s)\}$ is logconcave in n for every fixed $s > 0$; that is,*

$$a_n^2(s) \geq a_{n-1}(s)a_{n+1}(s), \qquad n = 0, 1, 2, \ldots. \tag{9.14}$$

Proof. The "only if" part is proved first.

(i) Suppose first that F belongs to the IFR class and has a density f on $(0, \infty)$. We will use a standard result on sign changes [see Lehmann (1986, page 85)]. Let $\delta > 0$. Using (9.11), we get

$$\begin{aligned}\int_0^\infty e^{-st}\frac{t^n}{n!}[f(t) - \delta\bar{F}(t)]dt &= b_n(s) - \delta a_n(s)\\ &= a_{n-1}(s) - (s+\delta)a_n(s)\\ &= a_n(s)\left[\frac{a_{n-1}(s)}{a_n(s)} - (s+\delta)\right].\end{aligned} \tag{9.15}$$

Since F has IFR, the function $f(t)/\bar{F}(t)$ is nondecreasing. Therefore $f(t) - \delta\bar{F}(t)$ changes sign at most once and, if it does, it does so from $-$ to $+$ values. Further the kernel $e^{-st}t^n/n!$ is TP_2 in t and n. Therefore, the right side of (9.15) also changes sign at most once as n increases from 0 to ∞. Moreover any change is from $-$ to $+$ values. Since δ is arbitrary and $[a_{n-1}(s)/a_n(s)] \geq s$ by (9.11), we conclude that $a_{n-1}(s)/a_n(s)$ is nondecreasing in n and so

$$\frac{a_{n-1}(s)}{a_n(s)} \leq \frac{a_n(s)}{a_{n+1}(s)},$$

which proves (9.14).

(ii) Suppose now that F has IFR but is not absolutely continuous. Let $\mathscr{E}_\lambda$ be the exponential distribution function with mean $(1/\lambda)$. Write $F_\lambda^* = F * \mathscr{E}_\lambda$. Since $\mathscr{E}_\lambda$ has a logconcave density, it follows easily from Theorems 2.16 and

9.6 that F_λ^* has IFR. If we write

$$a_n^*(s) = \int_0^\infty \left[\frac{e^{-st}t^n}{n!}\right]\bar{F}_\lambda^*(t)\,dt,$$

then it follows from part (i) of this proof that the sequence $\{a_n^*(s)\}$ is logconcave in n. But $a_n^*(s) \to a_n(s)$ as $\lambda \to \infty$. Therefore, $\{a_n(s)\}$ is also logconcave in n. The "only if" part of the theorem is thus proved.

(iii) To prove the "if" part of the theorem, we introduce the Laplace transforms:

$$\varphi(s) = \int_0^\infty e^{-st}\,dF(t),$$

and

$$\psi(s) = \int_0^\infty e^{-st}\bar{F}(t)\,dt.$$

Then

$$\psi(s) = \frac{1-\varphi(s)}{s}.$$

Assume for the moment that F has a bounded continuous density f. Then

$$\left(\frac{n}{s}\right)^{n+1} b_n\left(\frac{n}{s}\right) = \frac{(n/s)^{n+1}}{n!}\int_0^\infty e^{-nt/s}\,t^n f(t)\,dt$$
$$= \int_0^\infty f(t)\,dG_n(t),$$

where G_n is the gamma distribution function with mean $s(n+1)/n$ and variance $s^2(n+1)/n^2$. Since G_n converges to the degenerate distribution at s, we see that, as $n \to \infty$,

$$\left(\frac{n}{s}\right)^{n+1} b_n\left(\frac{n}{s}\right) \to f(s).$$

Similarly,

$$\left(\frac{n}{s}\right)^{n+1} a_n\left(\frac{n}{s}\right) \to \bar{F}(s).$$

Thus,

$$r(s) = \frac{f(s)}{\bar{F}(s)} = \lim_{n\to\infty}\frac{b_n(n/s)}{a_n(n/s)} = \lim_{n\to\infty} c_n\left(\frac{n}{s}\right). \tag{9.16}$$

If (9.14) holds, then (9.13) shows that $c_n(s)$ is nonincreasing in s for every fixed n. Therefore, it follows from (9.16) that $r(s)$ is nondecreasing in s.

(iv) To complete the proof of the "if" part, we need to remove the restriction that F has a bounded continuous density. Let $\mathscr{E}_\lambda$ be the exponential distribution function with mean $(1/\lambda)$ and write $F_2 = F * \mathscr{E}_\lambda$. Let $\{\beta_n(s)\}$ be the Laplace sequence for F_2. Define φ_2, ψ_2 from F_2 in the same way as φ, ψ are from F. Then

$$\varphi_2(s) = \frac{\lambda\varphi(s)}{s+\lambda}.$$

and

$$\begin{aligned}\psi_2(s) = \frac{1-\varphi_2(s)}{s} &= \frac{1}{s}\left[1 - \frac{\lambda\varphi(s)}{(s+\lambda)}\right] \\ &= \frac{1}{s}\left[1 - \frac{\lambda(1-s\psi(s))}{(s+\lambda)}\right] \\ &= \frac{1}{s+\lambda} + \frac{\lambda\psi(s)}{(s+\lambda)}.\end{aligned} \tag{9.17}$$

Now

$$\psi(s-z) = \sum_{n=0}^{\infty} a_n(s)z^n \quad \text{and} \quad \psi_2(s-z) = \sum_{n=0}^{\infty} \beta_n(s)z^n.$$

Therefore (9.17) shows that

$$\beta_n(s) = (s+\lambda)^{-(n+1)}\left[1 + \lambda \sum_{k=0}^{n} (s+\lambda)^k a_k(s)\right].$$

Write $d_k = \lambda(s+\lambda)^k a_k(s)$ and $D_n = \sum_{k=0}^{n} d_k$. Then $(s+\lambda)^{n+1}\beta_n(s) = 1 + D_n$ and so

$$\begin{aligned}(s+\lambda)^{2n+2}[\beta_n^2(s) - \beta_{n-1}(s)\beta_{n+1}(s)] &= (1+D_n)^2 - (1+D_{n-1})(1+D_{n+1}) \\ &= (D_n^2 - D_{n-1}D_{n+1}) \\ &\quad + (2D_n - D_{n-1} - D_{n+1}) \\ &= A + B, \text{ say.}\end{aligned}$$

Using the definition of D_n, we get

$$\begin{aligned}A &= D_n^2 - (D_n - d_n)(D_n + d_{n+1}) \\ &= D_n d_n - D_n d_{n+1} + d_n d_{n+1} \\ &= d_{-1}d_{n+1} + \sum_{k=0}^{n} (d_k d_n - d_{k-1}d_{n+1}) \\ &= d_{-1}d_{n+1} + C, \text{ say,}\end{aligned}$$

and

$$B = d_n - d_{n+1}.$$

Thus,

$$\begin{aligned} A + B &= C + d_{-1}d_{n+1} + d_n - d_{n+1} \\ &= C + \lambda(s+\lambda)^n[\lambda a_{n+1} + a_n - (s+\lambda)a_{n+1}] \\ &= C + \lambda(s+\lambda)^n[a_n - sa_{n+1}]. \end{aligned} \tag{9.18}$$

The logconcavity of $\{a_n(s)\}$ implies the logconcavity of $\{d_n\}$. Therefore, $C \geq 0$. Further (9.11) shows that the second term of (9.18) is nonnegative. It follows that $\{\beta_n(s)\}$ is logconcave.

The distribution F_2 has a density which may not be continuous. Therefore, we need to take one more convolution. Let $F_3 = F_2 * \mathscr{E}_\lambda = F * \mathscr{E}_\lambda * \mathscr{E}_\lambda$. Write $\{\gamma_n(s)\}$ for the Laplace sequence of F_3. The preceding discussion shows that $\{\gamma_n(s)\}$ is logconcave. But F_3 has a bounded and continuous density. Therefore, by part (iii) of the proof, F_3 has IFR. Now $F_3 \to F$ weakly as $\lambda \to \infty$ and the IFR class is closed under weak limits. Therefore F has IFR if $\{a_n(s)\}$ is logconcave. The proof of Theorem 9.7 is now complete. ■

A Laplace transform characterization of the IFRA class is given in the next theorem.

Theorem 9.8. *A distribution function F on $(0, \infty)$ has IFRA if, and only if, for every $s > 0$, the sequence $\{[a_n(s)]^{1/(n+1)}\}$ is nonincreasing in $n \geq 0$.*

Proof. (i) Assume that F has IFRA. Define a new distribution function G by $\bar{G}(x) = e^{-sx}\bar{F}(x)$, $x > 0$. Let μ_n denote the nth moment of $\bar{G}$ and write $\gamma_n = \mu_n/(n!)$. Then

$$\begin{aligned} a_n(s) &= \int_0^\infty \left[\frac{e^{-st}t^n}{(n!)}\right]\bar{F}(t)\,dt \\ &= \int_0^\infty \left[\frac{t^n}{(n!)}\right]\bar{G}(t)\,dt \\ &= \frac{\mu_{n+1}}{(n+1)!} = \gamma_{n+1}. \end{aligned}$$

By Corollary 6.5, page 112 of Barlow and Proschan (1981), $(\gamma_{n+1})^{1/(n+1)}$ is nonincreasing in $n > -1$. Therefore $[a_n(s)]^{1/(n+1)}$ is nonincreasng in $n \geq 0$.

(ii) To prove the converse, note that we saw in the proof of Theorem 9.7

that, if F is continuous, then, as $n \to \infty$,

$$\left(\frac{n}{s}\right)^{n+1} a_n\left(\frac{n}{s}\right) \to \bar{F}(s).$$

More generally, one easily shows that

$$(\delta_n)^{n+1} a_n(\delta_n) \to \bar{F}(x),$$

whenever x is a continuity point of F and $(n/\delta_n) \to x$.

Suppose that $[a_n(s)]^{1/(n+1)}$ is nonincreasing in $n \geq 0$. Let $\alpha \in (0, 1)$ be a rational number and let x and αx both be continuity points of F. Write $\alpha = p/q$, where $0 < p < q$ and p, q are integers. For $m = 1, 2, \ldots$, let $n = mp$, $k = m(q-p)$ and take $\delta_n = n/(\alpha x)$. Now

$$[a_n(s)]^{1/(n+1)} \geq [a_{n+k}(s)]^{1/(n+k+1)}$$

Therefore,

$$(\delta_n)^{n+1} a_n(\delta_n) \geq [(\delta_n)^{n+k+1} a_{n+k}(\delta_n)]^{(n+1)/(n+k+1)} \tag{9.19}$$

Now $n/\delta_n = \alpha x$, $(n+k)/\delta_n = x$ and $(n+1)/(n+k+1) \to \alpha$. Therefore, letting $n \to \infty$ in (9.19) we get

$$\bar{F}(\alpha x) \geq [\bar{F}(x)]^{\alpha}. \tag{9.20}$$

By the right continuity of $\bar{F}$, we can extend (9.20) to all $\alpha \in (0, 1)$ and all $x > 0$. Now (9.9) shows that F has IFRA. The theorem is now proved. ∎

It is known that the IFR and IFRA classes are closed under convolutions. The proof is given for the IFR class by Barlow and Proschan (1981) (see Theorem 4.2 on page 100). An alternative proof, to be given in the next section, is valid for the multivariate IFR case also. For the IFRA class, the proof was given by Block and Savits (1976) and is presented below. We need a lemma.

Lemma 9.1. *A distribution function F on $(0, \infty)$ has IFRA if, and only if,*

$$\int_0^\infty h(t)\, dF(t) \leq \left\{\int h^{\alpha}\left(\frac{t}{\alpha}\right) dF(t)\right\}^{1/\alpha}, \tag{9.21}$$

for all $\alpha \in (0, 1)$ and for all nonnegative, nondecreasing functions h on $(0, \infty)$.

Proof. (i) Suppose (9.21) holds and take $h(t) = I_{(x, \infty)}(t)$. We then get

$$\bar{F}(x) \leq [\bar{F}(\alpha x)]^{1/\alpha}, \tag{9.22}$$

which is equivalent to (9.9) and so F has IFRA. Thus the "if" part is trivial.

(ii) To prove the "only if" part, assume that F has IFRA. Then (9.9) or, equivalently, (9.22) holds. Now let

$$h(t) = \sum_{i=1}^{n} a_i I_{(x_i,\,\infty)}(t), \tag{9.23}$$

where $a_i \geq 0$ and $0 \leq x_1 \leq \cdots \leq x_n < \infty$. Then

$$\begin{aligned}\int_0^\infty h(t)\,dF(t) &= \sum_{i=1}^{n} a_i \bar{F}(x_i)\\ &\leq \sum_{i=1}^{n} a_i \{\bar{F}(\alpha x_i)\}^{1/\alpha}, \qquad \text{by} \quad (9.22)\\ &= \sum_{i=1}^{n} \{a_i^\alpha \bar{F}(\alpha x_i)\}^{1/\alpha}\\ &= \sum_{i=1}^{n} \left\{\int_0^\infty a_i^\alpha I^\alpha_{(x_i,\,\infty)}\left(\frac{t}{\alpha}\right) dF(t)\right\}^{1/\alpha}\\ &= \sum_{i=1}^{n} \left\| a_i I_{(x_i,\,\infty)}\left(\frac{\cdot}{\alpha}\right)\right\|_\alpha,\end{aligned} \tag{9.24}$$

where, for a given nonnegative function g,

$$\| g \|_\alpha = \left\{\int_0^\infty g^\alpha(t)\,dF(t)\right\}^{1/\alpha}.$$

Now by the Minkowski inequality for $\alpha \in (0, 1)$ [see Marshall and Olkin (1979), p. 460],

$$\begin{aligned}\sum_{i=1}^{n} \left\| a_i I_{(x_i,\,\infty)}\left(\frac{\cdot}{\alpha}\right)\right\|_\alpha &\leq \left\| \sum_{i=1}^{n} a_i I_{(x_i,\,\infty)}\left(\frac{\cdot}{\alpha}\right)\right\|_\alpha\\ &= \left\| h\left(\frac{\cdot}{\alpha}\right)\right\|_\alpha\\ &= \left\{\int_0^\infty h^\alpha\left(\frac{t}{\alpha}\right) dF(t)\right\}^{1/\alpha}.\end{aligned} \tag{9.25}$$

Combining (9.24) and (9.25), we see that (9.21) holds for functions h of the form (9.23). The "only if" part of the lemma is now a direct consequence of the monotone convergence theorem. The lemma is thus proved. ■

Theorem 9.9. *The class of IFRA distributions is closed under convolutions.*

Proof. Let F and G have IFRA and write $H = F * G$. Let h be a nonnegative nondecreasing function. Then

$$\int h(z)\, dH(z) = \int \int h(x+y)\, dF(x)\, dG(y).$$

Since F has IFRA and $h(x+y)$ is nondecreasing in x for every fixed y,

$$\int h(x+y)\, dF(x) \le \left\{ \int h^{\alpha}\left[\left(\frac{x}{\alpha}\right) + y\right] dF(x) \right\}^{1/\alpha}$$
$$= \psi(y), \text{ say.}$$

Therefore,

$$\int h(z)\, dH(z) \le \int \psi(y)\, dG(y). \tag{9.26}$$

Since G has IFRA and ψ is nondecreasing, Lemma 9.1 implies that

$$\int \psi(y)\, dG(y) \le \left\{ \int \psi^{\alpha}\left(\frac{y}{\alpha}\right) dG(y) \right\}^{1/\alpha}$$
$$= \left\{ \int \int h^{\alpha}\left[\left(\frac{x}{\alpha}\right) + \left(\frac{y}{\alpha}\right)\right] dF(x)\, dG(y) \right\}^{1/\alpha}$$
$$= \left\{ \int h^{\alpha}\left(\frac{z}{\alpha}\right) dH(z) \right\}^{1/\alpha}. \tag{9.27}$$

Combining (9.26) and (9.27), we get

$$\int h(z)\, dH(z) \le \left\{ \int h^{\alpha}\left(\frac{z}{\alpha}\right) dH(z) \right\}^{1/\alpha}.$$

Lemma 9.1 again shows that H has IFRA. The proof of the theorem is thus complete. ■

9.3. Multivariate IFR and IFRA Classes

This section is devoted to a brief discussion of the multivariate versions of the IFR and IFRA concepts. The results we present are due to Block and Savits (1980b) and Savits (1985).

The definition of the multivariate IFR class (MIFR) given by Savits (1985) is based on the following lemma.

Lemma 9.2. *A nonnegative random variable X has IFR if, and only if, X has the same distribution as $\psi(S)$, where S has the standrd exponential distribution and ψ is a continuous nonnegative nondecreasing concave function on* $[0, \infty)$.

Proof. (i) Suppose X has IFR and let $H(x) = -\log \bar{F}(x)$ be the hazard function of X. If X is degenerate at some point x_0, then we can take $\psi(x) \equiv x_0$. Suppose X is not degenerate. Let $b = \inf\{x \geq 0 : \bar{F}(x) = 0\}$ and $a = \sup\{x \geq 0 : \bar{F}(x) = 1\}$. Then $0 \leq a < b \leq \infty$. Now H is convex and continuous on $(-\infty, b)$ and also strictly increasing on $[a, b)$. If $A = H(b-)$ and φ is the restriction of H to $[a, b)$, then the inverse function φ^{-1} is continuous, strictly increasing and concave on $[0, A)$. Now define

$$\psi(s) = \inf\{x \geq 0 \colon H(x) > s\}, \qquad s \geq 0.$$

Then $\psi(s) = \varphi^{-1}(s)$ for $s \in [0, A)$ and $\psi(s) = b$ for $s \geq A$. Now ψ is clearly concave on $[0, \infty)$. We claim that, if S is a standard exponential random variable, then $\psi(S)$ has the same distribution as X. To see this, observe that if $0 \leq x < b$, then $0 \leq H(x) < A$ and so

$$\bar{F}(x) = e^{-H(x)} = P[S > H(x)] = P[\psi(S) > x].$$

On the other hand, if $x \geq b$, then

$$\bar{F}(x) = 0 = P[\psi(S) > x].$$

The lemma is thus proved. ■

This lemma immediately gives an interesting characterization of the IFR class.

Theorem 9.10 *A nonnegative random variable X has IFR if, and only if,* $E[h(X, \mathbf{y})]$ *is logconcave in* $\mathbf{y} \in R^m$ *for every function* $h(x, \mathbf{y})$ *which is logconcave in* $(x, \mathbf{y})$ *and is nondecreasing in x for every fixed* $\mathbf{y}$.

Proof. Suppose X has IFR. By Lemma 9.2, there is a continuous nondecreasing nonnegative concave function ψ on $[0, \infty)$ such that, if S is a standard exponential random variable, then $\psi(S)$ has the same distribution as X. Let

$h(x, \mathbf{y})$ be logconcave in $(x, \mathbf{y})$ and nondecreasing in x for every fixed $\mathbf{y}$. Then

$$E[h(X, \mathbf{y})] = E[h(\psi(S), \mathbf{y})] = \int_0^\infty h[\psi(s), \mathbf{y}]e^{-s}\, ds.$$

$$= \int_0^\infty g(s, \mathbf{y})e^{-s}\, ds,$$

where $g(s, \mathbf{y}) = h[\psi(s), \mathbf{y}]$. Since ψ is concave and $h(x, \mathbf{y})$ is logconcave and nondecreasing in x, it is easily checked that $g(s, \mathbf{y})$ is logconcave in $(s, \mathbf{y})$. The logconcavity of $E[h(X, \mathbf{y})]$ now follows by Theorem 2.16. This proves the "only if" part. To prove the converse, let $h(x, y) = I_{(y, \infty)}(x)$, where $y \in R.$. Then h has the required properties. Further $E[h(X, y)] = \bar{F}(y)$. Therefore, the logconcavity of $E[h(X, y)]$ implies that F has IFR. The theorem is thus proved. ■

Theorem 9.10 motivates a definition of the multivariate IFR (or MIFR) class.

Definition 9.7. A nonnegative random vector $\mathbf{X}$ is said to have MIFR if $E[h(\mathbf{X}, \mathbf{y})]$ is logconcave in $\mathbf{y}$ for all functions $h(\mathbf{x}, \mathbf{y})$ which are logconcave in $(\mathbf{x}, \mathbf{y})$ and nondecreasing in $\mathbf{x}$ for every fixed $\mathbf{y}$.

Remark. The function $h(\mathbf{x}, \mathbf{y})$ in the above definition can be restricted to be continuous and/or bounded.

With the above definition, the class of MIFR distributions enjoys many desirable properties. All of these follow from the theorems on logconcave distributions proved in Chapter 2 and by suitable choices of the function h in Definition 9.7. Two of these properties are isolated in the next theorem. We list a few more properties after the proof of the theorem.

Theorem 9.11

(a) *If* $\mathbf{X}$ *has a logconcave density, then* $\mathbf{X}$ *has MIFR.*
(b) *The MIFR class is closed under convolutions.*

Proof. (i) Suppose $\mathbf{X}$ has a logconcave density f. Let $h(\mathbf{x}, \mathbf{y})$ be nondecreasing in $\mathbf{x}$ and logconcave in $(\mathbf{x}, \mathbf{y})$. Then

$$E[h(\mathbf{x}, \mathbf{y})] = \int h(\mathbf{x}, \mathbf{y})f(\mathbf{x})\, d\mathbf{x}. \tag{9.28}$$

Now $h(\mathbf{x}, \mathbf{y})f(\mathbf{x})$ is logconcave in $(\mathbf{x}, \mathbf{y})$. Therefore, by Theorem 2.16, the right side of (9.28) is logconcave in $\mathbf{y}$. Thus, $\mathbf{X}$ has MIFR.

(ii) Let $\mathbf{X}$, $\mathbf{Y}$ be independent random n-vectors in the MIFR class with distribution functions F and G. Write $\mathbf{Z} = \mathbf{X} + \mathbf{Y}$. Suppose $h(\mathbf{z}, \mathbf{w})$ is nondecreasing in $\mathbf{z}$ and logconcave in $(\mathbf{z}, \mathbf{w})$. Then

$$E[h(\mathbf{Z}, \mathbf{w})] = \int\int h(\mathbf{x} + \mathbf{y}, \mathbf{w})\, dG(\mathbf{y})\, dF(\mathbf{x})$$

$$= \int h_1(\mathbf{x}, \mathbf{w})\, dF(\mathbf{x}), \tag{9.29}$$

where

$$h_1(\mathbf{x}, \mathbf{w}) = \int h(\mathbf{x} + \mathbf{y}, \mathbf{w})\, dG(y).$$

Since $\mathbf{Y}$ has MIFR, $h_1(\mathbf{x}, \mathbf{w})$ is logconcave in $(\mathbf{x}, \mathbf{w})$. It is also nondecreasing in $\mathbf{x}$. Therefore, the MIFR property for $\mathbf{X}$ means that the right side of (9.29) is logconcave in $\mathbf{w}$. Thus $\mathbf{Z}$ has MIFR and the theorem is proved. ■

Remark. Theorem 9.10 and Definition 9.7 show that the univariate MIFR class is the same as the IFR class. Therefore, we see from Theorem 9.11 that the IFR class of distributions on $[0, \infty)$ is closed under convolutions.

The following is a list of some more properties of MIFR distributions. As observed before, the proofs of these properties are straightforward.

(a) The marginals of an MIFR distribution are MIFR.
(b) If $\mathbf{X}$, $\mathbf{Y}$ have MIFR distributions and $\mathbf{X}$, $\mathbf{Y}$ are independent, then $(\mathbf{X}, \mathbf{Y})$ has MIFR.
(c) The MIFR class is closed under weak limits.
(d) If $\mathbf{X}$ has MIFR and g_i, $i = 1, \ldots, k$ are nonnegative, nondecreasing and concave functions, then $(g_1(\mathbf{X}), \ldots, g_k(\mathbf{X}))$ has MIFR.

An important example of an MIFR distribution which is useful in certain shock models is as follows.

Example 9.1. Consider the multivariate exponential distribution of Marshall and Olkin (1967). This distribution has the following structure. Let $S_1, \ldots, S_n$ be independent exponentially distributed random variables. Let $\{A_1, \ldots, A_k\}$ be a class of subsets of $\{1, \ldots, n\}$. Write $T_i = \min\{S_j : j \in A_i\}$, $i = 1, \ldots, k$. The distribution of the vector $\mathbf{T} = (T_1, \ldots, T_k)$ is then called a *multivariate exponential distribution.* Now $\mathbf{S} = (S_1, \ldots, S_n)$ has a logconcave density and so $\mathbf{S}$ has MIFR. Let $g_i(\mathbf{s}) = \min\{s_j : j \in A_i\}$. The functions g_i satisfy the hypotheses of (d) above and $T_i = g_i(\mathbf{S})$. Therefore, $\mathbf{T}$ has MIFR.

A multivariate generalization of the IFRA class was given by Block and Savits (1980b). This is based on the characterization of the univariate IFRA distributions given in Lemma 9.1.

Definition 9.8. A nonnegative random vector $\mathbf{X} = (X_1, \ldots, X_n)$ is said to have MIFRA if for every nonnegative, nondecreasing function h and for every $\alpha \in (0, 1)$, we have

$$E[h(\mathbf{X})] \le \left\{ E\left[h^{\alpha}\left(\frac{\mathbf{X}}{\alpha}\right)\right]\right\}^{1/\alpha}. \tag{9.30}$$

Remark. The function h in the above definition can be restricted to be continuous and/or bounded.

It is of interest to see how (9.30) is related to the "star shaped" character of certain functions. Call a subset C of the nonnegative orthant in R^n an *upper set* if

$$(\mathbf{x} \in C \text{ and } \mathbf{y} \ge \mathbf{x}) \quad \Rightarrow \quad \mathbf{y} \in C.$$

Let Γ denote the class of all upper sets. Then (9.30) is equivalent to

$$P[\mathbf{X} \in C] \le \{P[\mathbf{X} \in \alpha C]\}^{1/\alpha}, \tag{9.31}$$

for every $C \in \Gamma$ and $\alpha \in (0, 1)$. The proof of the equivalence of (9.30) and (9.31) follows the same lines as the proof of Lemma 9.1. A function g defined on Γ can be called *star-shaped* if

$$g(\alpha C) \le \alpha g(C),$$

for all $C \in \Gamma$ and $\alpha \in (0, 1)$. Then (9.31) says that $\mathbf{X}$ has MIFRA if, and only if, $-\log P(\mathbf{X} \in C)$ is a star-shaped function on Γ.

The following lemma is useful in deriving some properties of the MIFRA class of distributions.

Lemma 9.3. *Let* $\mathbf{X}$ *have MIFRA and let* $Y_i = g_i(\mathbf{X})$, $i = 1, \ldots, m$, *where each* g_i *is a nondecreasing function satisfying*

$$\alpha g_i\left(\frac{\mathbf{x}}{\alpha}\right) \le g_i(\mathbf{x}),$$

for all $\alpha \in (0, 1)$ *and for all* $\mathbf{x} \ge \mathbf{0}$. *Then* $\mathbf{Y}$ *has MIFRA.*

Proof. Let h be a nondecreasing, nonnegative function on R^k. Write $\mathbf{g} = (g_1, \ldots, g_k)$. Then

$$\begin{aligned} E[h(\mathbf{Y})] &= E[h(\mathbf{g}(\mathbf{X}))] \\ &\le E^{1/\alpha}\{h^\alpha[\mathbf{g}(\mathbf{X}/\alpha)]\} \\ &\le E^{1/\alpha}[h^\alpha(\alpha^{-1}\mathbf{g}(\mathbf{X}))] \\ &= E^{1/\alpha}\left[h^\alpha\left(\frac{\mathbf{Y}}{\alpha}\right)\right], \end{aligned}$$

which proves the lemma. ■

We now mention some important closure properties of the MIFRA class. These can be proved by direct generalizations of the univariate proofs and by suitably choosing the functions g_i in Lemma 9.3. So we will omit the proofs and list the properties.

(a) The marginals of an MIFRA distribution are MIFRA.
(b) The MIFRA class of distribution in R^n is closed under weak limits.
(c) If $\mathbf{X}$ and $\mathbf{Y}$ have MIFRA and $\mathbf{X}, \mathbf{Y}$ are independent, then $(\mathbf{X}, \mathbf{Y})$ has MIFRA.
(d) The MIFRA class is closed under convolutions.

Using these closure properties and Lemma 9.3, Block and Savits (1980b) have constructed several examples of MIFRA distributions. Of these, we mention just two.

Example 9.2. Recall the multivariate exponential distribution of Marshall and Olkin (1967). This is the distribution of $\mathbf{T} = (T_1, \ldots, T_m)$ with $T_i = \min_{j \in A_i} S_j$, where $S_1, \ldots, S_n$ are independent exponentially distributed random variables and $A_1, \ldots, A_m$ are subsets of $\{1, \ldots, n\}$. Since each S_i has IFRA, property (c) above shows that $(S_1, \ldots, S_n)$ has MIFRA. Further, the functions $g_i(\mathbf{s}) = \min_{j \in A_i} s_j$ satisfy the hypotheses of Lemma 9.3. Therefore, $\mathbf{T}$ has MIFRA.

Example 9.3. A multivariate gamma distribution has been discussed by Johnson and Kotz (1977, pp. 216–219). This has the following structure. Let $Y_j = X_0 + X_j$, $j = 1, \ldots, n$, where $X_0, X_1, \ldots, X_n$ are independent and X_i has the gamma density

$$f_i(x) = [\Gamma(\theta_i)]^{-1} x^{\theta_i - 1} e^{-x}, \qquad x > 0.$$

Suppose $\theta_i \ge 1$ for all $i \ge 0$. Then each X_i has an IFRA distribution. Therefore,

$\mathbf{X} = (X_0, \ldots, X_n)$ has MIFRA. Further, the functions $g_j(\mathbf{x}) = x_0 + x_j$ satisfy the hypotheses of Lemma 9.3. Therefore $\mathbf{Y} = (Y_1, \ldots, Y_n)$ has MIFRA.

Block and Savits (1980b) have given another method of constructing MIFRA distributions through the use of the following lemma.

Lemma 9.4. *Let* $\mathbf{X}$ *be a random n-vector having MIFRA. Suppose that* Y *is a real random variable whose conditional survival function*

$$\bar{G}(y \mid \mathbf{x}) = P(Y > y \mid \mathbf{X} = \mathbf{x})$$

satisfies the conditions

(i) $\bar{G}(y \mid \mathbf{x})$ *is nondecreasing in* $\mathbf{x}$, *and*
(ii) $\bar{G}(y \mid \mathbf{x}) \leq [\bar{G}(\alpha y \mid \alpha \mathbf{x})]^{1/\alpha}$ *for* $\alpha \in (0, 1)$.

Then $(\mathbf{X}, Y)$ *has MIFRA.*

Proof. (A) Let u be a nonnegative and nondecreasing function on R. Then we claim that

(a) $E[u(Y) \mid \mathbf{X} = \mathbf{x}]$ is nondecreasing in $\mathbf{x}$, and
(b) $E[u(Y) \mid \mathbf{X} = \mathbf{x}] \leq E^{1/\alpha}[u^{\alpha}(Y/\alpha) \mid \mathbf{X} = \alpha \mathbf{x}]$.

To see (a), observe that if $u(y) = I_{(t,\infty)}(y)$, then (a) reduces to condition (i). Now (a) follows as usual by taking linear combinations and monotone limits. Similarly, if $u(y) = I_{(t,\infty)}(y)$, then (b) reduces to condition (ii). Now (b) can be proved by taking linear combinations and using Minkowski's inequality as in the proof of Lemma 9.1.

(B) Let $h(\mathbf{x}, y)$ be nonnegative and nondecreasing on R^{n+1}. For $\mathbf{x}, \mathbf{x}'$ in R^n, let

$$v_{\alpha}(\mathbf{x}', \mathbf{x}) = E\left[h^{\alpha}\left(\mathbf{x}', \frac{Y}{\alpha}\right) \middle| \mathbf{X} = \alpha \mathbf{x}\right],$$

where $\alpha \in (0, 1)$. From (a) above, $v_{\alpha}(\mathbf{x}', \mathbf{x})$ is nondecreasing in $(\mathbf{x}', \mathbf{x})$. Therefore, if we write $w_{\alpha}(\mathbf{x}) = v_{\alpha}(\mathbf{x}, \mathbf{x})$, then w_{α} is also nondecreasing in $\mathbf{x}$. Now by (b) above,

$$E[h(\mathbf{x}', Y) \mid \mathbf{X} = \mathbf{x}] \leqq E^{1/\alpha}\left[h^{\alpha}\left(\mathbf{x}', \frac{Y}{\alpha}\right) \middle| \mathbf{X} = \alpha \mathbf{x}\right].$$

That is,

$$v_1(\mathbf{x}', \mathbf{x}) \leqq [v_{\alpha}(\mathbf{x}', \mathbf{x})]^{1/\alpha}.$$

Therefore,

$$w_1(\mathbf{x}) \le [w_\alpha(\mathbf{x})]^{1/\alpha}.$$

The last inequality and the MIFRA character of **X** show that

$$E[w_1(\mathbf{X})] \le E\{[w_\alpha(\mathbf{X})]^{1/\alpha}\} \le E^{1/\alpha}\left[w_\alpha\left(\frac{\mathbf{X}}{\alpha}\right)\right]. \tag{9.32}$$

Now

$$E[w_\alpha(\mathbf{X}/\alpha)] = E \cdot E\left[h^\alpha\left(\frac{\mathbf{X}}{\alpha}, \frac{Y}{\alpha}\right)\middle| \mathbf{X}\right]$$

$$= E\left[h^\alpha\left(\frac{\mathbf{X}}{\alpha}, \frac{Y}{\alpha}\right)\right]. \tag{9.33}$$

Setting $\alpha = 1$ in (9.33), we get

$$E[w_1(\mathbf{X})] = E[h(\mathbf{X}, Y)]. \tag{9.34}$$

Combine (9.32), (9.33) and (9.34) to obtain

$$E[h(\mathbf{X}, Y)] \le E^{1/\alpha}\left[h^\alpha\left(\frac{\mathbf{X}}{\alpha}, \frac{Y}{\alpha}\right)\right].$$

This proves that $(\mathbf{X}, Y)$ has MIFRA and completes the proof of the theorem. ■

Block and Savits (1980b) have given the following example to illustrate the use of Lemma 9.4.

Example 9.4. Consider a two component system where the lifetimes X, Y of components 1, 2 satisfy the following conditions.

(a) As long as both components survive, their lifetimes behave as independent exponential random variables with parameters λ_1, λ_2, respectively.
(b) If component 1 fails first, the remaining lifetime of component 2 is exponential with parameter λ_2' where $\lambda_2' \ge \lambda_2$.
(c) If component 2 fails first, the remaining lifetime of component 1 continues to be exponential with parameter λ_1.

It follows from these assumptions that X has the exponential distribution with parameter λ_1 and so X has IFRA. Further, the conditional survival function of Y given $X = x$ is

$$\bar{G}(y \mid x) = \begin{cases} \exp(-\lambda_2 y), & y < x \\ \exp[-\lambda_2' y + (\lambda_2' - \lambda_2)x], & y \geq x. \end{cases}$$

It is trivial to see that $\bar{G}(y \mid x)$ satisfies conditions (i) and (ii) of Lemma 9.4. Therefore, (X, Y) has a bivariate IFRA distribution. We note that the model considered here is a special ase of a model given by Freund (1961).

The rest of this section discusses the relationship between the MIFR and MIFRA classes. In the univariate case, we have seen that the IFR class is contained in the IFRA class. However, the corresponding result in the multivariate case is still open. An explanation for this situation will be given after the next theorem of which the first part is a restatement of (9.31) and the second part was given by Savits (1985).

Theorem 9.12. *Let* $\mathbf{X}$ *be a nonnegative random n-vector and let* μ *be the probability measure induced by* $\mathbf{X}$ *on the Borel* σ*-algebra in* R^n*. Then*

(a) $\mathbf{X}$ *has MIFRA if, and only if*

$$\mu(\alpha C) \geq [\mu(C)]^{\alpha} \tag{9.35}$$

for all upper sets $C \subset R^n$ *and for all* $\alpha \in (0, 1)$.

(b) $\mathbf{X}$ *has MIFR, if, and only if,*

$$\mu[\alpha C + (1 - \alpha)D] \geq [\mu(C)]^{\alpha}[\mu(D)]^{1-\alpha}, \tag{9.36}$$

for all upper convex sets C *and* D *in* R^n *and for all* $\alpha \in (0, 1)$.

Proof. (i) As noted above, part (a) is a restatement of (9.31).

(ii) Suppose $\mathbf{X}$ has MIFR. Let C and D be upper convex sets. For $\mathbf{y} = (y_1, y_2)$ with $y_1 > 0$ and $y_2 > 0$, let

$$h(\mathbf{x}, \mathbf{y}) = I_{y_1 C + y_2 D}(\mathbf{x}).$$

Then $E[h(\mathbf{X}, \mathbf{y})] = \mu(y_1 C + y_2 D)$. Since C and D are upper sets, it is easily seen that $h(\mathbf{x}, \mathbf{y})$ is nondecreasing in $\mathbf{x}$. Further the convexity of C and D also implies the logconcavity of $h(\mathbf{x}, \mathbf{y})$ in $(\mathbf{x}, \mathbf{y})$. From Defnition 9.7, we conclude that $\mu(y_1 C + y_2 D)$ is logconcave in $\mathbf{y}$. From this, (9.36) follows easily if we set $y_1 = \alpha$ and $y_2 = (1 - \alpha)$.

Conversely, suppose (9.36) holds. Let $h(\mathbf{x}, \mathbf{y})$ be logconcave in $(\mathbf{x}, \mathbf{y})$ and nondecreasing in $\mathbf{x}$. For $t \in R$, let

$$A(t, \mathbf{y}) = \{\mathbf{x} : h(\mathbf{x}, \mathbf{y}) \geq e^{-t}\}.$$

The logconcavity of h shows that $A(t, \mathbf{y})$ is a convex subset of R^n. It is also an upper set because $h(\mathbf{x}, \mathbf{y})$ is nondecreasing in $\mathbf{x}$. Now fix $\alpha \in (0, 1)$ and let t, t_1, t_2 and $\mathbf{y}$, $\mathbf{y}_1$, $\mathbf{y}_2$ be such that

$$t = \alpha t_1 + (1 - \alpha)t_2 \quad \text{and} \quad \mathbf{y} = \alpha \mathbf{y}_1 + (1 - \alpha)\mathbf{y}_2.$$

Then the logconcavity of h again shows that

$$A(t, \mathbf{y}) \supset \alpha A(t_1, \mathbf{y}_1) + (1 - \alpha)A(t_2, \mathbf{y}_2).$$

Therefore (9.36) shows that

$$\begin{aligned} \mu[A(t, \mathbf{y})] &\geq \mu[\alpha A(t_1, \mathbf{y}_1) + (1 - \alpha)A(t_2, \mathbf{y}_2)] \\ &\geq \{\mu[A(t_1, \mathbf{y}_1)]\}^{\alpha}\{\mu[A(t_2, \mathbf{y}_2)]\}^{1-\alpha}. \end{aligned}$$

That is, $\mu[A(t, \mathbf{y})]$ is logconcave in $(t, \mathbf{y})$. Now

$$\begin{aligned} h(\mathbf{x}, \mathbf{y}) &= \int_{-\log h(\mathbf{x}, \mathbf{y})}^{\infty} e^{-t}\, dt \\ &= \int_{-\infty}^{\infty} I_{A(t, \mathbf{y})}(\mathbf{x}) e^{-t}\, dt. \end{aligned}$$

Therefore,

$$E[h(\mathbf{X}, \mathbf{y})] = \int_{-\infty}^{\infty} \mu[A(t, \mathbf{y})] e^{-t}\, dt. \tag{9.37}$$

The integrand in (9.37) is logconcave in $(t, \mathbf{y})$. Therefore, by Theorem 2.16, $E[h(\mathbf{X}, \mathbf{y})]$ is logconcave in $\mathbf{y}$. Thus $\mathbf{X}$ has MIFR. The proof of the theorem is now complete. ∎

To end the section, let us see why the implication MIFR ⇒ MIFRA is still open. In (9.36), put $D = Q_n$, the nonnegative orthant in R^n. We get

$$\mu[\alpha C + (1 - \alpha)Q_n] \geq [\mu(C)]^{\alpha}[\mu(Q_n)]^{1-\alpha}. \tag{9.38}$$

But $\alpha C + (1 - \alpha)Q_n = \alpha C$ because C is an upper set and $\mu(Q_n) = 1$ because $\mathbf{X}$ is nonnegative. Therefore, (9.38) reduces to

$$\mu[\alpha C] \geq [\mu(C)]^{\alpha},$$

for all upper *convex* sets C and for all $\alpha \in (0, 1)$. To claim that $\mathbf{X}$ has MIFRA, we must have (9.35) for *all* upper sets, not just upper *convex* sets. On the line, every upper set is convex and so IFR ⇒ IFRA. But in dimension $n \geq 2$, there are upper sets which are not convex and this fact is the source of the difficulty in proving the required implication. Indeed, it is not known whether a distribution having a logconcave density in R^n, $n \geq 2$, has MIFRA.

9.4. Unimodality and Other Convexity Properties of Order Statistics

Order statistics play an important role in statistical work. They arise naturally in reliability theory when one considers k-out-of-n systems. Therefore, it is of some interest to know whether the order statistics based on a vector observation $\mathbf{X}$ inherit any type of unimodality or some other convexity property from $\mathbf{X}$. This question is discussed in the present section. It will be seen that strong unimodality (logconcavity) of $\mathbf{X}$ is passed on to the order statistics calculated from $\mathbf{X}$. However, weaker properties of unimodality may not be inherited. Other convexity properties considered here include the IFR and IFRA properties. Some of the results are stated in terms of the partial orderings introduced in Section 9.1.

The first result we present considers individual order statistics. Let G be a distribution function on R and let $X_{(1)} < \cdots < X_{(n)}$ be the order statistics computed from a random sample of size n drawn from G. As before, the survival function $\bar{G}$ is given by $\bar{G}(x) = 1 - G(x)$. Whenever G is continuous and unimodal, G will have a density g which can be taken to be the right or the left derivative of G. The next theorem is due to Alam (1972).

Theorem 9.13. *Let G be continuous and unimodal. Suppose the density g of G is such that $1/g$ is convex. Then each order statistic $X_{(i)}$ has a unimodal distribution.*

Proof. The density f_i of $X_{(i)}$ is given by

$$f_i(x) = Cg(x)[G(x)]^{i-1}[\bar{G}(x)]^{n-i}, \tag{9.39}$$

where C is a constant depending on i and n. We see easily from (9.39) that

$$\frac{d}{dx} \log f_i(x) = g(x) \left[\frac{g'(x)}{g^2(x)} + \frac{(i-1)}{G(x)} - \frac{(n-i)}{\bar{G}(x)} \right]. \tag{9.40}$$

Here the derivatives can be taken to be the right derivatives. By hypothesis, $g'(x)/g^2(x)$ is nonincreasing. Since G is nondecreasing and $\bar{G}$ is nonincreasing, we see from (9.40) that the right derivative of log $f_i(x)$ changes sign at most once and any change of sign must be from positive to negative. Thus log $f_i(x)$ and hence $f_i(x)$ is unimodal. This proves the theorem. ■

Remark. If g is a density on R such that $1/g$ is convex, then g has to be unimodal. This is because $\{x: g(x) \geq c\} = \{x: [1/g(x)] \leq 1/c\}$ and the latter set is an interval for every $c > 0$.

Remark. Alam (1972) points out that the largest order statistic $X_{(n)}$ has a unimodal distribution under hypotheses weaker than those of Theorem 9.13. Since $i = n$ in this case, we see from (9.40) that $X_{(n)}$ would be unimodal as soon as G is unimodal with a mode M and a continuous density g for which $1/g$ is convex on the interval (M, ∞). Similarly, the convexity of $1/g$ on $(-\infty, M)$ would be sufficient to yield the unimodality of the smallest order statistic $X_{(1)}$.

The following theorem is a strengthened version of the result proved by Huang and Ghosh (1982).

Theorem 9.14. *Suppose* **X** *has a logconcave density* g *on* R^n *which has permutation symmetry. Then the joint distribution of* $X_{(1)}, \ldots, X_{(n)}$ *is logconcave and all linear combinations* $\sum a_i X_{(i)}$ *have strongly unimodal distributions.*

Proof. Write $Y_i = X_{(i)}$ and $\mathbf{Y} = (Y_1, \ldots, Y_n)$. The density f of $\mathbf{Y}$ is given by

$$f(\mathbf{y}) = n!\, g(\mathbf{y}), \qquad -\infty < y_1 < \cdots < y_n < \infty.$$

Thus f is the restriction of a logconcave function to a convex set and so f must be logconcave. The second assertion of the theorem follows because the marginals of a logconcave density are logconcave. ∎

Remark. The theorem given by Huang and Ghosh (1982) is a special case of Theorem 9.14 in that they only prove the strong unimodality of each order statistic and also assume that the components of **X** are independent.

We now present a few examples which answer some questions related to Theorems 9.13 and 9.14. These examples have been discussed by Huang and Ghosh.

Example 9.5. Suppose g is the Cauchy density given by $g(x) = [\pi(1 + x^2)]^{-1}$. Then $1/g$ is convex and so by Theorem 9.13, all the $X_{(i)}$ have unimodal distributions for all sample sizes n. However, these distributions are never strongly unimodal. To see this, observe that

$$\bar{G}(x) = \frac{[\arctan(1/x)]}{\pi} \qquad \text{for } x > 0$$

and

$$G(x) = -\frac{[\text{arc } \tan(1/x)]}{\pi} \qquad \text{for } x < 0.$$

Therefore (9.39) shows that we can find constants A and B and integcrs a and b such that $f_i(x) \sim Ax^a$ as $x \to \infty$ and $f_i(x) \sim B|x|^b$ as $x \to -\infty$. It follows that $E(|X_{(i)}|^k) = \infty$ for all sufficiently large k. Since logconcave densities have moment generating functions and, in particular, finite moments of all orders, we conclude that $X_{(i)}$ cannot have a strongly unimodal distribution. Thus the hypotheses of Theorem 9.13 are not strong enough to imply the strong unimodality of order statistics.

Example 9.6. Let g be given by

$$g(x) = \begin{cases} 1, & -\frac{1}{2} < x < 0 \\ \frac{1}{2}, & 0 < x < 1. \end{cases}$$

Then g is unimodal. The density f_n of the largest order statistic for a sample of size $n \geq 2$ is given by

$$f_n(x) = ng(x)[G(x)]^{n-1}.$$

Since G is strictly increasing on $[-\frac{1}{2}, 1]$ and g has a downward jump at 0, we see that f_n has two local modes at 0 and at 1. Thus f_n is not unimodal. Thus the unimodality of G alone is not sufficient to give unimodality for all order statistics.

Example 9.7. Suppose $a > 1$ and let g be given by $g(x) = Cx^{-1/2}$, $1 < x < a^2$. Here C is a normalizing constant. We have $G(x) = 2C(\sqrt{x} - 1)$ and $\bar{G}(x) = 2C(a - \sqrt{x})$ for $x \in (1, a^2)$. We easily get

$$(\log g)''(x) = (2x^2)^{-1}, \tag{9.41}$$

$$(\log G)''(x) = \frac{1 - 2\sqrt{x}}{4x^{3/2}(\sqrt{x} - 1)^2},$$

and

$$(\log \bar{G})''(x) = \frac{a - 2\sqrt{x}}{4x^{3/2}(a - \sqrt{x})^2}.$$

Suppose now that we want to check whether the density $f_i(x)$ of the ith order statistic for a sample of size $n \geq 2$ from g is strongly unimodal. Since either

$(i-1) \geq 1$ or $(n-i) \geq 1$, we see from (9.39) that f_i is logconcave as soon as

$$[\log g + \log G] \quad \text{and} \quad [\log g + \log \bar{G}]$$

are both concave. Routine algebra yields

$$[\log g + \log G]''(x) = \frac{2 - 3\sqrt{x}}{4x^2(\sqrt{x}-1)^2},$$

and

$$[\log g + \log \bar{G}]''(x) = \frac{2a^2 - 3a\sqrt{x}}{4x^2(a - \sqrt{x})^2}.$$

The last two expressions are nonpositive for $x \geq 1$ if $a \leq \frac{3}{2}$. Thus, if $a \leq \frac{3}{2}$ then the order statistics have strongly unimodal distributions for samples of size $n \geq 2$. On the other hand (9.41) shows that g is not logconcave. Thus the strong unimodality of G is not necessary for the strong unimodality of the order statistics for samples of size ≥ 2.

Let us now turn to some results on order statistics which involve the partial orderings considered in Section 9.1. As will be seen, any convex or star orderings among parent distributions are maintained by all the order statistics. This result has been given by Barlow and Proschan (1981, pp. 107–108). For the MR ordering, however, such a result can only be proved for the highest order statistic.

Theorem 9.15. *Let F_{1i} and F_{2i} denote the distribution functions of order statistics $X_{(1i)}$ and $X_{(2i)}$ for samples of size n drawn from two distribution functions G_1 and G_2 respectively. If $G_1 \leq^c G_2$, then $F_{1i} \leq^c F_{2i}$.*

Proof. Define the incomplete beta function $B_{i,n}(x)$ as follows.

$$B_{i,n}(x) = \int_0^x \frac{u^{i-1}(1-u)^{n-i}}{B(i, n-i+1)}\, du, \qquad 0 < x < 1. \tag{9.42}$$

By a well-known calculation

$$B_{i,n}(x) = \sum_{j=i}^{n} \binom{n}{j} x^j (1-x)^{n-j}. \tag{9.43}$$

Now let F_i be the distribution function of the ith order statistic for a sample of size n from a distribution function G. Then

$$F_i(x) = \sum_{j=i}^{n} \binom{n}{j} [G(x)]^j [\bar{G}(x)]^{n-j}$$
$$= B_{i,n}[G(x)], \qquad \text{using (9.43).}$$

We also note that $B_{i,n}$ is strictly increasing and thus has an inverse.

Now, in terms of the notation of the statement of the theorem,

$$F_{2i}^{-1}F_{1i} = [B_{i,n}(G_2)]^{-1}[B_{i,n}(G_1)]$$
$$= G_2^{-1}B_{i,n}^{-1}B_{i,n}G_1 = G_2^{-1}G_1.$$

The theorem now follows from Definition 9.2 because the hypothesis $G_1 \leq^c G_2$ means that $G_2^{-1}G_1$ is convex and the assertion $F_{1i} \leq^c F_{2i}$ means that $F_{2i}^{-1}F_{1i}$ is convex.

Theorem 9.15 can be used to prove that the IFR property of a distribution is inherited by the order statistics. Recall that a distribution function G on $[0, \infty)$ has IFR if, and only if, $G \leq^c \mathscr{E}_\lambda$, where $\mathscr{E}_\lambda$ is the exponential distribution function with mean $1/\lambda$.

Theorem 9.16. *The IFR property of a distribution function G is inherited by the order statistics constructed from random samples drawn from G.*

Proof. Let F_i denote the distribution function of the ith order statistic for a random sample of size n from G. We denote by $K_{\lambda,i}$ the distribution function of the ith order statistic for a sample of size n from the exponential distribution function $\mathscr{E}_\lambda$. If G has the IFR property, then $G \leq^c \mathscr{E}_\lambda$. Therefore, by Theorem 9.15, $F_i \leq^c K_{\lambda,i}$. Now $\mathscr{E}_\lambda$ is strongly unimodal. Therefore, by Theorem 9.14, $K_{\lambda,i}$ is also strongly unimodal. But then, by Theorem 9.6, $K_{\lambda,i}$ has the IFR property. That is $K_{\lambda,i} \leq^c \mathscr{E}_\lambda$. Since the relation $\leq^c$ is transitive, we see that $F_i \leq^c \mathscr{E}_\lambda$, which again means that F_i has the IFR property. The theorem is thus proved. ■

Let us now consider the IFRA property and the ordering $\leq^*$. Recall that $G_1 \leq^* G_2$ means that $G_2^{-1}G_1$ is star-shaped and the IFRA property for G means that $G \leq^* \mathscr{E}_\lambda$. The proofs of the following two theorems follow the same lines as the proofs of Theorems 9.15 and 9.16.

Theorem 9.17. *With the same notation as in Theorem* 9.15, $G_1 \leq^* G_2 \Rightarrow F_{1i} \leq^* F_{2i}$.

Theorem 9.18. *The IFRA property of a distribution function G is inherited by the order statistics constructed from random samples drawn from G.*

Finally, we present a result involving order statistics and the MR ordering.

Theorem 9.19. *Suppose F_{1n} and F_{2n} denote the distribution functions of the highest order statistics $X_{(1n)}$ and $X_{(2n)}$ for random samples of size n from distribution functions G_1 and G_2 respectively. If $G_1 \leq^{mr} G_2$, then $F_{1n} \leq^{mr} F_{2n}$.*

Proof. Let $\mathbf{X}_i = (X_{i1}, \ldots, X_{in})$ be a random sample from G_i. For $\mathbf{x} \in R^n$, let

$$u(\mathbf{x}) = \max(x_1, \ldots, x_n).$$

Then we can write $X_{(in)} = u(\mathbf{X}_i)$. Assume that v is a nondecreasing convex function on R. Write $h(\mathbf{x}) = v[u(\mathbf{x})]$. Then h is coordinatewise convex and nondecreasing. Therefore, by Theorem 9.5

$$E[h(\mathbf{X}_1)] \leq E[h(\mathbf{X}_2)]. \tag{9.44}$$

But $h(\mathbf{X}_i) = v[u(\mathbf{X}_i)] = v(X_{(in)})$. Therefore, (9.44) means that

$$E[v(X_{(1n)})] \leq E[v(X_{(2n)})], \tag{9.45}$$

for all convex nondecreasing functions v. Now (9.45) and Theorem 9.4 show that $X_{(1n)} \leq^{mr} X_{(2n)}$. This proves the theorem. ■

Remark. The conclusion of Theorem 9.19 cannot be extended to order statistics other than the maximum. The reason is that the mappings which define these other order statistics from the sample are not convex.

Appendix

The purpose of this Appendix is to bring together some background material with which the reader is expected to be familiar. To facilitate the presentation, the material is divided into a few sections. Proofs are generally omitted and references are mentioned from time to time.

A.1 Convex Sets in $\mathbf{R}^n$

A subset C of R^n is called *convex* if

$$\mathbf{x} \in C, \mathbf{y} \in C, \text{ and } \theta \in [0, 1] \quad \Rightarrow \quad \theta\mathbf{x} + (1 - \theta)\mathbf{y} \in C.$$

The *convex hull* (respectively, *closed convex hull*) of a set $B \subset R^n$ is defined to be the smallest convex (respectively, closed convex) set containing B. A set $A \subset R^n$ is called *affine* if

$$\mathbf{x} \in A, \mathbf{y} \in A, \text{ and } \theta \in R \quad \Rightarrow \quad \theta\mathbf{x} + (1 - \theta)\mathbf{y} \in A.$$

The *affine hull* of a set $B \subset R^n$ is the smallest affine set containing B. It should be noted that θ is unrestricted in the definition of an affine set whereas it is restricted to be in $[0, 1]$ in the definition of a convex set. As a result, an affine set is unbounded as soon as it has at least two points. In addition, as will be seen below, an affine subset of R^n is always closed. This is why there is no separate definition of the closed affine hull of a set.

The concept of linear independence of vectors in R^n is well known. For some applications, we need the concept of affine independence.

Definition A.1. Vectors $\mathbf{x}_1, \ldots, \mathbf{x}_k$ in R^n are called *affine independent* if

$$\sum_{i=1}^{k} \theta_i \mathbf{x}_i = \mathbf{0} \quad \text{and} \quad \sum_{i=1}^{k} \theta_i = 0 \quad \Rightarrow \quad \theta_i = 0 \quad \text{for all } i = 1, \ldots, k.$$

The following two observations follow easily from Definition A.1.

(i) $\mathbf{x}_1, \ldots, \mathbf{x}_k$ are affine independent if, and only if, none of the vectors is in the affine hull of the remaining $(k-1)$ vectors.
(ii) $\mathbf{x}_1, \ldots, \mathbf{x}_k$ are affine independent if, and only if, the $(k-1)$ vectors $\mathbf{x}_i - \mathbf{x}_k$, $i = 1, \ldots, k-1$ are linearly independent.

Consequently, an affine independent set of vectors in R^n has at most $(n+1)$ elements.

Theorem A.1. *The affine hull of a nonempty set $B \subset R^n$ is the affine hull of a finite set of affine independent vectors in B.*

Proof. Let $B \subset R^n$ be nonempty and let A be the affine hull of B. If B has only one element $\mathbf{x}$ then $A = B =$ the affine hull of $\{\mathbf{x}\}$ and the theorem holds trivially. So suppose that B has at least two elements. Choose $\mathbf{x}_1$, $\mathbf{x}_2$ in B such that $\mathbf{x}_1 \neq \mathbf{x}_2$. Then $\mathbf{x}_1$, $\mathbf{x}_2$ are affine independent. Write $B_2 = \{\mathbf{x}_1, \mathbf{x}_2\}$. If the affine hull A_2 of B_2 includes B then $A = A_2$ and again the assertion of the theorem is clear. So suppose A_2 does not include B. We can then find $\mathbf{x}_3 \in B$ such that $\mathbf{x}_3 \notin A_2$. The set $B_3 = \{\mathbf{x}_1, \mathbf{x}_2, \mathbf{x}_3\}$ is then a set of affine independent vectors. Proceeding in this way and using comment (ii) above, we see that we will reach an integer k, $2 \leq k \leq (n+1)$ and a set B_k of k affine independent vectors in B such that the affine hull A_k of B_k contains B. But then $A_k = A$ and so the theorem is proved. ■

Let A be an affine subset of R^n. Choose $\mathbf{x}_0 \in A$ and write

$$L = A - \mathbf{x}_0 = \{\mathbf{x} - \mathbf{x}_0 : \mathbf{x} \in A\}.$$

Then L is a subspace of R^n. Moreover, L depends only on A and not on the particular choice of $\mathbf{x}_0 \in A$. In other words, every affine set A is the translate of a subspace L. The subspace L is usually called the parallel subspace of A. Since a subspace of R^n is always closed, we see that an affine subset of R^n is also always closed. The dimension of an affine set A is defined to be the dimension of its parallel subspace L. By extension, the dimension of a convex set $C \subset R^n$ is defined to be the dimension of the affine hull of C. Thus C has dimension m if, and only if,

(a) C has a subset of $(m+1)$ affine independent vectors and
(b) Every subset of C of size $\geq (m+2)$ is affine dependent.

Let $C \subset R^n$ be a nonempty convex set and let A be the affine hull of C. In general, C may not have a nonempty interior relative to R^n. But C always has a nonempty interior relative to A. To see this, let m be the dimension of C. Then we can find $(m+1)$ affine independent vectors $\mathbf{x}_0, \ldots, x_m$ in C. By a change of origin, we can arrange to have $\mathbf{x}_0 = \mathbf{0}$. But then A is a subspace. Further, since $\mathbf{x}_0 = \mathbf{0}$, the vectors $\mathbf{x}_1, \ldots, \mathbf{x}_m$ are linearly independent. That is, $\{\mathbf{x}_1, \ldots, \mathbf{x}_m\}$ is a basis for A. So A is the set of all vectors $\mathbf{x}$ of the form

$$\mathbf{x} = \sum_{i=1}^{m} \theta_i(\mathbf{x}) \cdot \mathbf{x}_i,$$

where $\theta_i(\mathbf{x}) \in R$. Writing $\theta_0(\mathbf{x}) = 1 - \sum_{i=1}^{m} \theta_i(\mathbf{x})$, we get

$$\mathbf{x} = \sum_{i=0}^{m} \theta_i(\mathbf{x}) \cdot \mathbf{x}_i, \qquad \text{where } \sum_{i=0}^{m} \theta_i(\mathbf{x}) = 1.$$

Now let

$$V = \left\{ \sum_{i=0}^{m} \lambda_i \mathbf{x}_i \colon \lambda_i > 0 \text{ for all } i \text{ and } \sum_{i=0}^{m} \lambda_i + 1 \right\}.$$

Then $V \subset C$. Further the uniqueness and the continuity of the coefficients $\theta_i(\mathbf{x})$ in $\mathbf{x}$ easily implies that V is open in A. Thus C has a nonempty interior relative to its affine hull.

Definition A.2. A convex set $C \subset R^n$ is called a *body* if C has a nonempty interior relative to R^n.

The discusson preceding the above definition shows that a convex set is always a body relative to its affine hull.

Definition A.3. A point $\mathbf{x}$ in a convex set $C \subset R^n$ is said to be an *extreme point* of C if $\mathbf{x}$ is not in the relative interior of a line segment joining two distinct points of C.

Thus $\mathbf{x} \in C$ is an extreme point of C if, and only if,

$$\left.\begin{array}{l} \mathbf{x}_1 \in C, \quad \mathbf{x}_2 \in C, \quad \theta \in (0,1) \\ \text{and} \\ \mathbf{x} = \theta \mathbf{x}_1 + (1-\theta)\mathbf{x}_2 \end{array}\right\} \Rightarrow \mathbf{x}_1 = \mathbf{x}_2 = \mathbf{x}.$$

We use the symbol Ex(C) to denote the set of all extreme points of C. An extreme point of C is always a boundary point of C. The converse is easily shown to be false. Moreover, C may not have any extreme points. For instance, an open convex set has no extreme points. One of the most important results on convex sets is the *Krein–Milman Theorem* which takes the following form in R^n.

Theorem A.2. *Let C be a compact convex subset of R^n. Then C is the convex hull of* Ex(C).

The proof of Theorem A.2 is given in most books on convex sets; see, for instance, Rockafellar (1970, pp. 167–168) or Lay (1982, p. 43). In more general locally convex spaces, a compact convex set C is the closed convex hull of Ex(C); see, for instance, Kelly and Namioka (1963, p. 131).

Suppose B is a Borel subset of R^n and let C be the closed convex hull of B. If μ is a probability measure on the Borel subsets of B and

$$\mathbf{x}(\mu) = \int_B \mathbf{y}\, d\mu(\mathbf{y}),$$

then $\mathbf{x}(\mu)$ is called the *barycenter* of μ. It is easy to show that $\mathbf{x}(\mu) \in C$ for all μ. Sometimes, the converse is true. That is, given $\mathbf{x} \in C$, there is a probability measure $\mu_{\mathbf{x}}$ on the Borel subsets of B such that the barycenter of $\mu_{\mathbf{x}}$ is $\mathbf{x}$. In other words

$$\mathbf{x} = \int_B \mathbf{y}\, d\mu_{\mathbf{x}}(\mathbf{y}).$$

Such a representation is called a *Choquet representation.* Of course, one can look at such representations in spaces much more general than R^n. The standard reference on Choquet theory is the book by Phelps (1966). For the purposes of this monograph, we need to look at convex structures of sets of probability measures on R^n. In view of the topological nature of the conditions in the Krein–Milman theorem, we need a notion of convergence for such probability measures. One such notion is discussed in the next section.

A.2. Weak Convergence of Probability Measures

Suppose $(\mathscr{X}, d)$ is a complete separable metric space and let $\mathscr{B}$ be the Borel σ-algebra in $\mathscr{X}$, that is, the σ-algebra generated by the family of all open subsets of $\mathscr{X}$. Let $\mathscr{P}$ be the set of all probability measures on $\mathscr{B}$. The most widely used notion of convergence on $\mathscr{P}$ is as follows.

Definition A.4. A sequence $\{P_n\}$ of elements in $\mathscr{P}$ is said to converge *weakly* to an element P in $\mathscr{P}$ if

$$\int g\, dP_n \to \int g\, dP$$

for every bounded real-valued continuous function g on $\mathscr{X}$.

Two standard references on weak convergence are the books by Billingsley (1968) and Parthasarathy (1967). An important result on weak convergence is the following result due to Prohorov (1956).

Theorem A.3. *A subset A of $\mathscr{P}$ has compact closure under weak convergence if, and only if, for every $\varepsilon > 0$ there is a compact set $K_\varepsilon \subset \mathscr{X}$ such that*

$$P(K_\varepsilon) \geq 1 - \varepsilon$$

for all $P \in A$.

It is well known that the topology of weak convergence on $\mathscr{P}$ can be metrized so that $\mathscr{P}$ becomes a complete separable metric space. Therefore, if $\mathscr{P}^+$ denotes the set of all probability measures on the Borel σ-algebra in $\mathscr{P}$, then one can again consider the topology of weak convergence on $\mathscr{P}^+$. One might then want to know whether a subset of $\mathscr{P}^+$ has compact closure. A condition under which this holds is given in the next theorem. It was used by Kanter (1977); see Section 2.5. The proof given here was shown to us by Professor Anselm Iwanik (1983).

Theorem A.4. *Suppose $A^+ \subset \mathscr{P}^+$ is such that for every $\varepsilon > 0$ and $\delta > 0$, there is a compact set $K_{\varepsilon,\delta} \subset \mathscr{X}$ such that*

$$P^+\{P \in \mathscr{P} : P(K_{\varepsilon,d}) \geq 1 - \delta\} \geq 1 - \varepsilon$$

for all $P^+ \in A^+$. Then the closure of A^+ in $\mathscr{P}^+$ is compact.

Proof. (i) For a closed set $D \subset \mathscr{X}$ and $\delta > 0$, let

$$A(\delta, D) = \{P \in \mathscr{P} : P(D) \geq 1 - \delta\}.$$

Then $A(\delta, D)$ is a closed subset of $\mathscr{P}$. To see this, let $P_m \in A(\delta, D)$, $P_0 \in \mathscr{P}$ and $P_m \to P_0$ weakly. Then, by Theorem 2.1 of Billingsley (1968),

$$P_0(D) \geq \limsup P_m(D) \geq 1 - \delta.$$

Thus $P_0 \in A(\delta, D)$ and $A(\delta, D)$ is closed.

(ii) Choose two sequences $\{\delta_n\}$ and $\{\varepsilon_n\}$ of positive numbers such that $\delta_n \to 0$ and $\sum \varepsilon_n < \infty$. By hypothesis, we can choose a compact set $K_n \subset \mathscr{X}$ such that (in the notation of part (i) of the proof)

$$P^+[A(\delta_n, K_n)] \geq 1 - \varepsilon_n$$

for all $P^+ \in A^+$. For simplicity, write A_n for $A(\delta_n, K_n)$. We know that A_n is closed. Therefore, if we write $B_n = \bigcap_{k \geq n} A_k$, then B_n is also closed. We now show that B_n is in fact compact. We do this by showing that B_n has compact closure in $\mathscr{P}$. For this purpose, we use Theorem A.3.

Let $\eta > 0$ be given. Choose $m \geq n$ such that $\delta_m < \eta$. Then

$$\begin{aligned} P \in B_n &\Rightarrow P \in A_m \\ &\Rightarrow P(K_m) \geq 1 - \delta_m > 1 - \eta. \end{aligned}$$

Thus each B_n is a compact subset of $\mathscr{P}$.

(iii) Let $\varepsilon > 0$ be given. Choose n so large that $\sum_{k=n}^{\infty} \varepsilon_k \leq \varepsilon$. Then, for all $P^+ \in A^+$,

$$\begin{aligned} 1 - P^+(B_n) = P^+(B_n^c) &= P^+\left(\bigcup_{k=n}^{\infty} A_k^c\right) \\ &\leq \sum_{k=n}^{\infty} P^+(A_k^c) \leq \sum_{k=n}^{\infty} \varepsilon_k \leq \varepsilon. \end{aligned}$$

Since B_n is compact and n is determined by ε, we see that A^+ has compact closure in $\mathscr{P}^+$. ■

A.3. Convex Sets of Probability Measures

Let $\mathscr{P}_n$ be the set of all probablity measures on the Borel σ-algebra $\mathscr{B}_n$ in R^n. In order to convert $\mathscr{P}_n$ into a convex set, we consider the operation of taking mixtures. If P_1, P_2 are in $\mathscr{P}_n$ and $\theta \subset [0, 1]$, then the $(\theta, (1 - \theta))$-*mixture* of P_1, P_2 is the measure $\theta P_1 + (1 - \theta)P_2$ defined in a natural way by

$$[\theta P_1 + (1 - \theta)P_2](B) = \theta P_1(B) + (1 - \theta)P_2(B).$$

With this definition of mixtures, the set $\mathscr{P}_n$ is convex. Under the topology of weak convergence $\mathscr{P}_n$ is not compact. Nevertheless $\mathscr{P}_n$ has extreme points and $\mathscr{P}_n$ is the closed convex hull of the set of these extreme points. In addition, every $P \in \mathscr{P}_n$ has a Choquet-type representation in terms of the extreme points. These two and other similar results are discussed in the present section.

In order to identify the extreme points of $\mathscr{P}_n$, we define the support of a probability measure P.

Definition A.5. The *support* of a measure $P \in \mathscr{P}_n$ is defined by

$$\operatorname{supp} P = \{\mathbf{x} \in R^n : P(N) > 0 \text{ for every neighborhood } N \text{ of } \mathbf{x}\}.$$

Alternatively, supp P can be defined to be the smallest closed set B such that $P(B) = 1$.

Let $\mathscr{F}_n$ denote the set of all measures in $\mathscr{P}_n$ having finite support. Then it is not difficult to show that $\mathscr{F}_n$ is dense in $\mathscr{P}_n$ under weak convergence. Within $\mathscr{F}_n$ we can identify the subset $\mathscr{D}_n$ of those measures which have singleton supports. Elements of $\mathscr{D}_n$ are the *degenerate* distributions $D_{\mathbf{x}}$, $\mathbf{x} \in R^n$, where

$$D_{\mathbf{x}}(B) = \begin{cases} 1, & \text{if } \mathbf{x} \in B \\ 0, & \text{if } \mathbf{x} \notin B. \end{cases}$$

Since the convex hull of $\mathscr{D}_n$ is clearly $\mathscr{F}_n$, we see that $\mathscr{P}_n$ is the closed convex hull of $\mathscr{D}_n$. This gives the first part of the next theorem.

Theorem A.5. *Let $\mathscr{P}_n$ (respectively, $\mathscr{D}_n$) be the set of all (respectively, all degenerate) probability measures on R^n. Then*

(a) *$\mathscr{P}_n$ is the closed convex hull of $\mathscr{D}_n$, and*
(b) *$\mathscr{D}_n$ is the set of all extreme points of $\mathscr{P}_n$.*

Proof. Only part (b) needs to be proved. First suppose $P_0 \in \mathscr{P}_n$ and $P_0 \notin \mathscr{D}_n$. Then the support of P_0 has at least two points. Therefore, we can find $B_0 \in \mathscr{B}_n$ such that $0 < P(B_0) < 1$. Write $\theta = P_0(B_0)$, $P_1(B) = P_0(B \cap B_0)/\theta$ and $P_2(B) = P_0(B \cap B_0^c)/(1-\theta)$ where $B \in \mathscr{B}_n$. Then P_1, P_2 are distinct because $P_1(B_0) = 1$ and $P_2(B_0) = 0$. Further $P_0 = \theta P_1 + (1-\theta)P_2$ and so P_0 is not extreme in $\mathscr{P}_n$.

Next we show that each degenerate distribution $D_{\mathbf{x}}$ is extreme in $\mathscr{P}_n$. Suppose that $D_{\mathbf{x}} = \theta P_1 + (1-\theta)P_2$ for some P_1, P_2 in $\mathscr{P}_n$ and some $\theta \in (0, 1)$. Then

$$1 = D_{\mathbf{x}}(\{\mathbf{x}\}) = \theta P_1(\{\mathbf{x}\}) + (1-\theta)P_2(\{\mathbf{x}\}).$$

Since $P_i(\{\mathbf{x}\}) \le 1$ for $i = 1, 2$ and $0 < \theta < 1$, it follows that $P_i(\{\mathbf{x}\}) = 1$ for $i = 1, 2$. Thus $P_1 = P_2 = D_{\mathbf{x}}$ and so $D_{\mathbf{x}}$ is extreme. The theorem is thus proved. ■

The above theorem is a result of the Krein–Milman type for the convex set $\mathscr{P}_n$. It is easy to see that a Choquet-type theorem is also available in this case. Indeed, if $P \in \mathscr{P}_n$ and $B \in \mathscr{B}_n$, we have

$$P(B) = \int D_{\mathbf{x}}(B)\, dP(\mathbf{x}).$$

So we have the representation

$$P = \int D_{\mathbf{x}}\, dP(\mathbf{x})$$

which expressed P as a *generalized mixture* (that is, an integral) of the degenerate measures $D_{\mathbf{x}}$.

In probability theory, one frequently considers interesting convex sets of probability measures (like $\mathscr{P}_n$) and tries to look for Krein–Milman and Choquet-type theorems. We illustrate this observation by presenting a couple of examples.

Example A.1. Let $\{X_n\}$ be a sequence of real-valued random variables on some probability space. The sequence $\{X_n\}$ determines a probability measure P_∞ on the Borel subsets of R^∞. The sequence $\{X_n\}$ or the probability P_∞ is called *exchangeable* if for every $n \geq 2$ and for every choice of Borel sets $B_1, \ldots, B_n$ in R, the probability

$$P[X_1 \in B_{i_1}, \ldots, X_n \in B_{i_n}]$$

is the same for all the $n!$ permutations $(i_1, \ldots, i_n)$ of $(1, \ldots, n)$.

The set $\mathscr{E}_\infty$ of all exchangeable probabilities P_∞ is clearly convex under mixtures. Within $\mathscr{E}_\infty$ we can identify the subset $\mathscr{I}_\infty$ of those measures P_∞ for which the random variables X_n are independent and identically distributed. The well known theorem of de Finetti says that $\mathscr{I}_\infty$ is the set of all extreme points of $\mathscr{E}_\infty$ and that every measure in $\mathscr{E}_\infty$ can be expressed as a generalized mixture of measures in $\mathscr{I}_\infty$.

It is instructive to look at the simple special case where each X_n takes only two values, namely, 0 and 1 and $\{X_n\}$ is exchangeable. In this case, the probability

$$P(X_1 = x_1, \ldots, X_n = x_n)$$

is a function of n and $\sum_{i=1}^n x_i$ only. If $\{Y_n\}$ corresponds to an extreme point relevant for this case, then its distribution is completely determined by the point $\theta \in [0, 1]$, where $\theta = P(Y_1 = 1)$. The relevant extreme points can therefore be indexed by points in $[0, 1]$. de Finetti's theorem states that there is a probability measure μ on the Borel σ-algebra in $[0,1]$ such that

$$P(X_1 = x_1, \ldots, X_n = x_n) = \int \theta^{\Sigma x_i}(1-\theta)^{n-\Sigma x_i}\, d\mu(\theta).$$

For a proof of the last formula, see Feller (1971, p. 228).

Example A.2. Let $\mathscr{P}^*$ be the set of all probability measures on R having mean zero. The set $\mathscr{P}^*$ is clearly convex under mixtures. Once again, one can identify the extreme points of $\mathscr{P}^*$ and get a Choquet-type representation for measures in $\mathscr{P}^*$.

Suppose Q_1 is the open positive quadrant in R^2 and write $Q = Q_1 \cup \{(0,0)\}$. For $(u, v) \in Q_1$, let $P_{u,v}$ be the probability measure which puts mass $v/(u+v)$ at the point u and mass $u/(u+v)$ at the point $(-v)$. The measure $P_{0,0}$ is defined to be the degenerate distribution at 0. It is easily verified that all these measures $P_{u,v}$ have mean zero and so they belong to $\mathscr{P}^*$. The following convexity result arises in the context of the well known Skorohod representation; see, for instance, Freedman (1971, p. 68). The proof given here seems to be new.

Theorem A.6.

(a) *Given $P \in \mathscr{P}^*$, there is a probability measure μ on the Borel σ-algebra in Q such that*

$$P = \int_Q P_{u,v}\, d\mu(u, v). \tag{A.1}$$

(b) *The measures $P_{u,v}$ are the extreme points of $\mathscr{P}^*$.*

Proof. (i) We prove (a) first. Let $P \in \mathscr{P}^*$. Write $\theta_0 = P(\{0\})$, $\theta_1 = P[(0,\infty)]$ and $\theta_2 = P[(-\infty,0)]$. If $\theta_0 = 1$, then in (A.1) we can take μ to be the degenerate distribution at $(0,0)$. So, let $\theta_0 < 1$. Then $\theta_1 > 0$ and $\theta_2 > 0$. Let

$$\alpha = \int_0^\infty x\, dP(x).$$

Since P has mean zero, we can also write

$$\alpha = -\int_{-\infty}^0 x\, dP(x).$$

Further, $\alpha > 0$ because $\theta_1 > 0$. Define measures ν_1 and ν_2 on $(0,\infty)$ as follows.

For $0 < a < b < \infty$,

$$\nu_1[(a, b)] = \frac{P[(a, b)]}{\sqrt{\alpha}},$$

$$\nu_2[(a, b)] = \frac{P[(-b, -a)]}{\sqrt{\alpha}}.$$

Then $\nu_i[(0, \infty)] = \theta_i/\sqrt{\alpha}$ and

$$\int_0^\infty u \, d\nu_1(u) = \int_0^\infty v \, d\nu_2(v) = \sqrt{\alpha}.$$

We now describe the measure μ required by assertion (a). Under μ the mass at $(0, 0)$ will be θ_0 and on the open positive quadrant Q_1, we will set

$$\frac{d\mu}{d(\nu_1 \times \nu_2)}(u, v) = (u + v).$$

With this definition of μ, we easily get $\mu(Q_1) = (\theta_1 + \theta_2)$. Since $\theta_0 + \theta_1 + \theta_2 = 1$, we see that μ is a probability measure.

Now we claim that (A.1) holds. Suppose $x > 0$. We need to verify that the evaluation of the right side of (A.1) for the set (x, ∞) equals $P[(x, \infty)]$. But this evaluation equals

$$\begin{aligned}
\int_{v>0} \int_{u>x} \frac{v}{u+v} \, d\mu(u, v) &= \int_{v>0} \int_{u>x} v \, d\nu_1(u) \, d\nu_2(v) \\
&= \left(\int_0^\infty v \, d\nu_2(v) \right) \cdot \nu_1[(x, \infty)] \\
&= \sqrt{\alpha} \cdot \nu_1[(x, \infty)] = P[(x, \infty)].
\end{aligned}$$

A similar verification holds for the interval $(-\infty, -x)$. Thus (a) is proved.

(ii) The assertion that $P_{u,v}$ is extreme in $\mathscr{P}^*$ follows from the fact that $P_{u,v}$ is the only distribution with support contained in the set $\{u, -v\}$ and having mean zero. If $P \in \mathscr{P}^*$ and P is not one of the measures $P_{u,v}$ then the measure μ in the representation (A.1) cannot be degenerate. Therefore, we can express P as a convex combination of two distinct measures in $\mathscr{P}^*$. This proves (b) and completes the proof of the theorem. ■

Looking for Krein-Milman and Choquet-type theorems for convex sets of probability measures is one of the themes of this monograph. For instance, we study the convex structure of unimodal distributions on R in Chapter 1, the convex structure of star unimodal distributions in Chapter 2, etc.

A.4. The Brunn–Minkowski Inequality

Suppose λ_n denotes Lebesgue measure on R^n. If A and B are subsets of R^n and $\theta \in [0, 1]$, we write

$$\theta A + (1-\theta)B = \{\theta\mathbf{x} + (1-\theta)\mathbf{y} : \mathbf{x} \in A,\ \mathbf{y} \in B\}.$$

An important result on convex sets which has been used several times in this monograph is the *Brunn-Minkowski inequality* which states that for convex bodies A and B in R^n and $\theta \in [0, 1]$,

$$\lambda_n^{1/n}[\theta A + (1-\theta)B] \geq \theta\lambda_n^{1/n}(A) + (1-\theta)\lambda_n^{1/n}(B). \tag{A.2}$$

In terms of the terminology of Section 3.3, (A.2) states that the measure λ_n is $(1/n)$-concave. A scattered proof of (A.2) is contained in the results of Section 3.3. To make this observation a little clearer, note that λ_n has a constant density which is thus ∞-concave. By Theorem 3.16, the measure λ_n is therefore $(1/n)$-concave. Of course, one needs to ensure that a circular argument is avoided. So we note that in the proof of Theorem 3.15 (of which Theorem 3.16 is a corollary) the Brunn–Minkowski inequality is used only for the case $n = 1$, where it is trivial.

It is of some interest to note the inequality (A.2) is really a consequence of Hölder's inequality. To see this, suppose that A and B are rectangles with sides parallel to the axes. That is, let

$$A = (a_{11}, a_{12}) \times \cdots \times (a_{n1}, a_{n2}),$$

and

$$B = (b_{11}, b_{12}) \times \cdots \times (b_{n1}, b_{n2}).$$

Write $c_i = (a_{i2} - a_{i1})$ and $d_i = (b_{i2} - b_{i1})$. Then (A.2) becomes

$$\prod_{i=1}^{n} [\theta c_i + (1-\theta)d_i]^{1/n} \geq \theta \prod_{i=1}^{n} c_i^{1/n} + (1-\theta) \prod_{i=1}^{n} d_i^{1/n}. \tag{A.3}$$

Suppose now that a random variable X takes the value 0 and 1 with probabilities θ and $(1-\theta)$ respectively. Let f_i be the function on $\{0, 1\}$ defined by $f_i(0) = c_i^{1/n}$ and $f_i(1) = d_i^{1/n}$. Write $Y_i = f_i(X)$. Then the inequality (A.3) is equivalent to

$$\prod_{i=1}^{n} [E(Y_i^n)]^{1/n} \geq E\left[\prod_{i=1}^{n} Y_i\right],$$

which is a special case of Hölder's inequality. Now that (A.2) is established for rectangles with sides parallel to the axes, it can be extended to general convex bodies by using the Hadwiger–Ohman technique which has already been covered in the proof of Theorem 2.6. The discussion of this section should explain the important role played by Hölder's inequality in the proofs of Section 3.3.

References

Adriano, K., Gentle, J. E., and Sposito, V. A. (1977). On the asymptotic bias of Grenander's mode estimator. *Commun. Statist.*, A, **6**, 773–776.

Alam, K. (1972). Unimodality of the distribution of an order statistic. *Ann. Math. Statist.*, **43**, 2041–2044.

Anderson, T. W. (1955). The integral of a symmetric unimodal function over a symmetric convex set and some probability inequalities. *Proc. Amer. Math. Soc.*, **6**, 170–176.

Anderson, T. W. and Das Gupta, S. (1964). Monotonicity of the power functions of some tests of independence between two sets of variates. *Ann. Math Statist.*, **35**, 206–208.

Barlow, R. E. and Proschan, F. (1966). Inequalities for linear combinations of order statistics from restricted families. *Ann. Math. Statist.*, **37**, 1574–1592.

Barlow, R. E. and Proschan, F. (1981). *Statistical Theory of Reliability and Life Testing*, Second Printing. To Begin With, Silver Springs.

Barnett, V. D. (1966). Evaluation of the maximum likelihood estimator when the likelihood equation has multiple roots. *Biometrika*, **53**, 151–165.

Beckenbach, E. F. and Bellman, R. (1965). *Inequalities*, Second revised printing. Springer-Verlag, New York.

Berge, P. O. (1937). A note on a form of Tchebycheff's theorem for two variables. *Biometrika*, **29**, 405–406.

Berk, R. and Hwang, J. T. (1984). Optimality of the least squares estimator. Submitted for publication.

Bertin, E. M. J. and Theodorescu, R. (1984). Some characterizations of discrete unimodality. *Statistics and Probability Letters*, **2**, 23–30.

Bessler, S. A. and Veinott, A. F. (1966). Optimal policy for a dynamic multi-echelon inventory model. *Naval Research Logistics Quarterly*, **13**, 355–389.

Bickel, P. J. and Doksum, K. A. (1977). *Mathematical Statistics*. Holden Day, San Francisco.

Billingsley, P. (1968). *Weak Convergence of Probability Measures*. Wiley, New York.

Birnbaum, Z. W. (1948). On random variables with comparable peakedness. *Ann. Math. Statist.* **19**, 76–81.

Block,H. W. and Savits, T. H. (1976). The IFRA closure problem. *Ann. Probability*, **4**, 1030–1032.

Block, H. W. and Savits, T. H. (1980a). Laplace transforms for classes of life distributions. *Ann. Probability*, **8**, 465–474.

Block, H. W. and Savits, T. H. (1980b). Multivariate increasing failure rate average distributions. *Ann. Probability*, **8**, 793–801.

Bochner, S. (1955). *Harmonic Analysis and the Theory of Probability*. Univ. of California Press.

Bølviken, E. (1982). Probability inequalities for the multivariate normal with nonnegative partial correlation coefficients. *Scand. J. Statist.* **9**, 49–58.

Borell, C. (1975). Convex set functions in *d*-space. *Period Math. Hungar.*, **6**, 111–136.

Brascamp, H. J. and Lieb, E. H. (1975). Some inequalities for Gaussian measures, in *Functional Integration and its Applications*, edited by A. M. Arthurs. Clarendon Press, Oxford.

Brockett, P. L. and Kemperman, J. H. B. (1982). On the unimodality of high convolutions. *Ann. Probability*, **10**, 270–277.

Brown, G. W. (1939). On the power of the L_1 test for equality of several variances. *Ann. Math. Statist.*, **10**, 119–129.

Brunk, H. D. (1965). Conditional expectation given a σ-lattice and applications. *Ann. Math. Statist.*, **36**, 1339–1350.

Burridge, J. (1982). Some unimodality properties of likelihoods derived from grouped data. *Biometrika*, **69**, 145–151.

Camp, B. H. (1922). A new generalization of Tchebycheff's statistical inequality. *Bull. Amer. Math. Soc.*, **28**, 427–432.

Chernoff, H. (1964). Estimation of the mode. *Ann. Inst. Statist. Math.*, **16**, 31–41.

Chung, K. L. and Fuchs, J. (1951). On the distribution of values of sums of random variables. *Mem. Amer. Math. Soc.*, No. 6, 1–12.

Cochran, W. G. (1941). The distribution of the largest of a set of variances as a fraction of their total. *Ann. Eugenics*, **11**, 47–52.

Cohen, A. and Strawdermann, W. E. (1971). Unbiasedness of tests for homogeneity of variances. *Ann. Math. Statist.*, **42**, 355–360.

Conlon, J. C., Leon, R., Proschan, F., and Sethuraman, J. (1977). *G*-ordered functions with applications in statistics. Technical Report No. 73, Department of Statistics, Florida State University.

Converse, G. (1977). Sets with a mode. *Israel J. Math.*, **27**, 124–130.

Copas, J. B. (1975). On the unimodality of the likelihood for the Cauchy distribution. *Biometrika*, **62**, 701–704.

Dalenius, T. (1965). The mode—A neglected statistical parameter. *J. R. Statist. Soc.*, A, **128**, 110–117.

Darling, D. A. (1955). The Cramer–Smirnov test in the parametric case. *Ann. Math. Statist.*, **26**, 1–20.

Das Gupta, S. (1976a). A generalization of Anderson's theorem on unimodal functions. *Proc. Amer. Math. Soc.*, **60**, 85–91.

Das Gupta, S. (1976b). S-unimodal functions: related inequalities and statistical applications. *Sankhyā*, B, **38**, 301–314.

Das Gupta, S. (1976c). Personal communication.

Das Gupta, S. (1980). Brunn–Minkowski inequality and its aftermath. *J. Multiv. Anal.*, **10**, 296–318.

Das Gupta, S., Anderson, T. W., and Mudholkar, G. S. (1964). Montonicity of the power functions of some tests of the multivariate linear hypothesis. *Ann. Math. Statist.*, **35**, 200–205.

Das Gupta, S., Eaton, M. L., Olkin, I., Perlman, M. D., Savage, L. J., and Sobel, M. (1972). Inequalities on the probability content of convex regions for elliptically contoured distributions. *Proc. 6th Berkeley Symp.*, Vol. II, Univ. of California Press, 241–265.

Dharmadhikari, S. W. and Joag-dev, K. (1983a). Unimodality of symmetrized unimodal laws and related results. In *A Festschrift for Erich L. Lehmann*, edited by P. J. Bickel, K. Doksum, and J. L. Hodges, Jr. Wadsorth, Belmont, California, 131–139.

Dharmadhikari, S. W. and Joag-dev, K. (1983b). Reccurrence of symmetric random walks. *Amer. J. Math. Management Sci.*, **3**, 25–34.

Dharmadhikari, S. W. and Joag-dev, K. (1983c). Mean, median, mode III. *Statist. Neerl.*, **37**, 165–168.

Dharmadhikari, S. W. and Joag-dev, K. (1985a). Examples of nonunique maximum likelihood estimators. *Amer. Statistician*, **39**, 199–200.

Dharmadhikari, S. W. and Joag-dev, K. (1985b). The Gauss–Tchebyshev inequality for unimodal distributions. *Theor. Probability Appl.*, **30**, 817–820.

Dharmadhikari, S. W. and Joag-dev, K. (1986). Some results on generalized unimodality and an application to Chebyshev's inequality. In *Reliability and Quality Control*, edited by A. P. Basu, North-Holland, 127–132.

Dharmadhikari, S. W. and Jogdeo, K. (1973). A characterization of convexity and central symmetry for planar polygonal sets. *Israel J. Math.*, **15**, 356–366.

Dharmadhikari, S. W. and Jogdeo, K. (1974). Convolutions of α-modal distributions. *Z. Wahrscheinlichkeitsth.*, **30**, 203–208.

Dharmadhikari, S. W. and Jogdeo, K. (1976a). Multivariate Unimodality. *Ann. Statist.*, **4**, 607–613.

Dharmadhikari, S. W. and Jogdeo, K. (1976b). On characterizations of unimodality of discrete distributions. *Ann. Inst. Statist. Math.*, **28**, 9–18.

Dinghas, A. (1957). Uber eine Klasse superadditiver Mengenfuktionale von Brunn-Minkowski-Lusterniksсhen Typus. *Math. Z.* **68**, 111–129.

Dunn, O. J. (1958). Estimation of the means for dependent variables. *Ann. Math. Statist.*, **29**, 1095–1111.

Eaton, M. L. (1982). A review of selected topics in multivariate probability inequalities. *Ann. Statist.*, **10**, 11–43.

Eaton, M. L. and Perlman, M. D. (1974). A monotonicity property of the power functions of some invariant tests for MANOVA. *Ann. Statist.*, **2**, 1022–1028.

Eaton, M. L. and Perlman, M. D. (1977). Reflection groups, generalized Schur functions and the geometry of majorization. *Ann. Probability*, **6**, 829–860.

Esary, J., Proschan, F., and Walkup, W. (1967). Associated random variables with applications. *Ann. Math. Statist.*, **38**, 1466–1474.

Fáry, I. and Rédei, L. (1950). Der zentralsymmetrische Kern und die zentralsymmetrische Hülle von konvexen Körpern. *Math. Ann.*, **122**, 205–220.

Fefferman, C., Jodeit, M., and Perlman, M. D. (1972). A spherical surface measure inequality for convex sets. *Proc. Amer. Math. Soc.*, **33**, 114–119.

Feller, W. (1971). *An Introduction to Probability Theory and Its Applications*, Vol. II, Second Edition. Wiley, New York.

Fisz, M. (1963). On the orthogonality of measures induced by *L*-processes. *Trans. Amer. Math. Soc.*, **106,** 185–192.

Fisz, M. and Varadarajan, V. S. (1963). A condition for absolute continuity of infinitely divisible distribution functions. *Z. Wahrscheinbichkeitsth.*, **1**, 335–339.

Fortuin, C. M., Kasteleyn, P. W., and Ginibre, J. (1971). Correlation inequalities on some partially ordered sets. *Comm. Math. Phys.*, **22**, 89–103.

Freedman, D. A. (1970). *Brownian Motion and Diffusion.* Holden-Day, San Francisco.

Freund, J. E. (1961). A bivariate extension of the exponential distribution. *J. Amer. Statist. Assoc.*, **56**, 971–977.

Gauss, C. F. (1821). *Theoria Combinationis Observationum*, Art 10, Göttingen.

Ghosh, P. (1974). On generalized unimodality. *Comm. Statist.*, **3**, 567–580.

Gnedenko, B. V. and Kolmogorov, A. N. (1954). *Limit Distributions for Sums of Independent Random Variables.* Addision–Wesley, Cambridge, Massachusetts.

Good, I. J. (1965). *The Estimation of Probabilities: An Essay on Modern Bayesian Methods.* MIT Press, Cambridge, Massachusetts.

Grenander, U. (1965). Some direct estimators of the mode. *Ann. Math. Statist.*, **36**, 131–138.

Groenveld, R. A. and Meeden, G. (1977). The mode, median and mean inequality. *Amer. Statistician*, **31**, 120–121.

Guenther, W. C. (1969). Shortest Confidence Intervals. *Amer. Statistician*, **33**, 22–25.

Hadwiger, H. and Ohman, D. (1956). Brunn-Minkowskischer Satz und Isoperimetrie. *Math. Z.* **66**, 1–8.

Hartley, H. O. (1950). The maximum F ratio as a shortcut test for heterogeneity of variances. *Biometrika*, **37**, 308–312.

Hartman, P. and Wintner, A. (1942). On thc infinitesimal generators of integral convolutions. *Amer. J. Math.*, **64**, 273–298.

Hodges, J. L. and Lehmann, E. L. (1954). Matching in paired comparisons. *Ann. Math. Statist.*, **25**, 787–791.

Hoeffding, W. (1940). Masstabinivariante Korrelationstheorie. *Schriften Math. Inst. Univ. Berlin*, **5**, 181–233.

Hoeffding, W. (1955). The extrema of the expected value of a function of independent random variables. *Ann. Math. Statist.*, **26**, 268–275.

Hogg, R. V. and Tanis, E. A. (1983). *Probability and Statistical Inference*, Second Edition. Macmillan, New York.

Holgate, P. (1970). The modality of some compound Poisson distributions. *Biometrika*, **57**, 666–667.

Hotelling, H. and Pabst, M. R. (1936). Rank correlation and tests of significance involving no assumptions of normality *Ann. Math. Statist.*, **7**, 29–43.

Huang, J. S. and Ghosh, M. (1982). A note on strong unimodality of order statistics. *J. Amer. Statist. Assoc.*, **77**, 929–930.

Ibragimov, I. A. (1956). On the composition of unimodal distributions. *Theor. Probability Appl.*, **1**, 255–266.

Ibragimov, I. A. (1957). A remark on probability distributions of class *L*. *Theor. Probability Appl.*, **2**, 117–119.

Iwanik, A. (1983). Personal communication.

Jeyaratnam, S. (1985) Minimum volume confidence regions. *Statistics and Probability Letters*, **3**, 307–308.

Jogdeo K. (1970). A simple proof of an inequality for multivariate normal probabilities of rectangles. *Ann. Math. Statist.*, **41**, 1357–1359.

Jogdeo, K. (1977). Association and probability inequalities. *Ann. Statist.*, **3**, 495–504.

Johnson, N. L. and Kotz, S. (1977). *Distributions in Statistics: Continuous Multivariate Distributions*. Wiley, New York.

Johnson, N. L. and Rogers, C. A. (1951). The moment problem for unimodal distributions. *Ann. Math. Statist.*, **22**, 432–439.

Kanter, M. (1977). Unimodality and dominance for symmetric random vectors. *Trans. Amer. Math. Soc.*, **229**, 65–85.

Karlin, S. (1968) *Total Positivity*, Vol. I. Stanford University Press.

Karlin, S. and Rinott, Y. (1981). Total positivity properties of absolute value multinormal variables with applications to confidence interval estimates and related probability inequalities. *Ann. Statist*, **9**, 1035–1049.

Keilson, J. and Gerber, H. (1971). Some results for discrete unimodality. *J. Amer. Statist. Assoc.*, **66**, 386–389.

Kelly, J. L. and Namioka, I. (1963). *Linear Topological Spaces*. van Nostrand, Princeton.

Khintchine, A. Y. (1938). On unimodal distributions. *Izv. Nauchno-Isled. Inst. Mat. Mech. Tomsk. Gos. Univ.*, **2**, 1–7.

Lapin, A. I. (1947). On properties of stable laws. Ph.D. Dissertation.

Laue, R. V. (1965). Heterogeneity of variances in a one-way and two-way classification model. Ph.D. Dissertation, Rutgers University.

Lay, S. R. (1982). *Convex Sets and Their Applications*. Wiley, New York.

Lehmann, E. L. (1966). Some concepts of dependence. *Ann. Math. Statist*, **37**, 1137–1153.

Lehmann, E. L. (1986). *Testing Statistical Hypotheses*, Second Edition. Wiley, New York.

Levin, B. and Reeds, J. (1977). Compound multinomial likelihood functions are unimodal: proof of a conjecture of I. J. Good. *Ann. Statist.*, **5**, 79–87.

Lévy, P. (1954). *Théorie de L'addition des Variables Aléatoires*. Gauthier-Villars, Paris.

Loève, M. (1977). *Probability Theory*, Vol. I, Fourth Edition. Springer-Verlag, New York.

Lukacs, E. (1960). *Characteristic functions*. Griffin, London.

MacGillivray, H. L. (1981). The mean, median, mode inequality and skewness for a class of densities. *Austral. J. Statist.*, **23**, 247–250.

Marshall, A. W. and Olkin, I. (1967). A multivariate exponential distribution. *J. Amer. Statist. Assoc.*, **62**, 30–44.

Marshall, A. W. and Olkin, I. (1974). Majorization in multivariate distributions. *Ann. Statist.*, **2**, 1189–1200.

Marshall, A. W. and Olkin, I. (1979). *Inequalities: The Theory of Majorization with Applications to Combinatorics, Probability, Statistics, and Matrix Theory*. Academic Press, New York.

Medgyessy, P. (1972). On the unimodality of discrete distributions. *Period. Math. Hungar.*, **2**, 245–257.

Meidell, B. (1922). Sur un probléme du calcul des probabilités et les statistiques mathematiques. *C. R. Acad. Sci. Paris*, t, **173**, 806–808.

Mudholkar, G. S. (1966). The integral of an invariant unimodal function over an invariant convex set — an inequality and applications. *Proc. Amer. Math. Soc.*, **17**, 299–306.

Odlyzko, A. M. and Richmond, L. B. (1985). On the unimodality of high convolutions. *Ann. Probability*, **13**, 299–306.

Olkin, I. and Tong, Y. L. (1987). Peakedness in multivariate distributions. To appear in *Statistical Decision Theory and Related Topics* — IV, edited by S. S. Gupta and J. O. Berger, Springer-Verlag.

Olshen, R. A. and Savage, L. J. (1970). A generalized unimodality. *J. Appl. Probability*, **7**, 21–34.

Parthasarathy, K. R. (1967). *Probability Measures on Metric Spaces*. Academic Press, New York.

Parzen, E. (1962). On estimation of a probability density function and mode. *Ann. Math. Statist.*, **33**, 1065–1076.

Perlman, M. D. (1976). Personal communication.

Perlman, M. D. and Olkin, I. (1980). Unbiasedness of invariant tests for MANOVA and other multivariate problems. *Ann. Statist.*, **8**, 1326–1341.

Phelps, R. R. (1966). *Lectures on Choquet Theory*. van Nostrand, Princeton.

Pitman, E. J. G. (1939). Tests of hypotheses concerning location and scale parameters. *Biometrika*, **31**, 200–215.

Pitt, L. D. (1977). A Gaussian correlation inequality for symmetric convex sets. *Ann. Probability*, **5**, 470–474.

Plackett, R. L. (1954). A reduction formula for normal multivariate integrals. *Biometrika*, **41**, 351–360.

Pratt, J. W. (1981). Concavity of the log likelihood. *J. Amer. Statist. Assoc.*, **76**, 103–106.

Prékopa, A. (1971). Logarithmic concave measures with application to stochastic programming. *Acta. Sci. Mat.*, **32**, 301–306.

Prékopa, A. (1973). On logarithmic concave measures and functions. *Acta Sci. Mat.*, **34**, 335–343.

Prohorov, Yu, V. (1956). Convergence of random processes and limit theorems in probability theory. *Theor. Probability Appl.*, **1**, 159-214.

Proschan, F. (1965). Peakedness of distributions of convex combinations. *Ann. Math. Statist.*, **36**, 1703–1706.

Ramachandran, K. V. (1956). On the Tukey test for the equality of means and the Hartley test for equality of variances. *Ann. Math. Statist.*, **27**, 825–831.

Reeds, J. A. (1985). Asymptotic number of roots of Cauchy location likelihood equations. *Ann. Statist.*, **13**, 775–784.

Rinott, Y. (1976). On convexity of measures. *Ann. Probability*, **4**, 1020–1026.

Robertson, T. (1967). On estimating a density which is measurable with respect to a σ-lattice. *Ann. Math. Statist.*, **38**, 482–493.

Rockafellar, R. T. (1970). *Convex Analysis*. Princeton Univ. Press, Princeton.

Ross, S. M. (1983) *Stochastic Processes*. Wiley, New York.

Royden, H. L. (1968). *Real Analysis*, Second Edition. Macmillan, New York.

Rudin, W. (1987). *Real and Complex Analysis*, Third Edition. McGraw Hill, New York.

Runnenburg, J. Th. (1978). Mean median, mode. *Statist. Neerl.*, **32**, 73–80.

Sager, T. W. (1975). Consistency in nonparametric estimation of the mode. *Ann. Statist.*, **3**, 698–706.

Sager, T. W. (1978). Estimation of a multivariate mode. *Ann. Statist.*, **6**, 802–812.

Sager, T. W. (1979). An interative procedure for estimating a multivariate mode and isopleth. *J. Amer. Statist. Assoc.*, **74**, 329–339.

Sager, T. W. (1983). Estimating modes and isopleths. *Commun. Statist. Theor. Method.*, **12**, 529–557.

Savits, T. H. (1985). A multivariate IFR class. *J. Appl. Probability*, **22**, 197–204.

Schwartz, R. (1967). Admissible tests in multivariate analysis of variance. *Ann. Math. Statist.*, **38**, 698–710.

Shepp, L. A. (1962). Symmetric random walk. *Trans. Amer. Math. Soc.*, **104**, 144–153.

Sherman, S. (1955). A theorem on convex sets with applications. *Ann. Math. Statist.*, **26**, 763–766.

Šidák, Z. (1967). Rectangular confidence regions for the means of multivariate normal distributions. *J. Amer. Statist. Assoc.*, **62**, 626–633.

Šidák, Z. (1968). On multivariate normal probabilities of rectangles: their dependence on correlations. *Ann. Math. Statist.*, **39**, 1425–1434.

Slepian, D. (1962). The one-sided barrier problem for Gaussian noise. *Bell System Tech. J.*, **41**, 463–501.

Stoyan, D. (1983) *Comparison Methods for Queues and Other Stochastic Models* (edited and revised by D. J. Daley). Wiley, New York.

Sun, T. C. (1967). A note on the unimodality of distributions of class L. *Ann. Math. Statist.*, **38**, 1296–1299.

Tate, R. F. and Klett, G. W. (1959). Optimum confidence intervals for the variance of a normal distribution. *J. Amer. Statist. Assoc.*, **54**, 674–682.

Tong, Y. L. (1980). *Probability Inequalities in Multivariate Distributions.* Academic Press, New York.

Tucker, H. G. (1964). On continuous singular infinitely divisible distribution functions. *Ann. Math. Statist.*, **35**, 330–335.

Uhrin, B. (1984). Some remarks about the convolutions of unimodal functions. *Ann. Probability*, **12**, 640–645.

Ulin, B. (1953). An extremal problem in mathematical statistics. *Skand. Akt.* **36**, 158–167.

Urbanik, K. (1972). Lévy probability measures on Euclidean spaces. *Studia Math.*, **44**, 119–148.

Ushakov, N. G. (1982). On a problem of Rényi. *Theor. Probability Appl.*, **27**, 361–362.

van Zwet, W. R. (1964). *Convex Transformations of Random Variables.* Math. Centrum, Amsterdam.

van Zwet, W. R. (1979). Mean, median, mode, II. *Statist. Neerl.*, **33**, 1–5.

Venter, J. H. (1967). On estimation of the mode. *Ann. Math. Statist.*, **38**, 15–29.

Vinogradov, O. P. (1973). The definition of distribution functions with increasing hazard rate in terms of the Laplace transform. *Theor. Probability Appl.*, **18**, 811–814.

Vysochanskii, D. F. and Petunin, Yu., (1979). Justification of the three-sigma rule for unimodal distributions. *Theor. Probability and Math. Statist.*, **21**, 25–36.

Vysochanskii, D. F. and Petunin, Yu. I., (1980). On a Gauss inequality for unimodal distributions. *Theor. Probability Appl.*, **27**, 359–361.

Wegman, E. J. (1969). A note on estimating a unimodal density. *Ann. Math. Statist.*, **40**, 1661–1667.

Wegman, E. J. (1970a). Maximum lielihood estimation of a unimodal density function. *Ann. Math. Statist.*, **41**, 457–471.

Wegman, E. J. (1970b). Maximum likelihood estimation of a unimodal density function. *Ann. Math. Statist.*, **41**, 2196–2174.

Wells, D. R. (1978). A result in multivariate unimodality. *Ann. Statist.*, **6**, 926–931.

Wintner, A. (1936). On a class of Fourier transforms. *Amer. J. Math.*, **58**, 45–90.

Wintner, A. (1938). *Asymptotic Distributions and Infinite Convolutions*. Edwards Brothers, Ann Arbor, Michigan.

Wintner, A. (1956). Cauchy's stable distributions and an "explicit formula" of Mellin. *Amer. J. Math.*, **78**, 819–861.

Wolfe, S. J. (1971). On the unimodaity of *L* functions. *Ann. Math. Statist.*, 42, 912–918.

Wolfe, S. J. (1975). On the unimodality of spherically symmetric stable distribution functions. *J. Multiv. Anal.*, **5**, 236–241.

Wolfe, S. J. (1978a). On the unimodality of infinitely divisible distribution functions. *Z. Wahrscheinlichkeitsth.*, **45**, 329–335.

Wolfe, S. J. (1978b). On the unimodality of multivariate symmetric distribution functions of class *L*. *J. Multiv. Anal.*, **8**, 141–145.

Wolfe, S. J. (1980). A characterization of Lévy probability distribution functions on Euclidean spaces. *J. Multiv. Anal.*, **10**, 379–384.

Yamazato, M. (1978). Unimodality of infinitely divisible distribution functions of class *L*. *Ann. Probability*, **6**, 523–531.

Author Index

Subject Index

PROBABILITY AND MATHEMATICAL STATISTICS

Thomas Ferguson, *Mathematical Statistics: A Decision Theoretic Approach*
Howard Tucker, *A Graduate Course in Probability*
K. R. Parthasarathy, *Probability Measures on Metric Spaces*
P. Révész, *The Laws of Large Numbers*
H. P. McKean, Jr., *Stochastic Integrals*
B. V. Gnedenko, Yu. K. Belyayev, and A. D. Solovyev, *Mathematical Methods of Reliability Theory*
Demetrios A. Kappos, *Probability Algebras and Stochastic Spaces*
Ivan N. Pesin, *Classical and Modern Integration Theories*
S. Vajda, *Probabilistic Programming*
Robert B. Ash, *Real Analysis and Probability*
V. V. Fedorov, *Theory of Optimal Experiments*
K. V. Mardia, *Statistics of Directional Data*
H. Dym and H. P. McKean, *Fourier Series and Integrals*
Tatsuo Kawata, *Fourier Analysis in Probability Theory*
Fritz Oberhettinger, *Fourier Transforms of Distributions and Their Inverses: A Collection of Tables*
Paul Erdös and Joel Spencer, *Probabilistic Methods of Statistical Quality Control*
K. Sarkadi and I. Vincze, *Mathematical Methods of Statistical Quality Control*
Michael R. Anderberg, *Cluster Analysis for Applications*
W. Hengartner and R. Theodorescu, *Concentration Functions*
Kai Lai Chung, *A Course in Probability Theory, Second Edition*
L. H. Koopmans, *The Spectral Analysis of Time Series*
L. E. Maistrov, *Probability Theory: A Historical Sketch*
William F. Stout, *Almost Sure Convergence*
E. J. McShane, *Stochastic Calculus and Stochastic Models*
Robert B. Ash and Melvin F. Gardner, *Topics in Stochastic Processes*
Avner Friedman, *Stochastic Differential Equations and Applications, Volume 1, Volume 2*
Roger Cuppens, *Decomposition of Multivariate Probabilities*
Eugene Lukacs, *Stochastic Convergence, Second Edition*
H. Dym and H. P. McKean, *Gaussian Processes, Function Theory, and the Inverse Spectral Problem*
N. C. Giri, *Multivariate Statistical Inference*
Lloyd Fisher and John McDonald, *Fixed Effects Analysis of Variance*
Sidney C. Port, and Charles J. Stone, *Brownian Motion and Classical Potential Theory*

Konrad Jacobs, *Measure and Integral*
K. V. Mardia, J. T. Kent, and J. M. Biddy, *Multivariate Analysis*
Sri Gopal Mohanty, *Lattice Path Counting and Applications*
Y. L. Tong, *Probability Inequalities in Multivariate Distributions*
Michel Metivier and J. Pellaumail, *Stochastic Integration*
M. B. Priestly, *Spectral Analysis and Time Series*
Ishwar V. Basawa and B. L. S. Prakasa Rao, *Statistical Inference for Stochastic Processes*
M. Csörgö and P. Révész, *Strong Approximations in Probability and Statistics*
Sheldon Ross, *Introduction to Probability Models, Second Edition*
P. Hall and C. C. Heyde, *Martingale Limit Theory and Its Application*
Imre Csiszár and János Körner, *Information Theory: Coding Theorems for Discrete Memoryless Systems*
A. Hald, *Statistical Theory of Sampling Inspection by Attributes*
H. Bauer, *Probability Theory and Elements of Measure Theory*
M. M. Rao, *Foundations of Stochastic Analysis*
Jean-Rene Barra, *Mathematical Basis of Statistics*
Harald Bergström, *Weak Convergence of Measures*
Sheldon Ross, *Introduction to Stochastic Dynamic Programming*
B. L. S. Prakasa Rao, *Nonparametric Functional Estimation*
M. M. Rao, *Probability Theory with Applications*
A. T. Bharucha-Reid and M. Sambandham, *Random Polynomials*
Sudhakar Dharmadhikari and Kumar Joag-dev, *Unimodality, Convexity, and Applications*